HUMAN AND MACHINE PERCEPTION 3

THINKING, DECIDING, AND ACTING

HUMAN AND MACHINE PERCEPTION 3

THINKING, DECIDING, AND ACTING

Edited by

Virginio Cantoni
University of Pavia
Pavia, Italy

Vito Di Gesù
University of Palermo
Palermo, Italy

Alessandra Setti
University of Pavia
Pavia, Italy

and

Domenico Tegolo
University of Palermo
Palermo, Italy

Kluwer Academic / Plenum Publishers
New York, Boston, Dordrecht, London, Moscow

Proceedings of the Fourth International Workshop on Human and Machine Perception: Thinking, Deciding, and Acting, held June 20–23, 2000, in Palermo, Italy

ISBN: 0-306-46673-2

233 Spring Street, New York, New York 10013

10 9 8 7 6 5 4 3 2 1

A C.I.P. record for this book is available from the Library of Congress.

Printed in the United States of America

PREFACE

The following are the proceedings of the Fourth International Workshop on Human and Machine Perception held in Palermo, Italy, on June 20 - 23, 2000, under the auspices of three Institutions: the Cybernetic and Biophysics Group (GNCB) of the Italian National Research Council (CNR) and the two Inter-Department Centers of Cognitive Sciences of Palermo and Pavia University respectively.

A broad spectrum of topics are covered in this series, ranging from computer perception to psychology and physiology of perception. The theme of this workshop on Human and Machine Perception was focused on Thinking, Deciding, and Acting. As in the past editions the final goal has been the analysis and the comparison of biological and artificial solutions.

The focus of the lectures has been on presenting the state-of-the-art and outlining open questions. In particular, they sought to stress links, suggesting possible synergies between the different cultural areas. The panel discussion has been conceived as a forum for an open debate, briefly introduced by each panelist, and mainly aimed at deeper investigation of the different approaches to perception and strictly related topics. The panelists were asked to prepare a few statements on hot-points as a guide for discussion. These statements were delivered to the participants together with the final program, for a more qualified discussion.

The number of participants to the workshop was limited to 50. Besides the 30 invited lecturers and panelists, 20 more participants were admitted. Priority for these positions was given to young researchers who made significant contributions to the open discussions. Both the lectures and the contents of the panel discussions are included in these proceedings.

The workshop structure consisted of five modules. The first one composed of five talks and a first general panel discussion to give the foundations. Four subsequent modules, each one consisting of two themes with invited lectures (corresponding to approaches in nature and machines respectively) and a panel discussion, dealt with the basic sequential subtopics on *thinking, deciding, planning and acting*. In this edition we included, for the first time in the series, also a few solicited presentations from the audience.

Table 1 details the subjects of the foundations. Table 2 shows the topics as they were originally conceived, with the title of the lectures divided by area and the planned title of the common panels.

Table 1. Workshop Foundation

Nature	Artificial
Processing of biological signals: a bridge between structure and function	Developing rational agents
Overview on decision making	Architectures for artificial mentation and behavior
Toward a perception-based theory of probabilistic theory	"Is thinking necessary to do actions" *(Panel discussion)*

Table 2. Structure of the Workshop in Four Subjects

Subject	Nature	Artificial	Panel
Thinking	Learning from mistake (the psychologist view-point)	Artificial thinking: learning from mistakes	On the edge of thinking
Deciding	Decision making in evolving systems	Cognitive Robotics	Rational and not rational behaviors in decision-making
Planning	The human brain between seriality and parallelism	Attention strategies in visual systems	Planning in natural and artificial systems
Acting	Action-reaction and cooperation-competition	Notion formation in artificial intelligence	Artificial and natural systems integration/interaction

The description of both natural and artificial structures and basic approaches represents a natural starting point for a workshop which intends to analyze the processes of strategy formation, planning and acting from different viewpoints. The lectures were organized by alternating the functional descriptions of natural and artificial information management and decision making related to perception, learning, knowledge acquisition and storage. Further inquiries concern the human ability to exploit spatial information for reasoning and communication purposes and the new technologies and facilities on this filed.

Thinking is based on those "processes" that starting from a collection of stimuli and acquisitions (perception, knowledge, experience, intention) produce chains of deductions and abductions which lead to a set of possible conclusions. This definition is vague, the term "process" need further investigation. Fields involved are very complicated: how reasoning is accomplished in brains (from the simplest to the most sophisticated)? how logic is embedded in biological systems? how intuition takes part in artificial reasoning? how perception is implanted in human brain and how can be effectively implemented in artificial systems?

Deciding goes beyond purely statistical approaches by attempting to add judgement values and requiring more capabilities to foreseeing the resulting effects of actions. In order to achieve this, a sophisticated *planning* faculty is needed. Often, subjective issues may result in a more appropriate outcome. Decision Theory provides a methodology for dealing with such circumstances. Each node of a decision tree may represents not only possible outcomes, but also the "cost" of alternatives and sophisticated 'figure of merits'. Nevertheless, is decision a simple consequence of reasoning on facts? This simple question put several drawbacks concerning the decision level (hierarchical view), its complexity and representation. Natural systems generate large collections of interacting and competing opportunities, and genetic algorithms could be at the basis of their survival and winning. Is there a general control rule for the optimization of fitness conditions (both for natural & artificial systems) or there are a priori evaluations and constraints?

Finally, *actions* result from the reasoning-planning-decision chain, in which the environment is involved through an action/reaction mechanism. Actions are strongly influenced by the system morphology and physiology. In this sense action and planning-decision can't be considered fully separated, and they depend on the system anatomy/hardware. Automatism (for example conditioned reflexes) are part of the biological system, however, in both natural and artificial systems learning by training influence the execution time of an action so that repeated conscious actions become automatic. For example in biology, postural or balance control is the foundation underlying performance of many other movements such as walking and eye-hand co-ordination skills. Further problems arise when more than one actor is present in the environment; the coordination is matter of critical investigation because of the wide interaction modality, ranging from cooperativeness to competitiveness.

The workshop dealt with most of these problems and the results of the presentations and discussions are herewith included even if the chapters of the book vary somewhat from the scheduled program, to take care of the positions that emerged from the debates and to the audience reactions.

Acknowledgments

The workshop, and thus indirectly this book, was made possible through the generous financial support of the universities involved, and the research and industrial organizations that are listed separately. Their support is gratefully acknowledged.

The editors would also like to express their appreciation to the Program Committee of the workshop for their advice: Ornella Andreani-Dentici, Jörg D. Becker, Roberto Bordogna, Luciano Da Fontoura Costa, Ivo De Lotto, Salvatore Gaglio, Edzard Gelsema , Giorgio Grisanti, Roland Köberle, Eugenio Mira, Vito Roberto, Silvana Vallerga, Yehezkel· Yeshurun, Bertrand Zavidovique.

Special thanks should go to Mrs. Clara Randazzo, for her precious help and patience in organizing the workshop.

Virginio Cantoni, Vito Di Gesù, Alessandra Setti, and Domenico Tegolo

SPONSORING INSTITUTIONS

The following institutions are gratefully acknowledged for their contributions and support to the Workshop:

- Gruppo Nazionale di Cibernetica e Biofisica del C.N.R.
- Palermo University
- Pavia University
- Regione Siciliana
- Centro Interdipartimentale di Scienze Cognitive of Pavia University
- Centro Interdipartimentale di Tecnologie della Conoscenza of Palermo University

CONTENTS

GENERAL TALKS

From Computing with Numbers to Computing with Words – From Manipulation of Measurements to Manipulation of Perceptions 1
L.A. Zadeh

Processing of Biological Signals: A Bridge between Structure and Function 27
S. Cerutti, A.M. Bianchi, and M.G. Signorini

Architectures for Artificial Mentation and Behavior 41
I. Aleksander

Developing Rational Agents 51
M. Colombetti and P.L. Lanzi

Overview on Decision Making 67
G. Mosconi and L. Macchi

Panel Summary – Think about This: Is Thinking Necessary to Do Actions 81
J.D. Becker, M. Semenza, and A. Setti

THINKING

Learning from Mistakes 89
A. Giordana and A. Serra

Panel Summary – The Parallel between Acting and Thinking: Using it for Better Process Understanding 103
B. Zavidovique

Time-Effective Detection of Objects of Interest in Images by Means of a Visual Attention Mechanism 113
R.M. Palenichka and P. Zinterhof

Experiments on Concurrent Artificial Environment 123
V. Di Gesù, G. Lo Bosco, and D. Tegolo

Rational Decision: A Matter of Common Sense 131
M. Semenza

A Hierarchical Entropy Based Representation for Medical Signals............................ 137
C. Di Ruberto, R. Distasi, and S. Vitulano

Adaptation and Learning for Pattern Recognition:
A Comparison between Neural and Evolutionary Computation............................ 149
C. De Stefano, A. Della Cioppa, and A. Marcelli

DECIDING

Decision Making in Evolving Artificial Systems.. 159
R. Walker and M. Cardaci

Case Studies in Cognitive Robotics.. 167
L. Carlucci Aiello, D. Nardi, and F. Pirri

Panel Summary – Rational and Not Rational Behaviors in Decision-Making 183
R. Bordogna, V. Di Gesù, B. Zavidovique, and V. Cantoni

PLANNING

Biological Brain and Mind.. 189
E. Boncinelli

Visual Attention Systems... 197
V. Cantoni and M. Cellario

Panel Summary – Planning and Plasticity in Artificial and Natural Systems.............. 209
V. Di Gesù and S. Termini

ACTING

Notion Formation in Machine Learning.. 215
L. Saitta and R. Esposito

Towards Artificial Action: Teaching by Showing...................................... 225

T. Drummond and R. Cipolla

Panel Summary – Frontiers of Human-Machine Interaction 237
G. Rigamonti, N. Intrator, and H. Yeshurun

Index... 241

FROM COMPUTING WITH NUMBERS TO COMPUTING WITH WORDS - FROM MANIPULATION OF MEASUREMENTS TO MANIPULATION OF PERCEPTIONS*

Lotfi A. Zadeh

Berkeley Initiative in Soft Computing (BISC)
Computer Science Division and the Electronics Research Laboratory
Department of EECS, University of California
Berkeley, CA 94720-1776 USA
Telephone: 510-642-4959; Fax: 510-642-1712
e-mail: zadeh@cs.berkeley.edu
http://http.cs.berkeley.edu/People/Faculty/Homepages/zadeh.html

ABSTRACT

Computing, in its usual sense, is centered on manipulation of numbers and symbols. In contrast, computing with words, or CW for short, is a methodology in which the objects of computation are words and propositions drawn from a natural language, e.g., *small, large, far, heavy, not very likely, the price of gas is low and declining, Berkeley is near San Francisco, it is very unlikely that there will be a significant increase in the price of oil in the near future*, etc. Computing with words is inspired by the remarkable human capability to perform a wide variety of physical and mental tasks without any measurements and any computations. Familiar examples of such tasks are parking a car, driving in heavy traffic, playing golf, riding a bicycle, understanding speech and summarizing a story. Underlying this remarkable capability is the brain's crucial ability to manipulate perceptions - perceptions of distance, size, weight, color, speed, time, direction, force, number, truth, likelihood and other characteristics of physical and mental objects. Manipulation of perceptions plays a key role in human recognition, decision and execution processes. As a methodology, computing with words provides a foundation for a computational theory of perceptions - a theory which may have an important bearing on how humans make - and machines might make - perception-based rational decisions in an environment of imprecision, uncertainty and partial truth.

A basic difference between perceptions and measurements is that, in general, measurements are crisp whereas perceptions are fuzzy. One of the fundamental aims of

Human and Machine Perception 3: Thinking, Deciding, and Acting, **edited by Cantoni** ***et al.***
Kluwer Academic/Plenum Publishers, New York, 2001.

science has been and continues to be that of progressing from perceptions to measurements. Pursuit of this aim has led to brilliant successes. We have sent men to the moon; we can build computers that are capable of performing billions of computations per second; we have constructed telescopes that can explore the far reaches of the universe; and we can date the age of rocks that are millions of years old. But alongside the brilliant successes stand conspicuous underachievements and outright failures. We cannot build robots which can move with the agility of animals or humans; we cannot automate driving in heavy traffic; we cannot translate from one language to another at the level of a human interpreter; we cannot create programs which can summarize non-trivial stories; our ability to model the behavior of economic systems leaves much to be desired; and we cannot build machines that can compete with children in the performance of a wide variety of physical and cognitive tasks.

It may be argued that underlying the underachievements and failures is the unavailability of a methodology for reasoning and computing with perceptions rather than measurements. An outline of such a methodology - referred to as a computational theory of perceptions - is presented in this paper. The computational theory of perceptions, or CTP for short, is based on the methodology of computing with words (CW). In CTP, words play the role of labels of perceptions and, more generally, perceptions are expressed as propositions in a natural language. CW-based techniques are employed to translate propositions expressed in a natural language into what is called the Generalized Constraint Language (GCL). In this language, the meaning of a proposition is expressed as a generalized constraint, *X* isr R, where *X* is the constrained variable, *R* is the constraining relation and isr is a variable copula in which *r* is a variable whose value defines the way in which *R* constrains X. Among the basic types of constraints are: possibilistic, veristic, probabilistic, random set, Pawlak set, fuzzy graph and usuality. The wide variety of constraints in GCL makes GCL a much more expressive language than the language of predicate logic.

In CW, the initial and terminal data sets, IDS and TDS, are assumed to consist of propositions expressed in a natural language. These propositions are translated, respectively, into antecedent and consequent constraints. Consequent constraints are derived from antecedent constraints through the use of rules of constraint propagation. The principal constraint propagation rule is the generalized extension principle. The derived constraints are retranslated into a natural language, yielding the terminal data set (TDS). The rules of constraint propagation in CW coincide with the rules of inference in fuzzy logic. A basic problem in CW is that of explicitation of X, *R* and *r* in a generalized constraint, *X* isr R, which represents the meaning of a proposition, p, in a natural language.

There are two major imperatives for computing with words. First, computing with words is a necessity when the available information is too imprecise to justify the use of numbers; and second, when there is a tolerance for imprecision which can be exploited to achieve tractability, robustness, low solution cost and better rapport with reality. Exploitation of the tolerance for imprecision is an issue of central importance in CW and CTP. At this juncture, the computational theory of perceptions - which is based on CW - is in its initial stages of development. In time, it may come to play an important role in the conception, design and utilization of information/intelligent systems. The role model for CW and CTP is the human mind.

INTRODUCTION

In the fifties, and especially late fifties, circuit theory was at the height of importance and visibility. It played a pivotal role in the conception and design of electronic circuits and was enriched by basic contributions of Darlington, Bode, McMillan, Guillemin, Carlin, Youla, Kuh, Desoer, Sandberg and other pioneers.

However, what could be discerned at that time was that circuit theory was evolving into a more general theory - system theory - a theory in which the physical identity of the elements of a system is subordinated to a mathematical characterization of their input/output relations. This evolution was a step in the direction of greater generality and, like most generalizations, it was driven by a quest for models which make it possible to reduce the distance between an object that is modeled - the modelizand - and its model in a specified class of systems.

In a paper published in 1961 entitled "From Circuit Theory to System Theory,"[1] I discussed the evolution of circuit theory into system theory and observed that the high effectiveness of system theory in dealing with mechanistic systems stood in sharp contrast to its low effectiveness in the realm of humanistic systems - systems exemplified by economic systems, biological systems, social systems, political systems and, more generally, man-machine systems of various types. In more specific terms, I wrote:

<< There is a fairly wide gap between what might be regarded as "animate" system theorists and 'inanimate' system theorists at the present time, and it is not at all certain that this gap will be narrowed, much less closed, in the near future. There are some who feel that this gap reflects the fundamental inadequacy of conventional mathematics - the mathematics of precisely-defined points, functions, sets, probability measures, etc. - for coping with the analysis of biological systems, and that to deal effectively with such systems, which are generally orders of magnitude more complex than man-made systems, we need a radically different kind of mathematics, the mathematics of fuzzy or cloudy quantities which are not describable in terms of probability distributions. Indeed, the need for such mathematics is becoming increasingly apparent even in the realm of inanimate systems, for in most practical cases the *a priori* data as well as the criteria by which the performance of a man-made system are judged are far from being precisely specified or having accurately-known probability distributions. >>

It was this observation that motivated my development of the theory of fuzzy sets, starting with the 1965 paper "Fuzzy Sets",[2] which was published in *Information and Control*.

Subsequently, in a paper published in 1973, "Outline of a New Approach to the Analysis of Complex Systems and Decision Processes,"[3] I introduced the concept of a linguistic variable, that is, a variable whose values are words rather than numbers. The concept of a linguistic variable has played and is continuing to play a pivotal role in the development of fuzzy logic and its applications.

The initial reception of the concept of a linguistic variable was far from positive, largely because my advocacy of the use of words in systems and decision analysis clashed with the deep-seated tradition of respect for numbers and disrespect for words. The essence of this tradition was succinctly stated in 1883 by Lord Kelvin:

<< In physical science the first essential step in the direction of learning any subject is to find principles of numerical reckoning and practicable methods for measuring some quality connected with it. I often say that when you can measure what you are speaking about and express it in numbers, you know something about it; but when you cannot measure it, when you cannot express it in numbers, your knowledge is of a meagre and unsatisfactory kind: it may be the beginning of knowledge but you have scarcely, in your thoughts, advanced to the state of science, whatever the matter may be. >>

The depth of scientific tradition of respect for numbers and derision for words was reflected in the intensity of hostile reaction to my ideas by some of the prominent members of the scientific elite. In commenting on my first exposition of the concept of a linguistic variable in 1972, Rudolph Kalman had this to say:

<< I would like to comment briefly on Professor Zadeh's presentation. His proposals could be severely, ferociously, even brutally criticized from a technical point of view. This would be out of place here. But a blunt question remains: Is Professor Zadeh presenting important ideas or is he indulging in wishful thinking? No doubt Professor Zadeh's

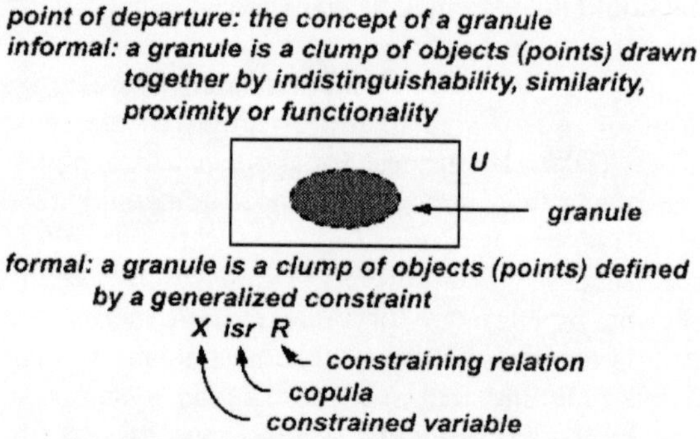

Figure 1. Informal and formal definitions of a granule.

enthusiasm for fuzziness has been reinforced by the prevailing climate in the U.S. - one of unprecedented permissiveness. 'Fuzzification' is a kind of scientific permissiveness; it tends to result in socially appealing slogans unaccompanied by the discipline of hard scientific work and patient observation. >>

In a similar vein, my esteemed colleague Professor William Kahan - a man with a brilliant mind - offered this assessment in 1975:

<< "Fuzzy theory is wrong, wrong, and pernicious." says William Kahan, a professor of computer sciences and mathematics at Cal whose Evans Hall office is a few doors from Zadeh's. "I can not think of any problem that could not be solved better by ordinary logic." What Zadeh is saying is the same sort of things 'Technology got us into this mess and now it can't get us out.' Well, technology did not get us into this mess. Greed and weakness and ambivalence got us into this mess. What we need is more logical thinking, not less. The danger of fuzzy theory is that it will encourage the sort of imprecise thinking that has brought us so much trouble." >>

What Lord Kelvin, Rudolph Kalman, William Kahan and many other brilliant minds did not appreciate is the fundamental importance of the remarkable human capability to perform a wide variety of physical and mental tasks without any measurements and any computations. Familiar examples of such tasks are parking a car; driving in heavy traffic; playing golf; understanding speech and summarizing a story.

Underlying this remarkable ability is the brain's crucial ability to manipulate perceptions - perceptions of size, distance, weight, speed, time, direction, smell, color, shape, force, likelihood, truth and intent, among others. A fundamental difference between measurements and perceptions is that, in general, measurements are crisp numbers whereas perceptions are fuzzy numbers or, more generally, fuzzy granules, that is, clumps of objects in which the transition from membership to nonmembership is gradual rather than abrupt.

The fuzziness of perceptions reflects finite ability of sensory organs and the brain to resolve detail and store information. A concomitant of fuzziness of perceptions is the preponderant partiality of human concepts in the sense that the validity of most human concepts is a matter of degree. For example, we have partial knowledge, partial understanding, partial certainty, partial belief and accept partial solutions, partial truth and partial causality. Furthermore, most human concepts have a granular structure and are context-dependent.

In essence, a granule is a clump of physical or mental objects (points) drawn together by indistinguishability, similarity, proximity or functionality (Fig. 1). A granule may be crisp or fuzzy, depending on whether its boundaries are or are not sharply defined. For example, age may be granulated crisply into years and granulated fuzzily into fuzzy intervals labeled very young, young, middle-aged, old and very old (Fig. 2). A partial taxonomy of granulation is shown in Figures 3a and 3b.

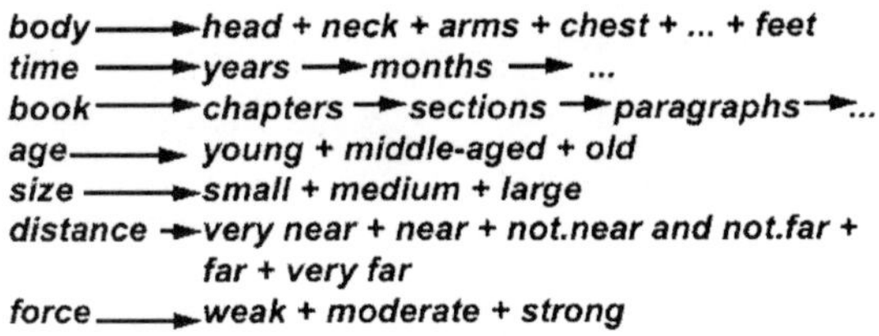

Figure 2. Examples of crisp and fuzzy granulation.

In a very broad sense, granulation involves a partitioning of whole into parts. Modes of information granulation (IG) in which granules are crisp play important roles in a wide variety of methods, approaches and techniques. Among them are: interval analysis, quantization, chunking, rough set theory, diakoptics, divide and conquer, Dempster-Shafer theory, machine learning from examples, qualitative process theory, decision trees, semantic networks, analog-to-digital conversion, constraint programming, image segmentation, cluster analysis and many others.

Important though it is, crisp IG has a major blind spot. More specifically, it fails to reflect the fact that most human perceptions are fuzzy rather than crisp. For example, when we mentally granulate the human body into fuzzy granules labeled head, neck, chest, arms, legs, etc., the length of neck is a fuzzy attribute whose value is a fuzzy number. Fuzziness of granules, their attributes and their values is characteristic of ways in which human concepts are formed, organized and manipulated. In effect, fuzzy information granulation (fuzzy IG) may be viewed as a human way of employing data compression for reasoning and, more particularly, making rational decisions in an environment of imprecision, uncertainty and partial truth.

The tradition of pursuit of crispness and precision in scientific theories can be credited with brilliant successes. We have sent men to the moon; we can build computers that are capable of performing billions of computations per second; we have constructed telescopes that can explore the far reaches of the universe; and we can date the age of rocks that are millions of years old. But alongside the brilliant successes stand conspicuous underachievements and outright failures. We cannot build robots which can move with the agility of animals or humans; we cannot automate driving in heavy traffic; we cannot translate from one language to another at the level of a human interpreter; we cannot create

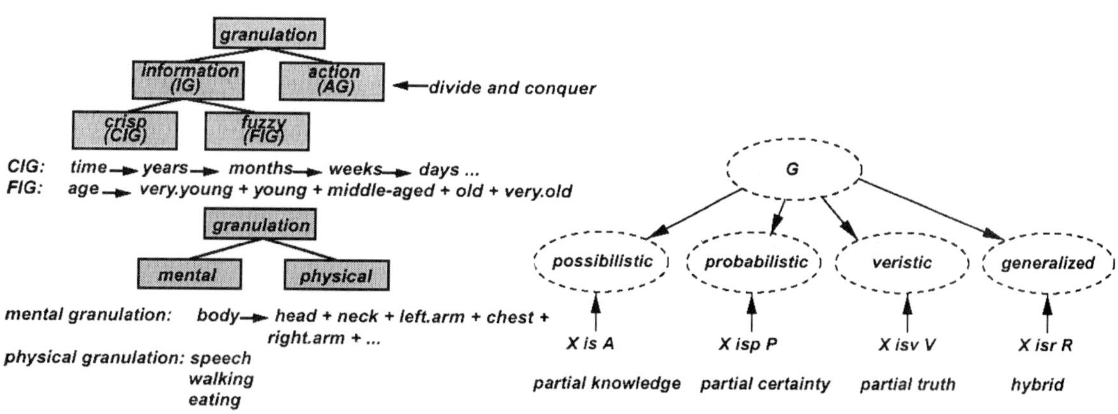

Figure 3. (a) Partial taxonomy of granulation. **(b)** Principal types of granules.

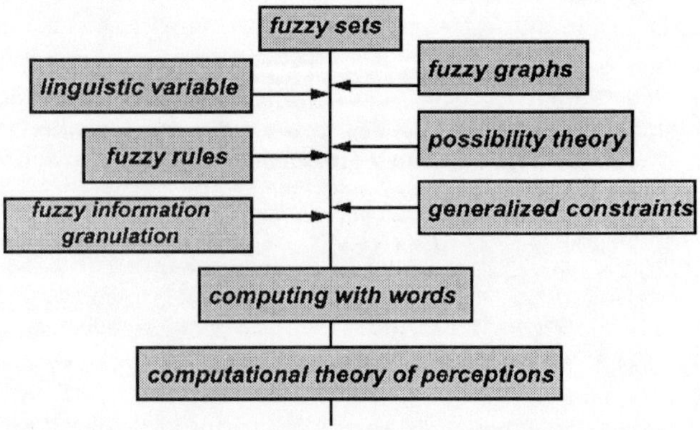

Figure 4. Conceptual structure of computational theory of perceptions.

programs which can summarize non-trivial stories; our ability to model the behavior of economic systems leaves much to be desired; and we cannot build machines that can compete with children in the performance of a wide variety of physical and cognitive tasks.

What is the explanation for the disparity between the successes and failures? What can be done to advance the frontiers of science and technology beyond where they are today, especially in the realms of machine intelligence and automation of decision processes? In my view, the failures are conspicuous in those areas in which the objects of manipulation are, in the main, perceptions rather than measurements. Thus, what we need are ways of dealing with perceptions, in addition to the many tools which we have for dealing with measurements. In essence, it is this need that motivated the development of the methodology of computing with words (CW) - a methodology in which words play the role of labels of perceptions.

Computing with words provides a methodology for what may be called a *computational theory of perceptions* (CTP) (Fig. 4). However, the potential impact of the methodology of computing with words is much broader. Basically, there are four principal rationales for the use of CW.

1) *The don't know rationale*. In this case, the values of variables and/or parameters are not known with sufficient precision to justify the use of conventional methods of numerical computing. An example is decision-making with poorly defined probabilities and utilities.

2) *The don't need rationale*. In this case, there is a tolerance for imprecision which can exploited to achieve tractability, robustness, low solution cost and better rapport with reality. An example is the problem of parking a car.

3) *The can't solve rationale*. In this case, the problem cannot be solved through the use of numerical computing. An example is the problem of automation of driving in city traffic.

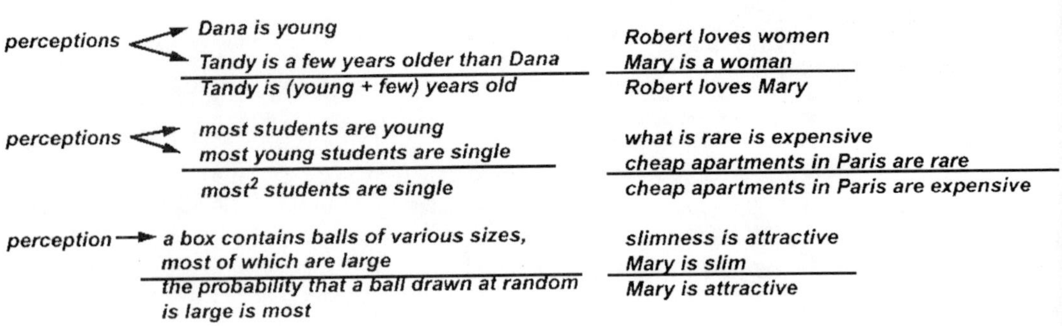

Figure 5. (a) Examples of reasoning with perceptions. **(b)** Examples of incorrect reasoning.

4) *The can't define rationale*. In this case, a concept that we wish to define is too complex to admit of definition in terms of a set of numerical criteria. A case in point is concept of causality. Causality is an instance of what may be called an amorphic concept.

The basic idea underlying the relationship between CW and CTP is conceptually simple. More specifically, in CTP perceptions and queries are expressed as propositions in a natural language. Then, propositions and queries are processed by CW-based methods to yield answers to queries. Simple examples of linguistic characterization of perceptions drawn from everyday experiences are:

Robert is highly intelligent
Carol is very attractive
Hans loves wine
Overeating causes obesity
Most Swedes are tall
Berkeley is more lively than Palo Alto
It is likely to rain tomorrow
It is very unlikely that there will be a significant increase in the price of oil in the near future

Examples of correct conclusions drawn from perceptions through the use of CW-based methods are shown in Figure 5a. Examples of incorrect conclusions are shown in Figure 5b.

Perceptions have long been an object of study in psychology. However, the idea of linking perceptions to computing with words is in a different spirit. An interesting system-theoretic approach to perceptions is described in a recent work of R. Vallée.[4] A logic of perceptions has been described by H. Rasiowa.[5] These approaches are not related to the approach described in our paper.

An important point that should be noted is that classical logical systems such as propositional logic, predical logic and modal logic, as well as AI-based techniques for natural language processing and knowledge representation, are concerned in a fundamental way with propositions expressed in a natural language. The main difference between such approaches and CW is that the methodology of CW - which is based on fuzzy logic - provides a much more expressive language for knowledge representation and much more versatile machinery for reasoning and computation.

In the final analysis, the role model for computing with words is the human mind and its remarkable ability to manipulate both measurements and perceptions. What should be stressed, however, is that although words are less precise than numbers, the methodology of computing with words rests on a mathematical foundation. An exposition of the basic concepts and techniques of computing with words is presented in the following sections. The linkage of CW and CTP is discussed very briefly because the computational theory of perceptions is still in its early stages of development.

WHAT IS CW?

In its traditional sense, computing involves for the most part manipulation of numbers and symbols. By contrast, humans employ mostly words in computing and reasoning, arriving at conclusions expressed as words from premises expressed in a natural language or having the form of mental perceptions. As used by humans, words have fuzzy denotations. The same applies to the role played by words in CW.

The concept of CW is rooted in several papers starting with my 1973 paper "Outline of a New Approach to the Analysis of Complex Systems and Decision Processes,"[3] in which the concepts of a linguistic variable and granulation were introduced. The concepts of a fuzzy constraint and fuzzy constraint propagation were introduced in "Calculus of Fuzzy Restrictions,"[6] and developed more fully in "A Theory of Approximate Reasoning,"[7] and "Outline of a Computational Approach to Meaning and Knowledge Representation Based

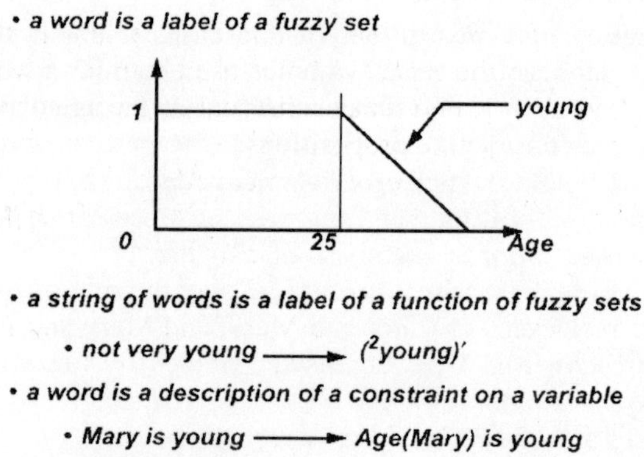

Figure 6. Words as labels of fuzzy sets.

on a Concept of a Generalized Assignment Statement."[8] Application of fuzzy logic to meaning representation and its role in test-score semantics are discussed in "PRUF - A Meaning Representation Language for Natural Languages,"[9] and "Test-Score Semantics for Natural Languages and Meaning Representation via PRUF."[10] The close relationship between CW and fuzzy information granulation is discussed in "Toward a Theory of Fuzzy Information Granulation and its Centrality in Human Reasoning and Fuzzy Logic[11]."

Although the foundations of computing with words were laid some time ago, its evolution into a distinct methodology in its own right reflects many advances in our understanding of fuzzy logic and soft computing - advances which took place within the past few years. (See References and Related Papers.) A key aspect of CW is that it involves a fusion of natural languages and computation with fuzzy variables. It is this fusion that is likely to result in an evolution of CW into a basic methodology in its own right, with wide-ranging ramifications and applications.

We begin our exposition of CW with a few definitions. It should be understood that the definitions are dispositional, that is, admit of exceptions.

As was stated earlier, a concept which plays a pivotal role in CW is that of a granule. Typically, a granule is a fuzzy set of points drawn together by similarity (Fig. 1). A word may be atomic, as in *young*, or composite, as in *not very young* (Fig. 6). Unless stated to the contrary, a word will be assumed to be composite. The denotation of a word may be a higher order predicate, as in Montague grammar.[12,13]

In CW, a granule, g, which is the denotation of a word, w, is viewed as a fuzzy constraint on a variable. A pivotal role in CW is played by fuzzy constraint propagation from premises to conclusions. It should be noted that, as a basic technique, constraint propagation plays important roles in many methodologies, especially in mathematical programming, constraint programming and logic programming. (See References and Related Papers.)

As a simple illustration, consider the proposition *Mary is young*, which may be a

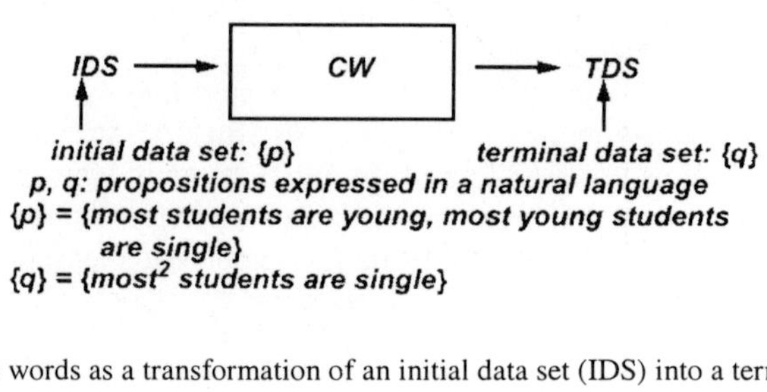

Figure 7. Computing with words as a transformation of an initial data set (IDS) into a terminal data set (TDS).

linguistic characterization of a perception. In this case, *young* is the label of a granule young. (Note that for simplicity the same symbol is used both for a word and its denotation.) The fuzzy set *young* plays the role of a fuzzy constraint on the age of Mary (Fig. 6).

As a further example consider the propositions

$$p1 = Carol\ lives\ near\ Mary$$

and

$$p2 = Mary\ lives\ near\ Pat.$$

In this case, the words *lives near* in *p1* and *p2* play the role of fuzzy constraints on the distances between the residences of Carol and Mary, and Mary and Pat, respectively. If the query is: How far is Carol from Pat?, an answer yielded by fuzzy constraint propagation might be expressed as *p*3, where

$$p3 = Carol\ lives\ not\ far\ from\ Pat.$$

More about fuzzy constraint propagation will be said at a later point.

A basic assumption in CW is that information is conveyed by constraining the values of variables. Furthermore, information is assumed to consist of a collection of propositions expressed in natural or synthetic language. Typically, such propositions play the role of linguistic characterization of perceptions.

A basic generic problem in CW is the following.

We are given a collection of propositions expressed in a natural language which constitute the *initial data se*t, or IDS for short.

From the initial data set we wish to infer an answer to a query expressed in a natural language. The answer, also expressed in a natural language, is referred to as the *terminal data se*t, or TDS for short. The problem is to derive TDS from IDS (Fig. 7).

A few problems will serve to illustrate these concepts. At this juncture, the problems will be formulated by not solved.

1) Assume that a function *f*, $f : U \rightarrow V, X \in U, Y \in V,$ is described in words by the fuzzy if-then rules

f: if *X* is *small* then *Y* is *small*
if *X* is *medium* then *Y* is *large*
if *X* is *large* then *Y* is *small*

What this implies is that *f* is approximated to by a fuzzy graph *f** (Fig. 8), where

$$f^* = small \times small + medium \times large + large \times small$$

In *f**, + and × denote respectively, the disjunction and cartesian product. An expression of the form $A \times B$, where *A* and *B* are words, will be referred to as a *Cartesian granule*. In this sense, a fuzzy graph may be viewed as a disjunction of cartesian granules. In essence, a

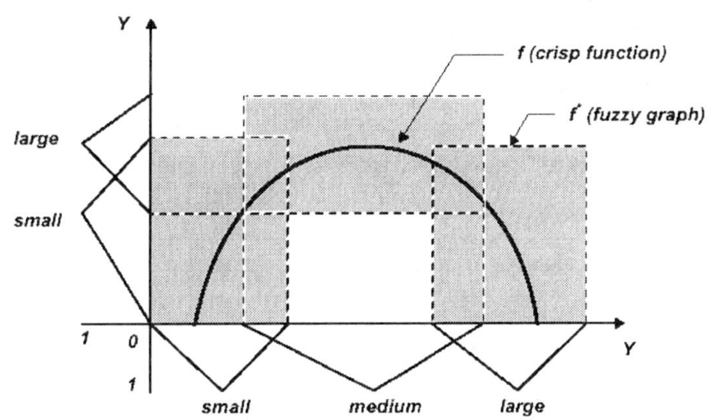

Figure 8. Fuzzy graph of a function.

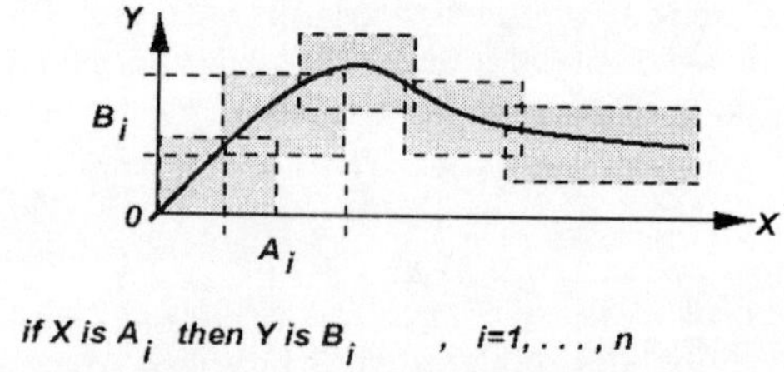

Figure 9. A fuzzy graph of a function represented by a rule-set.

fuzzy graph serves as an approximation to a function or a relation.[14,15] Equivalently, it may be viewed as a linguistic characterization of a perception of (Fig. 9).

In the example under consideration, the IDS consists of the fuzzy rule-set f. The query is: What is the maximum value of f (Fig. 10)? More broadly, the problem is: How can one compute an attribute of a function, f, e.g., its maximum value or its area or its roots if is described in words as a collection of fuzzy if-then rules? Determination of the maximum value will be discussed in greater detail at a later point.

2) A box contains ten balls of various sizes of which several are large and a few are small. What is the probability that a ball drawn at random is neither large nor small? In this case, the IDS is a verbal description of the contents of the box; the TDS is the desired probability.

3) A less simple example of computing with words is the following.

Let X and Y be independent random variables taking values in a finite set $V = \{v_1,\dots,v_n\}$ with probabilities $p_1,\dots p_n$ and $q_1,\dots q_n$, respectively. For simplicity of notation, the same symbols will be used to denote X and Y and their generic values, with p and q denoting the probabilities of X and Y, respectively.

Assume that the probability distributions of X and Y are described in words through the fuzzy if-then rules (Fig. 11):

P: if X is *small* then p is *small*
if X is *medium* then p is *large*
if X is *large* then p is *small*

and

Q: if Y is *small* then q is *large*
if Y is *medium* then q is *small*
if Y is *large* then q is *large*

where the granules *small*, *medium* and *large* are values of linguistic variables X and Y in their respective universes of discourse. In the example under consideration, these rule-sets

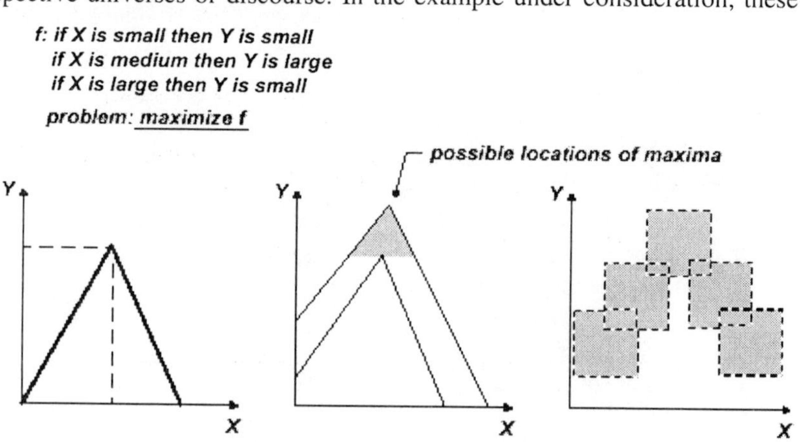

Figure 10. Fuzzy graph of a function defined by a fuzzy rule-set.

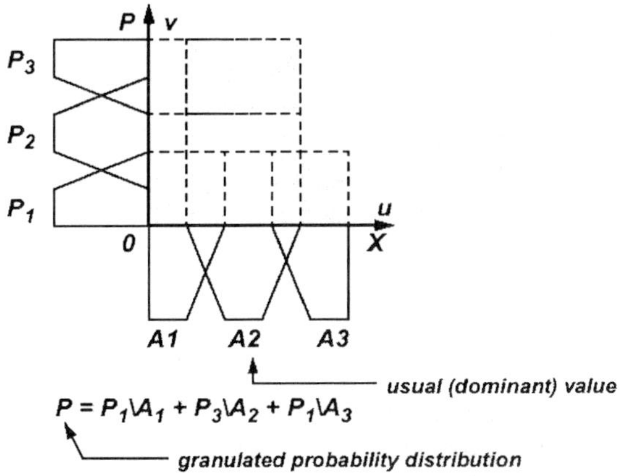

Figure 11. A fuzzy graph representation of a granulated probability distribution.

constitute the IDS. Note that *small* in *P* need not have the same meaning as *small* in Q, and likewise for *medium* and *large*.

The query is: How can we describe in words the joint probability distribution of *X* and *Y*? This probability distribution is the TDS.

For convenience, the probability distributions of *X* and *Y* may be represented as fuzzy graphs:

P: *small* × *small* + *medium* × *large* + *large* × *small*
Q: *small* × *large* + *medium* × *small* + *large* × *large*

with the understanding that the underlying numerical probabilities must add up to unity.

Since *X* and *Y* are independent random variables, their joint probability distribution (*P*,Q) is the product of *P* and Q. In words, the product may be expressed as:[15]

(*P*,Q): *small* × *small* × (*small* ∗ *large*) + *small* × *medium* × (*small* ∗ *small*) + *small* × *large* × (*small* ∗ *large*) + ...+ *large* × *large large* × (*small* ∗ *large*)

where ∗ is the arithmetic product in fuzzy arithmetic.[16] In this example, what we have done, in effect, amounts to a derivation of a linguistic characterization of the joint probability distribution of *X* and *Y* starting with linguistic characterizations of the probability distribution of *X* and the probability distribution of Y.

A few comments are in order. In linguistic characterizations of variables and their dependencies, words serve as values of variables and play the role of fuzzy constraints. In this perspective, the use of words may be viewed as a form of granulation, which in turn may be regarded as a form of fuzzy quantization.

Granulation plays a key role in human cognition. For humans, it serves as a way of achieving data compression. This is one of the pivotal advantages accruing through the use of words in human, machine and man-machine communication.

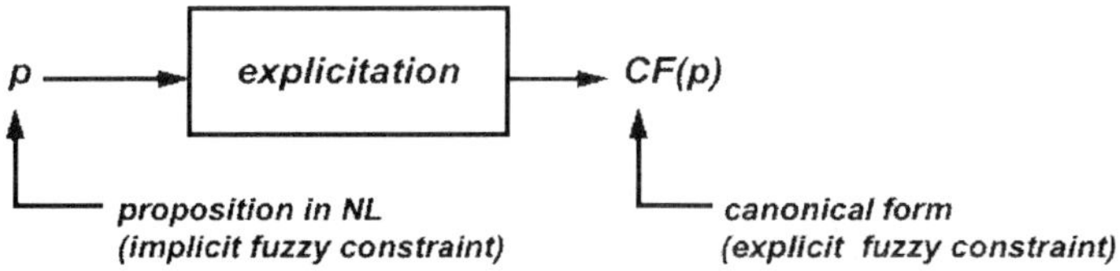

Figure 12. Canonical form of a proposition.

The point of departure in CW is the premise that the meaning of a proposition, p, in a natural language may be represented as an implicit constraint on an implicit variable. Such a representation is referred to as a *canonical form* of p, denoted as *CF* (Fig. 12). Thus, a canonical form serves to make explicit the implicit constraint which resides in p. The concept of a canonical form is described in greater detail in the following section.

As a first step in the derivation of TDS from IDS, propositions in IDS are translated into their canonical forms, which collectively represent *antecedent* constraints. Through the use of rules for constraint propagation, antecedent constraints are transformed into *consequent* constraints. Finally, consequent constraints are translated into a natural language through the use of *linguistic approximation*,[17,18] yielding the terminal data set TDS. This process is schematized in Fig. 13.

In essence, the rationale for computing with words rests on two major imperatives. First, computing with words is a necessity when the available information is too imprecise to justify the use of numbers. And second, when there is a tolerance for imprecision which can be exploited to achieve tractability, robustness, low solution cost and better rapport with reality.

In computing with words, there are two core issues that arise. First is the issue of representation of fuzzy constraints. More specifically, the question is: How can the fuzzy constraints which are implicit in propositions expressed in a natural language be made explicit. And second is the issue of fuzzy constraint propagation, that is, the question of how can fuzzy constraints in premises, i.e., antecedent constraints, be propagated to conclusions, i.e., consequent constraints.

These are the issues which are addressed in the following.

REPRESENTATION OF FUZZY CONSTRAINTS, CANONICAL FORMS, AND GENERALIZED CONSTRAINTS

Our approach to the representation of fuzzy constraints is based on test-score semantics.[10,19] In outline, in this semantics, a proposition, p, in a natural language is viewed as a network of fuzzy (elastic) constraints. Upon aggregation, the constraints which are embodied in *p* result in an overall fuzzy constraint which can be represented as an expression of the form

$$X \text{ is } R$$

where *R* is a constraining fuzzy relation and *X* is the constrained variable. The expression in question is the canonical form of p. Basically, the function of a canonical form is to place in evidence the fuzzy constraint which is implicit in p. This is represented schematically as

$$P \rightarrow X \text{ is } R$$

in which the arrow → denotes explicitation. The variable *X* may be vector-valued and/or conditioned.

In this perspective, the meaning of *p* is defined by two procedures. The first procedure acts on a so-called explanatory database, ED, and returns the constrained variable, X. The second procedure acts on ED and returns the constraining relation, R.

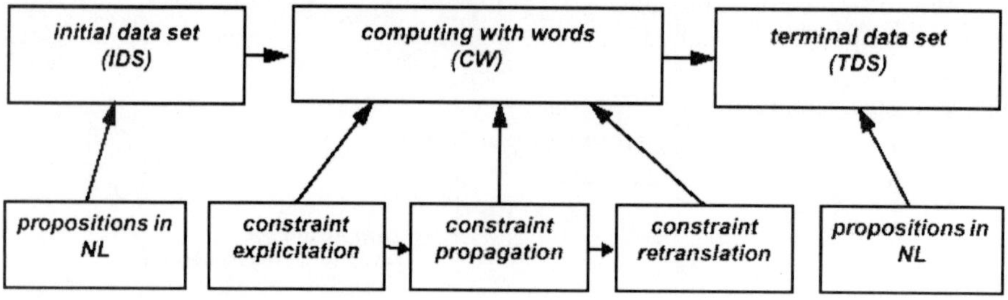

Figure 13. Conceptual structure of computing with words.

An explanatory database is a collection of relations in terms of which the meaning of p is defined. The relations are empty, that is, they consist of relation names, relations attributes and attribute domains, with no entries in the relations. When there are entries in ED, ED is said to be *instantiated* and is denoted EDI. EDI may be viewed as a description of a possible world in possible world semantics,[20] while ED defines a collection of possible worlds, with each possible world in the collection corresponding to a particular instantiation of ED.[19]

As a simple illustration, consider the proposition

$$p = \textit{Mary is not young}.$$

Assume that the explanatory database is chosen to be

$$ED = \text{POPULATION}[\text{ Name; Age }] + \text{YOUNG}[\text{ Age; } \mu\ ]$$

in which POPULATION is a relation with arguments Name and Age; YOUNG is a relation with arguments Age and μ; and + is the disjunction. In this case, the constrained variable is the age of Mary, which in terms of ED may be expressed as

$$X = \text{Age(Mary)} =_{\text{Age}} \text{POPULATION}[\text{ Name=Mary }].$$

This expression specifies the procedure which acts on ED and returns X. More specifically, in this procedure, Name is instantiated to Mary and the resulting relation is projected on Age, yielding the age of Mary.

The constraining relation, R, is given by

$$R = ({}^2\text{YOUNG})'$$

which implies that the intensifier *very* is interpreted as a squaring operation, and the negation *not* as the operation of complementation.[21]

Equivalently, R may be expressed as

$$R = \text{YOUNG [Age; } 1 - \mu^2\text{]}.$$

As a further example, consider the proposition

$$p = \textit{Carol lives in a small city near San Francisco}$$

and assume that the explanatory database is:

$$ED = \text{POPULATION[Name;Residence]} + \text{SMALL[City;}\mu] + \text{NEAR[City1;City2;}\mu]$$

In this case,

$$X = \text{Residence(Carol)} =_{\text{Residence}} \text{POPULATION[Name=Carol]}$$

and

$$R = \textit{SMALL}\ [\ \textit{City},\ \mu\ ] \cap_{\text{City1}} \text{NEAR}\ [\textit{City}2 = \text{San\_Francisco}]$$

In R, the first constituent is the fuzzy set of small cities; the second constituent is the fuzzy set of cities which are near San Francisco; and $\cap$ denotes the intersection of these sets.

So far we have confined our attention to constraints of the form

$$X \text{ is } R$$

In fact, constraints can have a variety of forms. In particular, a constraint - expressed as a canonical form – may be conditional, that is, of the form

$$\text{if } X \text{ is } R \text{ then } Y \text{ is S}$$

which may also be written as

$$Y \text{ is } S \text{ if } X \text{ is R}.$$

The constraints in question will be referred to as *basic*.

For purposes of meaning representation, the richness of natural languages necessitates a wide variety of constraints in relation to which the basic constraints form an important though special class. The so-called generalized constraints[8] contain the basic constraints as a special case and are defined as follows. The need for generalized constraints becomes obvious when one attempts to represent the meaning of simple propositions such as

Robert loves women
John is very honest
checkout time is 11 am
slimness is attractive

in the language of standard logical systems.

A generalized constraint is represented as

X isr R

where isr, pronounced "ezar", is a variable copula which defines the way in which R constrains X. More specifically, the role of R in relation to X is defined by the value of the discrete variable r. The values of r and their interpretations are defined below:

e : equal (abbreviated to =);
d : disjunctive (possibilistic) (abbreviated to blank);
ν: veristic;
p : probabilistic;
γ: probability value;
u : usuality;
rs : random set;
rfs: random fuzzy set;
fg: fuzzy graph;
ps: rough set (Pawlak set);

As an illustration, when r = e, the constraint is an equality constraint and is abbreviated to =. When r takes the value d, the constraint is *disjunctive* (possibilistic) and isd abbreviated to is, leading to the expression

X is R

in which R is a fuzzy relation which constrains X by playing the role of the possibility distribution of X. More specifically, if X takes values in a universe of discourse, $U = \{ u \}$, then $Poss\{ X = u \} = \mu_R(u)$, where μ_R is the membership function of R, and Π_X is the possibility distribution of X, that is, the fuzzy set of its possible values.[22] In schematic form:

$$XisR \begin{cases} \Pi_X = R \\ Poss\{X = u\} = \mu_R(u) \end{cases}$$

Similarly, when r takes the value ν, the constraint is *veristic*. In the case,

X isv R

means that if the grade of membership of u in R is μ, then X=u has truth value μ. For example, a canonical form of the proposition

p = *John is proficient in English, French and German*

may be expressed as

Proficiency (John) isv (1|English + 0.7|French + 0.6|German)

in which 1.0, 0.7 and 0.6 represent, respectively, the truth values of the propositions *John is proficient in English, John is proficient in French and John is proficient in German*. In a similar vein, the veristic constraint

Ethnicity (John) isv (0.5|German + 0.25|French + 0.25|Italian)

represents the meaning of the proposition *John is half German, quarter French and quarter Italian*.

When r = p, the constraint is *probabilistic*. In this case,

X isp R

means that R is the probability distribution of X. For example

X isp N(m, σ^2)

means that X is normally distributed with mean m and variance σ^2. Similarly,

X isp (0.2\\$a$ + 0.5\\$b$ + 0.3\\c)

means that X is a random variable which takes the values, a, b and c with respective probabilities 0.2, 0.5 and 0.3.

The constraint

X isu R

is an abbreviation for

usually(X is R)

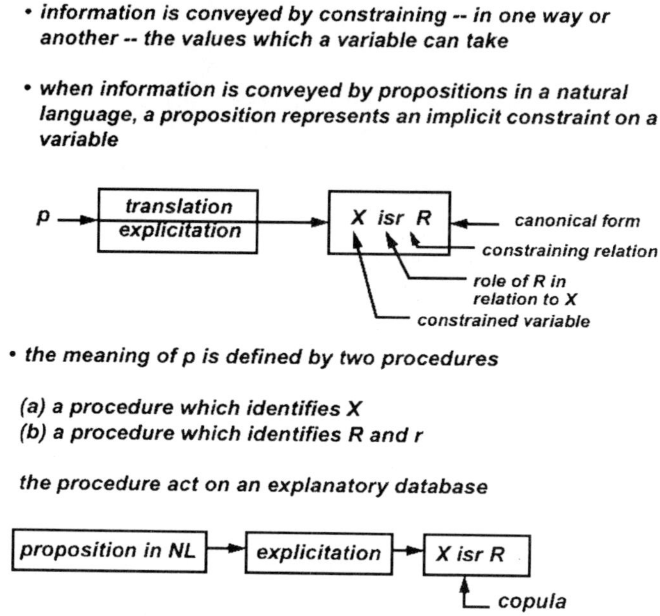

Figure 14. Representation of meaning in test-score semantics.

which in turn means that

$$Prob\{X \text{ is R}\} \text{ is } usually.$$

In this expression *X* is *R* is a fuzzy event and *usually* is its fuzzy probability, that is, the possibility distribution of its crisp probability.

The constraint

$$X \text{ isrs } P$$

is a random set constraint. This constraint is a combination of probabilistic and possibilistic constraints. More specifically, in a schematic form, it is expressed as

$$\begin{array}{c} X \text{ isp } P \\ (X,Y) \text{ is } Q \\ \hline Y \text{ isrs } R \end{array}$$

where *Q* is a joint possibilitistic constraint on *X* and Y, and *R* is a random set. It is of interest to note that the Dempster-Shafer theory of evidence[23] is, in essence, a theory of random set constraints.

In computing with words, the starting point is a collection of propositions which play the role of premises. In many cases, the canonical forms of these propositions are constraints of the basic, possibilistic type. In a more general setting, the constraints are of the generalized type, implying that explicitation of a proposition, p, may be represented as

$$p \rightarrow X \text{ isr } R$$

where *X* isr *R* is the canonical form of *p* (Fig. 14).

As in the case of basic constraints, the canonical form of a proposition may be derived through the use of test-score semantics. In this context, the depth of *p* is, roughly, a measure of the effort that is needed to explicitate p, that is, to translate *p* into its canonical form. In this sense, the proposition *X* isr *R* is a surface constraint (depth=zero), with the depth of explicitation increasing in the downward direction (Fig. 15). Thus a proposition such as *Mary is young* is shallow, whereas *it is unlikely that there will be a substantial increase in the price of oil in the near future*, is not.

Once the propositions in the initial data set are expressed in their canonical forms, the groundwork is laid for fuzzy constraint propagation. This is a basic part of CW which is discussed in the following section.

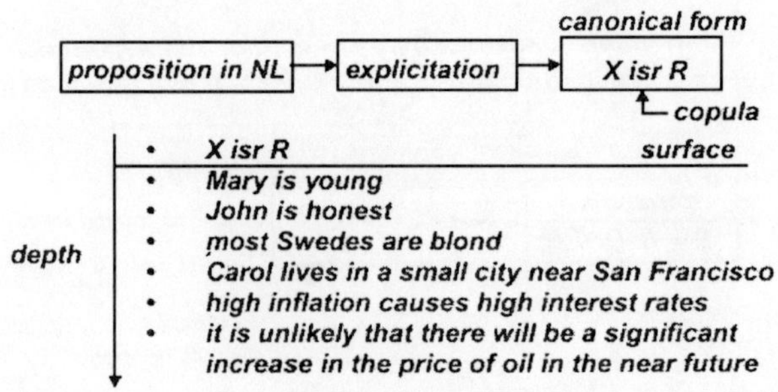

Figure 15. Depth of explicitation.

FUZZY CONSTRAINT PROPAGATION AND THE RULES OF INFERENCE IN FUZZY LOGIC

The rules governing fuzzy constraint propagation are, in effect, the rules of inference in fuzzy logic. In addition to these rules, it is helpful to have rules governing fuzzy constraint modification, The latter rules will be discussed at a later point in this section.

In a summarized form, the rules governing fuzzy constraint propagation are the following.[15] (*A* and *B* are fuzzy relations. Disjunction and conjunction are defined, respectively, as max and min, with the understanding that, more generally, they could be defined via t-norms and *s*-norms.[24,25] The antecedent and consequent constraints are separated by a horizontal line.)

Conjunctive Rule 1:

$$\frac{X \text{ is } A \qquad X \text{ is } B}{X \text{ is } A \cap B}$$

Conjunctive Rule 2: $(X \in \mathrm{U}, Y \in \mathrm{B}, A \subset \mathrm{U}, B \subset \mathrm{V})$

$$\frac{X \text{ is } A \qquad Y \text{ is } B}{(X,Y) \text{ is } A \times B}$$

Disjunctive Rule 1:

$$\frac{X \text{ is } A \quad \text{or} \quad X \text{ is } B}{X \text{ is } A \cup B}$$

Disjunctive Rule 2: $(A \subset \mathrm{U}, B \subset \mathrm{V})$

$$\frac{A \text{ is } A \qquad Y \text{ is } B}{(X,Y) \text{ is } A \times V \cup U \times B}$$

where $A \times V$ and $U \times B$ are cylindrical extensions of *A* and B, respectively.

Conjunctive Rule for isv:

$$\frac{X \text{ isv } A \qquad X \text{ isv } B}{X \text{ isv } A \cup B}$$

Projective Rule:

$$\frac{(X,Y) \text{ is } A}{Y \text{ is } proj_V\, A}$$

where $proj_V\, A = \sup_u A$

Surjective Rule:

$$\frac{X \text{ is } A}{(X,Y) \text{ is } A \times V}$$

Derived Rules:

Compositional Rule:

$$\frac{X \text{ is } A \quad (X,Y) \text{ is } B}{Y \text{ is } A \circ B}$$

where $A \circ B$ denotes the composition of A and B.

Extension Principle (mapping rule)[2,26]:

$$\frac{X \text{ is } A}{f(X) \text{ is } f(A)}$$

where $f: U \rightarrow V$, and $f(A)$ is defined by $\mu_{f(A)}(v) = \sup_{u \mid v = f(u)} \mu_A(u)$

Inverse Mapping Rule:

$$\frac{f(X) \text{ is } A}{X \text{ is } f^{-1}(A)}$$

where $\mu_{f^{-1}(A)}(u) = \mu_{A(f(u))}$

Generalized modus ponens:

$$\frac{X \text{ is } A \quad \text{if } X \text{ is } B \text{ then } Y \text{ is } C}{Y \text{ is } A \circ ((\neg B) \oplus C)}$$

where the bounded sum $\neg B \oplus C$ represents Lukasiewicz's definition of implication.

Generalized Extension Principle :

$$\frac{f(X) \text{ is } A}{q(X) \text{ is } q(f^{1}(A))}$$

where $\mu_q(v) = \sup_{u \mid v = q(u)} \mu_A(f(u))$.

The generalized extension principle plays a pivotal role in fuzzy constraint propagation. However, what is used most frequently in practical applications of fuzzy logic is the *basic interpolative rule*, which is a special case of the compositional rule of inference applied to a function which is defined by a fuzzy graph.[14,15] More specifically, if f is defined by a fuzzy rule set

$$f\text{: if } X \text{ is } A_i \text{ then } X \text{ is } B_i,\ i=1, \ldots, n$$

or equivalently, by a fuzzy graph

$$f \text{ is } \sum_i A_i \times B_i$$

and its argument, X, is defined by the antecedent constraint

$$X \text{ is } A$$

then the consequent constraint on Y may be expressed as

$$Y \text{ is } \sum_i m_i \wedge B_i$$

where m_i is a matching coefficient,

$$mi = \sup(A_i \cap A)$$

which serves as a measure of the degree to which A matches A_i.

Syllogistic Rule:[27]

$$\frac{\begin{array}{l} Q_1\ A\text{'s are } B\text{'s} \\ Q_2\ (A \text{ and } B)\text{'s are } C\text{'s} \end{array}}{(Q_1 \otimes Q_2)\ A\text{'s are } (B \text{ and } C)\text{'s}}$$

where Q_1 and Q_2 are fuzzy quantifiers; A, B and C are fuzzy relations; and $Q_1 \otimes Q_2$ is the product of Q_1 and Q_2 in fuzzy arithmetic.

Constraint Modification Rules:[21,9]

$$X \text{ is } mA \rightarrow X \text{ is } f(A)$$

where m is a modifier such as *not, very, more or less*, and $f(A)$ defines the way in which m modifies A. Specifically,

$$\text{if } m = \textit{not} \text{ then } f(A) = A' \text{ (complement)}$$
$$\text{if } m = \textit{very} \text{ then } f(A) = {}^2A \text{ (left square)}$$

where $\mu_{{}^2A}(u) = (\mu_A(u))^2$. This rule is a convention and should not be construed as a realistic approximation to the way in which the modifier *very* functions in a natural language.

Probability Qualification Rule:[7]

$$(X \text{ is A}) \text{ is } \Lambda \rightarrow P \text{ is } \Lambda$$

where X is a random variable taking values in U with probability density $p(u)$; Λ is a linguistic probability expressed in words like *likely, not very likely*, etc.; and P is the probability of the fuzzy event X, expressed as

$$P = \int_U \mu_A(u)p(u)du$$

The primary purpose of this summary is to underscore the coincidence of the principal rules governing fuzzy constraint propagation with the principal rules of inference in fuzzy logic. Of necessity, the summary is not complete and there are many specialized rules which are not included. Furthermore, most of the rules in the summary apply to constraints which are of the basic, possibilistic type. Further development of the rules governing fuzzy constraint propagation will require an extension of the rules of inference to generalized constraints.

As was alluded to in the summary, the principal rule governing constraint propagation is the generalized extension principle which in a schematic form may be represented as

$$\frac{f(X_1,\ldots,X_n) \text{ is } A}{q(X_1,\ldots,X_n) \text{ is } q(f^{-1}(\text{A}))}$$

In this expression, $X_1,\ldots,X_n$ are database variables; the term above the line represents the constraint induced by the IDS; and the term below the line is the TDS expressed as a constraint on the query $q(X_1,\ldots,X_n)$. In the latter constraint, f^{-1}(A) denotes the preimage of the fuzzy relation A under the mapping $f: U \rightarrow$ V, where A is a fuzzy subset of V and U is the domain of $f(X_1,\ldots,X_n)$.

Expressed in terms of the membership functions of A and $q(f^{-1}(\text{A}))$, the generalized extension principle reduces the derivation of the TDS to the solution of the constrained maximization problem

$$\mu_q(X_1,\ldots,X_n)(v) = \sup_{(u_1,\ldots u_n)} (\mu_A(f(u_1,\ldots u_n))).$$

in which $u_1,\ldots,u_n$ are constrained by

$$v = q(u_1,\ldots,u_n).$$

The generalized extension principle is simpler than it appears. An illustration of its use is provided by the following example.

The IDS is:

most Swedes are *tall*

The query is: *What is the average height of Swedes?*

The explanatory database consists of a population of N Swedes, $Name_1, \ldots, Name_N$. The database variables are $h_1, \ldots, h_N$, where h_i is the height of $Name_i$, and the grade of membership of $Name_i$ in *tall* is $\mu_{tall}(h_i)$, $i = 1, \ldots, n$.

The proportion of Swedes who are tall is given by the sigma-count[9]

$$\Sigma\, Count(tall - Swedes\,/\,Swedes) = \frac{1}{N}\sum_i \mu_{tall}(h_i)$$

from which it follows that the constraint on the database variables induced by the IDS is

$$\frac{1}{N}\sum_i \mu_{tall}(h_i) \text{ is } most$$

In terms of the database variables $h_1, \ldots, h_N$, the average height of Swedes is given by

$$h_{ave} = \frac{1}{N}\sum_i h_i$$

Since the IDS is a fuzzy proposition, h_{ave} is a fuzzy set whose determination reduces to the constrained maximization problem

$$\mu_{h_{ave}}(v) = \sup_{(h_1, \ldots, h_n)} \left(\mu_{most}\left(\frac{1}{N}\sum_i \mu_{tall} h_i\right)\right)$$

subject to the constraint

$$v = \frac{1}{N}\sum_i h_i$$

It is possible that approximate solutions to problems of this type might be obtainable through the use of neurocomputing or evolutionary-computing-based methods.

As a further example, we will return to a problem stated in an earlier section, namely, maximization of a function, f, which is described in words by its fuzzy graph, f^* (Fig. 10). More specifically, consider the standard problem of maximization of an objective function in decision analysis. Let us assume - as is frequently the case in real-world problems - that the objective function, f, is not well-defined and that what we know about can be expressed as a fuzzy rule-set

f: *if* X *is* A_1 *then* Y *is* B_1
if X *is* A_2 *then* Y *is* B_2
....
if X *is* A_n *then* Y *is* B_n

or, equivalently, as a fuzzy graph

$$f \text{ is } \sum_i A_i \times B_i$$

The question is: What is the point or, more generally, the maximizing set[28] at which f is maximized, and what is the maximum value of f?

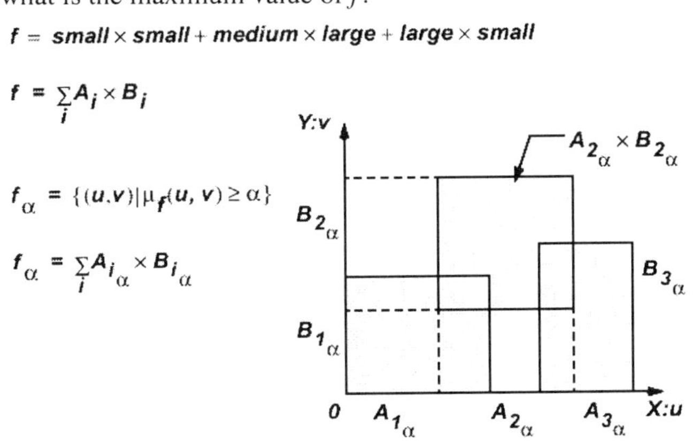

Figure 16. α-Cuts of a function described by a fuzzy graph.

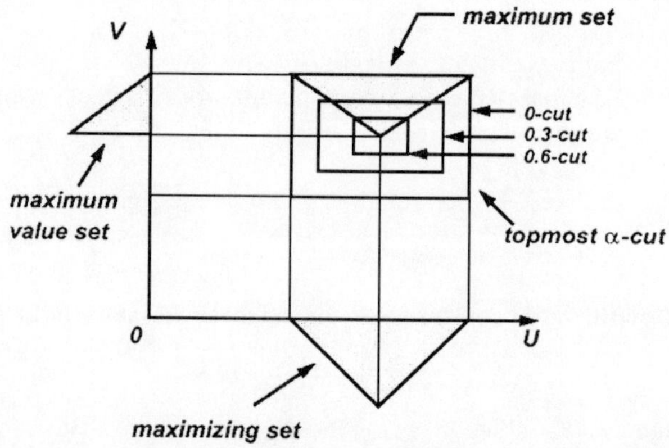

Figure 17. Computation of maximizing set, maximum set and maximum value set.

The problem can be solved by employing the technique of *α-cuts*.[2,26] With reference to Fig. 16, if A_{i_α} and B_{i_α} are *α-cuts* of A_i and B_i, respectively, then the corresponding *α-cut* of f^* is given by

$$f_\alpha^* = \sum_i A_{i_\alpha} \times B_{i_\alpha}$$

From this expression, the maximizing fuzzy set, the maximum fuzzy set and maximum value fuzzy set can readily be derived, as shown in Figs. 16 and 17.

A key point which is brought out by these examples and the preceding discussion is that explicitation and constraint propagation play pivotal roles in CW. This role can be concretized by viewing explicitation and constraint propagation as translation of propositions expressed in a natural language into what might be called the *generalized constraint language* (GCL) and applying rules of constraint propagation to expressions in this language - expressions which are typically canonical forms of propositions expressed in a natural language. This process is schematized in Fig. 18.

The conceptual framework of GCL is substantively differently from that of conventional logical systems, e.g., predicate logic. But what matters most is that the expressive power of GCL - which is based on fuzzy logic - is much greater than that of standard logical calculi. As an illustration of this point, consider the following problem.

A box contains ten balls of various sizes of which several are large and a few are small. What is the probability that a ball drawn at random is neither large nor small?

To be able to answer this question it is necessary to be able to define the meanings *of large, small, several large balls, few small balls* and *neither large nor small*. This is a problem in semantics which falls outside of probability theory, neurocomputing and other methodologies.

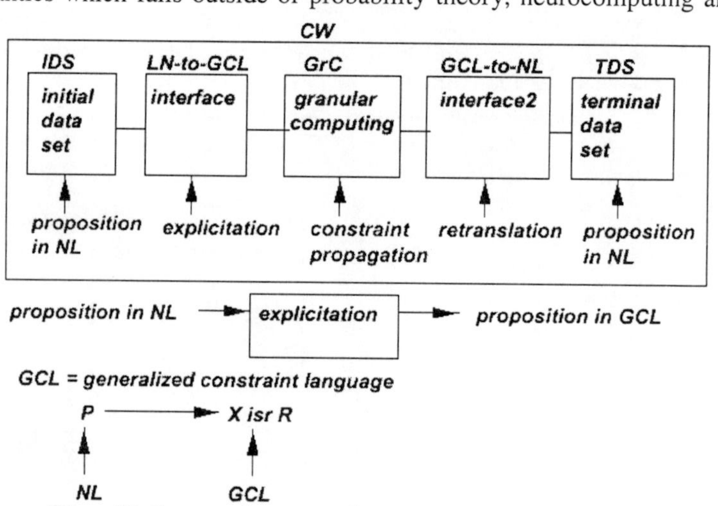

Figure 18. Conceptual structure of computing with words.

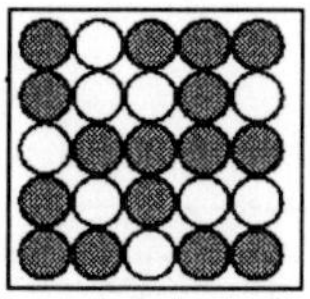

Figure 19. A box with black and white balls.

An important application area for computing with words and manipulation of perceptions is decision analysis since in most realistic settings the underlying probabilities and utilities are not known with sufficient precision to justify the use of numerical valuations. There exists an extensive literature on the use of fuzzy probabilities and fuzzy utilities in decision analysis. In what follows, we shall restrict our discussion to two very simple examples which illustrate the use of perceptions.

First, consider a box which contains black balls and white balls (Fig. 19). If we could count the number of black balls and white balls, the probability of picking a black ball at random would be equal to the proportion, r, of black balls in the box.

Now suppose that we cannot count the number of black balls in the box but our perception is that most of the balls are black. What, then, is the probability, p, that a ball drawn at random is black?

Assume that *most* is characterized by its possibility distribution (Fig. 20). In this case, p is a fuzzy number whose possibility distribution is *most*, that is,

$$p \text{ is } most.$$

Next, assume that there is a reward of a dollars if the ball drawn at random is black and a penalty of b dollars if the ball is white. In this case, if p were known as a number, the expected value of the gain would be:

$$e = ap - b(1 - p)$$

Since we know not p but its possibility distribution, the problem is to compute the value of e when p is *most*. For this purpose, we can employ the extension principle,[2,26] which implies that the possibility distribution, E, of e is a fuzzy number which may be expressed as

$$E = a\ most - b(1 - most)$$

For simplicity, assume that *most* has a trapezoidal possibility distribution (Fig. 20). In this case, the trapezoidal possibility distribution of E can be computed as shown in Fig. 21.

It is of interest to observe that if the support of E is an interval $[\alpha, \beta]$ which straddles O (Fig. 22), then there is no non-controversial decision principle which can be employed to answer the question: Would it be advantageous to play a game in which a ball is picked at random from a box in which most balls are black, and a and b are such that the support of E contains O.

Next, consider a box in which the balls $b_1, \ldots, b_n$ have the same color but vary in size, with b_i, $i = 1, \ldots, n$ having the grade of membership μ_i in the fuzzy set of large balls (Fig. 23). The question is: What is the probability that a ball drawn at random is large, given the perception that most balls are large?

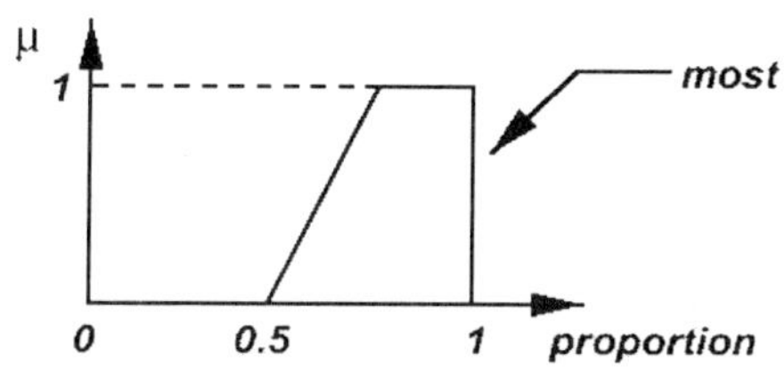

Figure 20. Membership function of *most*.

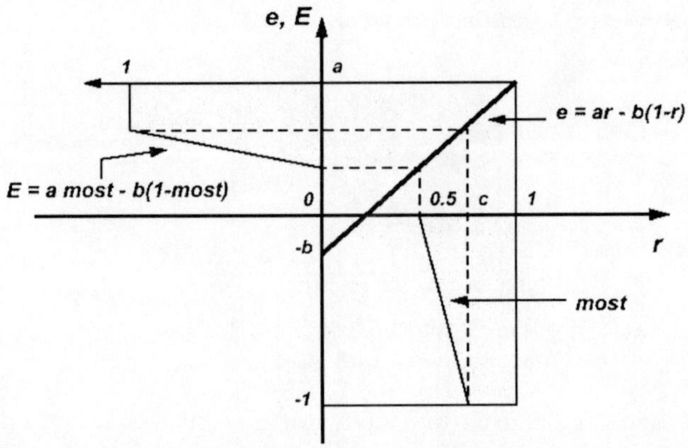

Figure 21. Computation of expectation through use of the extension principle.

The difference between this example and the preceding one is that the event *the ball drawn at random is large* is a fuzzy event, in contrast to the crisp event *the ball drawn at random is black*.

The probability of drawing b_i is $1/n$. Since the grade of membership of b_i in the fuzzy set of large balls is μ_i, the probability of the fuzzy event *the ball drawn at random is large* is given by[29]

$$P = \frac{1}{n}\sum \mu_i$$

On the other hand, the proportion of large balls in the box is given by the relative sigma-count[26,9]

$$\Sigma\text{Count}\ (large.balls\ /\ balls.in.box) = \frac{1}{n}\sum \mu_i$$

Consequently, the canonical form of the perception *most balls are large* may be expressed as

$$\frac{1}{n}\sum \mu_i \text{ is } most$$

which leads to the conclusion that

$$P \text{ is } most$$

It is of interest to observe that the possibility distribution of P is the same as in the preceding example.

If the question were: What is the probability that a ball drawn at random is *small*, the answer would be

$$P \text{ is } \frac{1}{n}\sum \nu_i$$

where ν_i, $i = 1, \ldots, n$, is the grade of membership of b_i in the fuzzy set of small balls, given that

$$\frac{1}{n}\sum \mu_i \text{ is } most$$

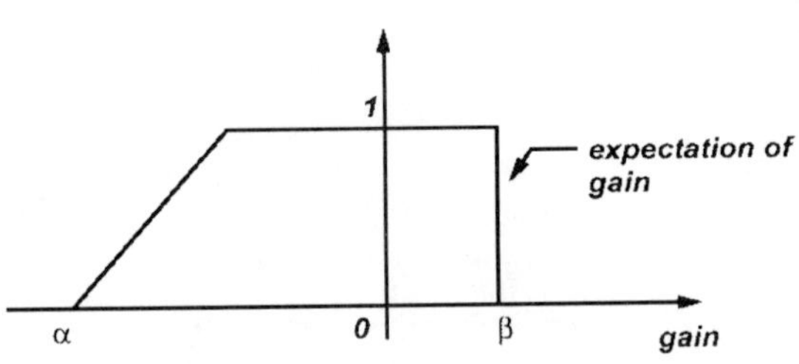

Figure 22. Expectation of gain.

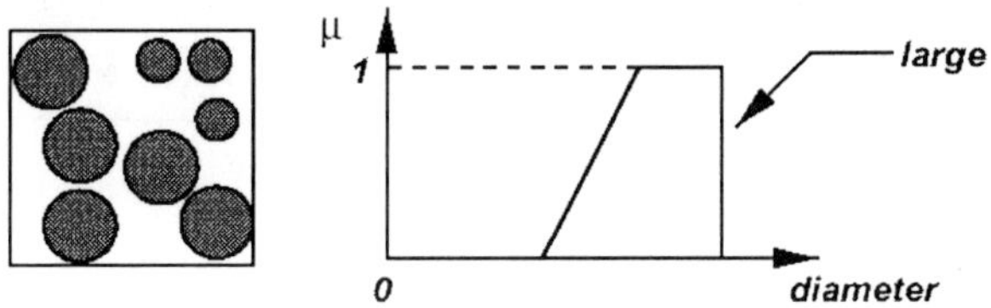

Figure 23. A box with balls of various sizes and a definition of large ball.

What is involved in this case is constraint propagation from the antecedent constraint on the í *i* to a consequent constraint on the ν_i. This problem reduces to the solution of a nonlinear program.

What this example points to is that in using fuzzy constraint propagation rules, application of the extension principle reduces, in general, to the solution of a nonlinear program. What we need -and do not have at present –are approximate methods of solving such programs which are capable of exploiting the tolerance for imprecision.

Without such methods, the cost of solutions may be excessive in relation to the imprecision which is intrinsic in the use of words. In this connection, an intriguing possibility is to use neurocomputing and evolutionary computing techniques to arrive at approximate solutions to constrained maximization problems. The use of such techniques may provide a closer approximation to the ways in which human manipulate perceptions.

CONCLUDING REMARKS

In our quest for machines which have a high degree of machine intelligence (high MIQ), we are developing a better understanding of the fundamental importance of the remarkable human capacity to perform a wide variety of physical and mental tasks without any measurements and any computations. Underlying this remarkable capability is the brain's crucial ability to manipulate perceptions -perceptions of distance, size, weight, force, color, numbers, likelihood, truth and other characteristics of physical and mental objects. A basic difference between perceptions and measurements is that, in general, measurements are crisp whereas perceptions are fuzzy. In a fundamental way, this is the reason why to deal with perceptions it is necessary to employ a logical system that is fuzzy rather than crisp.

Humans employ words to describe perceptions. It is this obvious observation that is the point of departure for the theory outlined in the preceding sections.

When perceptions are described in words, manipulation of perceptions is reduced to computing with words (CW). In CW, the objects of computation are words or, more generally, propositions drawn from a natural language. A basic premise in CW is that the meaning of a proposition, p, may be expressed as a generalized constraint in which the constrained variable and the constraining relation are, in general, implicit in p.

In coming years, computing with words and perceptions is likely to emerge as an important direction in science and technology. In a reversal of long-standing attitudes, manipulation of perceptions and words which describe them is destined to gain in respectability. This is certain to happen because it is becoming increasingly clear that in dealing with real-world problems there is much to be gained by exploiting the tolerance for imprecision, uncertainty and partial truth. This is the primary motivation for the methodology of computing with words (CW) and the computational theory of perceptions (CTP) which are outlined in this paper.

ACKNOWLEDGMENT

The author acknowledges Prof. Michio Sugeno, who has contributed so much and in so many ways to the development of fuzzy logic and its applications.

REFERENCES

1. L.A. Zadeh, From circuit theory to system theory, *Proc. IRE,* 50, pp. 856-865, 1961.
2. L.A. Zadeh, Fuzzy Sets, *Inf. Contro*l, 8, pp. 338-353, 1965.
3. L.A. Zadeh, Outline of a New Approach to the Analysis of Complex System and Decision Processes, *IEEE Trans. On Systems, Man, and Cybernetic*s, SMC-3, pp.28-44, 1973.
4. R. Vallée, *Cognition et Systè*me, Paris: l'Interdisciplinaire Systeme(s), 1995.
5. H. Rasiowa and M. Marek, On reaching consensus by groups of intelligent agents, in Z. W. Ras ed., *Methodologies for Intelligent Syste*ms, Amsterdam: North-Holland, pp. 234-243, 1989.
6. L.A. Zadeh, Calculus of Fuzzy Restrictions, in L.A. Zadeh, K.S. Fu, M. Shimura eds., *Fuzzy Sets and Their Applications to Cognitive and Decision Processe*s, New York: Academic Press, pp. 1-39, 1975.
7. L.A. Zadeh, A Theory of Approximate Reasoning, J. Hayes, D. Michie, and L.I. Mikulich eds., *Machine Intelligence* 9, New York: Halstead Press, pp. 149-194, 1979.
8. L.A. Zadeh, Outline of a Computational Approach to Meaning and Knowledge Representation. Based on a Concept of a Generalized Assignment Statement, M. Thoma and A. Wyner eds., *Proc. of the Int. Seminar on Artificial Intelligence and Man-Machine Syste*ms, Heidelberg: Springer-Verlag, pp. 198-211, 1986.
9. L.A. Zadeh, PRUF - A Meaning Representation Language for Natural Languages, *Int. J. Man-Machines Studi*es, 10, pp. 395-460, 1978.
10. L.A. Zadeh, Test-Score Semantics for Natural Languages and Meaning Representation via PRUF, B. Rieger ed., *Empirical Semanti*cs, W. Germany: Brockmeyer, pp. 281-349, also *Technical Report Memorandum* 246, AI Center, SRI International, Menlo Park, CA, 1981.
11. L.A. Zadeh, Toward a Theory of Fuzzy Information Granulation and its Centrality in Human Reasoning and Fuzzy Logic, *Fuzzy Sets and Systems* 90, pp. 111-127, 1997.
12. J.R. Hobbs, Making Computation Sense of Montague's Intensional Logic, *Artificial Intelligence* 9, pp. 287-306, 1978.
13. B. Partee, *Montague Gramm*ar, New York: Academic Press, 1976.
14. L.A. Zadeh, On the Analysis of Large Scale Systems, H. Gottinger ed., *Systems Approaches and Environment Problems*, Gottingen: Vandenhoeck and Ruprecht, pp. 23-37, 1974.
15. L.A. Zadeh, Fuzzy Logic and the Calculi of Fuzzy Rules and Fuzzy Graphs: A Precis, *Multiple Valued Logic* 1, Gordon and Breach Science Publishers, pp. 1-38, 1996.
16. A. Kaufmann and M.M. Gupta, *Introduction to Fuzzy Arithmetic: Theory and Applicatio*ns, New York: Von Nostrand, 1985.
17. E.C. Freuder and P. Snow, Improved Relaxation and Search Methods for Approximate Constraint Satisfaction with a Maximin Criterion, *Proc. of the 8th Biennial Conf. on the Canadian Society for Computational Studies of Intelligen*ce, Ontario, pp. 227-230, 1990.
18. E.H. Mamdani and B.R. Gaines eds., *Fuzzy Reasoning and its Applicatio*ns, London, 1981.
19. L.A. Zadeh, Test-score semantics for natural languages, *Proc. of the Ninth International Conference on Computational Linguisti*cs, Prague, pp. 425-430, 1982.
20. M.J. Cresswell, *Logic and Languag*es, London: Methuen, 1973.
21. L.A. Zadeh, A fuzzy-set-theoretic interpretation of linguistic hedges, *J. of Cybernetics* 2, pp. 4-34, 1972.
22. L.A. Zadeh, Fuzzy sets as a basis for a theory of possibility, *Fuzzy Sets and Systems* 1, pp. 3-28, 1978.
23. G. Shafer, *A Mathematical Theory of Eviden*ce, Princeton: Princeton University Press, 1976.
24. G. Klir and B. Yuan, *Fuzzy Sets and Fuzzy Log*ic, New Jersey: Prentice Hall, 1995.
25. W. Pedrycz and F. Gomide, *Introduction to Fuzzy Se*ts, Cambridge: MIT Press, page 38 of 40,August 24, 1998.
26. L.A. Zadeh, The concept of a linguistic variable and its application to approximate reasoning, Part I: *Inf. Sci.* 8, pp. 199-249; Part II: *Inf. Sci.* 8, pp. 301-357; Part III: *Inf. Sci.* 9, pp. 43-80, 1975.
27. L.A. Zadeh, Syllogistic reasoning in fuzzy logic and its application to reasoning with dispositions, *Proceedings of the 1984 International Symposium on Multiple-Valued Log*ic, Winnipeg, Canada, pp. 148-153, 1984.
28. L.A. Zadeh, Maximizing Sets and Fuzzy Markoff Algorithms, *IEEE Transactions on Systems, Man, and Cybernetics-Part C: Applications and Reviews,* Vol. 28, pp. 9-15, August 24, 1998.
29. L.A. Zadeh, Probability measures of fuzzy events, *Jour. Math. Analysis and Appl.* 23, pp. 421-427, 1968.

RELATED PUBLICATIONS

30. H.R. Berenji, Fuzzy Reinforcement Learning and Dynamic Programming, in A.L. Ralescu ed., *Fuzzy Logic in Artificial Intelligence, Proc. IJCAI '93 Workshop*, Berlin: Springer-Verlag, pp. 1-9, 1994.
31. M. Black, Reasoning with Loose Concepts, *Dialog* 2, pp. 1-12, 1963.
32. P. Bosch, Vagueness, Ambiguity and All the Rest, in M. Van de Velde and W. Vandeweghe eds., *Sprachstruktur, Individuum und Gesselschaft*, Tubingen: Niemeyer, 1978.
33. J. Bowen, R. Lai, and D. Bahler, Fuzzy Semantics and Fuzzy Constraint Networks, *Proc. of the 1st IEEE Conf. on Fuzzy Systems*, San Francisco, pp. 1009-1016, 1992.
34. J. Bowen, R. Lai, and D. Bahler, Lexical Imprecision in Fuzzy Constraint Networks, *Proc. of the National Conf. On Artificial Intelligence*, pp. 616-620, 1992.
35. D. Dubois, H. Fargier, and H. Prade, Propagation and Satisfaction of Flexible Constraints, in R.R. Yager, L.A. Zadeh eds., *Fuzzy Sets, Neural Networks, and Soft Computing*, New York: Von Nostrand Reinhold, pp. 166-187, 1994.
36. D. Dubois, H. Fargier, and H. Prade, Possibility Theory in Constraint Satisfaction Problems: Handling Priority, Preference and Uncertainty, to appear in *Applied Intelligence Journal*.
37. D. Dubois, H. Fargier, and H. Prade, The Calculus of Fuzzy Restrictions as a Basis for Flexible Constraint Satisfaction, *Proc. of the 2nd IEEE Int. Conf. on Fuzzy Systems*, San Francisco, pp. 1131-1136, 1993.
38. J.A. Goguen, The Logic of Inexact Concepts, *Synthese* 19, pp. 325-373, 1969.
39. O. Katai, S. Matsubara, H. Masuichi, M. Ida, et. al., Synergetic Computation for Constraint Satisfaction Problems Involving Continuous and Fuzzy Variables by Using Occam, in S. Noguchi and H. Umeo eds., *Transputer/Occam, Proc. of the 4th Transputer/Occam Int. Conf.*, Amsterdam: IOS Press, pp. 146-160, 1992.
40. K. Lano, A Constraint-Based Fuzzy Inference System, in P. Barahona, L.M. Pereira, and A. Porto eds., *EPIA* 91, *5th Portuguese Conf. on Artificial Intelligence*, Berlin: Springer-Verlag, pp. 45-59, 1991.
41. W.A. Lodwick, Analysis of Structure in Fuzzy Linear Programs, *Fuzzy Sets and Systems*, *38(*1), pp. 15-26, 1990.
42. M. Mares, *Computation Over Fuzzy Quantities*, Boca Raton: CRC Press, 1994.
43. V. Novak, Fuzzy Logic, Fuzzy Sets, and Natural Languages, *Int. J. of General Systems 20(*1), pp. 83-97, 1991.
44. V. Novak, M. Ramik, M. Cerny and J. Nekola eds., *Fuzzy Approach to Reasoning and Decision-Making*, Boston: Kluwer, 1992.
45. M.S. Oshan, O.M. Saad and A.G. Hassan, On the Solution of Fuzzy Multiobjective Integer Linear Programming Problems with a Parametric Study, *Advances in Modelling & Analysis* A, *24(*2), pp. 49-64, 1995.
46. G. Qi and G. Friedrich, Extending Constraint Satisfaction Problem Solving in Structural Design, in F. Belli and F.J. Radermacher eds., *Industrial and Engineering Applications of Artificial Intelligence and Expert Systems, 5th Int. Conf., IEA/AIE*-92, Berlin: Springer-Verlag, pp. 341-350, 1992.
47. A. Rosenfeld, R.A. Hummel and S.W. Zucker, Scene Labeling by Relaxation Operations, *IEEE Trans. on Systems, Man and Cybernetics,* 6, pp. 420-433, 1976.
48. M. Sakawa, K. Sawada, and M. Inuiguchi, A Fuzzy Satisficing Method for Large-Scale Linear Programming Problems with Block Angular Structure, *European Journal of Operational Research*, *81(*2), pp. 399-409, 1995.
49. S.C. Tong, Interval Number and Fuzzy Number Linear Programming, *Advances in Modelling & Analysis* A, *20(*2), pp. 51-56, 1994.
50. R.R. Yager, Some extensions of constraint propagation of label sets, *Int. J. of Approximate Reasoning,* 3, pp. 417-435, 1989.
51. L.A. Zadeh, A fuzzy-algorithmic approach to the definition of complex or imprecise concepts, *Int. Jour. Man-Machine Studies* 8, pp. 249-291, 1976.
52. L.A. Zadeh, Fuzzy Sets and Information Granularity, in M. Gupta, R.Ragade and R. Yager eds., *Advances in Fuzzy Set Theory and Applications*, Amsterdam: North-Holland, pp. 3-18, 1979.
53. L.A. Zadeh, Fuzzy Logic, Neural Networks and Soft Computing, *Communications of the* ACM, 37(3), pp. 77-84, 1994.
54. L.A. Zadeh, Fuzzy Logic = Computing with Words, *IEEE Transactions on Fuzzy Systems*, Vol. 4, pp. 103-111, 1996.

PROCESSING OF BIOLOGICAL SIGNALS: A BRIDGE BETWEEN STRUCTURE AND FUNCTION

Sergio Cerutti, Anna M. Bianchi, and Maria G. Signorini

Department of Biomedical Engineering
Polytechnic University
Milano, ITALY
e-mail: cerutti@biomed.polimi.it

INTRODUCTION

A biological structure is the source of a large number of signals of different nature, among which we may recall the electroencephalogram, the evoked and event related brain potentials, the electrocardiogram, the electromyogram, blood pressures (recorded in different body sites), fluid flows, biochemical concentrations, and many others. Most of them can be obtained non invasively and all of them carry information about the considered body organ or, from a more general point of view, about the whole signal generating process, that sometimes involves complex structures and systems. Due to this complexity the informative contents have sometimes to be enhanced through advanced signal processing techniques able to eliminate the superimposed noise, to extract new parameters, to quantitatively relate different signals. In particular, multiparametric approaches may lead to the definition of mathematical models of the underlying structure, for facilitating in the interpretation both in physiological and pathological conditions.

Even in the same subject, biological signals are characterized by high variability, which is related to different states and in general to a dynamical behavior of the underlying process.

Furthermore, the use of computational modeling to elucidate the relationships between the structural complexity of the brain or the heart and their functional resultants is an intriguing and fascinating point. The integration of models of various physiological coupled processes as a large-scale computational challenge, generate the need to develop hierarchical models from the level of the ion channel, to the cell, tissue, organ, up to organ systems.

In the recent past, someone theorized that if we know the structure and the mechanical and electrical properties of the tissues of a living organ, then by the principles of physics we should be able to predict the functions of that organ. It was a vision, as the complex relationships in the living systems do not allow the knowledge to be so complete.

Human and Machine Perception 3: Thinking, Deciding, and Acting, edited by Cantoni *et al.*
Kluwer Academic/Plenum Publishers, New York, 2001.

The brain capability in acquiring complex data, in elaborating and integrating them, also with previous information, for controlling different automatic and voluntary functions, is still inimitable.

The mammalian heart is a masterpiece of engineering design, it is highly reliable, exquisitely adaptive, and remarkably energy efficient.

Recent researches propose the use of a wide array of techniques from the engineering and biomedical sciences to investigate relationships between the cellular and tissue structure of organs and their mechanical and electrical function both in normal condition and in pathology.

These multi-disciplinary approaches involve a close interaction between engineering, life sciences and medicine.

Within this framework, the challenge of developing integrated, hierarchical models of physiological complexity certainly has to consider, among the others, the fractal and the multiparametric approaches. Quantification of complex temporal and geometric relationships of living structures introduces new interpretive tools to formulate and test quantitative hypotheses on structure-function relations of complex systems.

Procedures of parametric modeling are applied in the dynamical evaluation of the brain functions during the learning process. In particular the autoregressive filter with exogenous input (ARX) model allows the extraction of the single responses of event related potentials during a learning task, and the dynamic time warping (DTW) procedure allows the automatic tracking of specific brain activity points along the recording session. In particular, parameters related to the P300 wave show modifications during the learning phase. Time-variant spectral analysis in the bivariate form shows a modulation in the interrelationships between cortical areas which seem to be related to a learning process: coherence between the signals recorded in correspondence of the premotor and motor cortex follows a dynamical pattern linked to cognitive, attentive and decisional processes.

The aim of this chapter is also to show the structure of the physiological rhythms of the cardiac system involving nonlinear dynamics. We want to highlight how the presence of nonlinear deterministic phenomena could affect the Heart Rate Variability (HRV) signal both in short and long temporal windows. In the following we will present results in HRV signal analysis by measuring long period correlation and self-similarity properties. Methods for the Hurst self-similarity exponent H computation are compared in different groups of subjects with cardiovascular disease conditions. Results show that the H parameter could be useful to significantly differentiate healthy and diseased patients.

INFORMATION ENHANCEMENT FROM BIOLOGICAL SIGNALS OF CNS

ARX Single Sweep Filtering

The ARX was introduced for the evaluation of single sweep visual evoked potentials by Cerutti[1] in 1987. It provides a separation of the recorded signals in evoked response and background EEG on a sweep-by-sweep basis, and allows the evaluation of the dynamical changes of the central responses during a session of a few hundred stimuli. The model was then applied in different protocols both physiological and clinical, in which the interest was focused on the time evolution of relevant parameters[2,3] and was further developed as ARXX, with a double exogenous input, for the suppression of eye movement artifacts.[4] The filter is based on a model of interaction between the evoked response and the background EEG. They are hypothesized to be independent and summed in the recorded signal. The EEG contribution, not related to the stimulus, is modeled as an autoregressive process, according to various authors,[5,6] while the evoked response, related to the stimulus, is modeled as the filtered version of a reference signal. According to Cerutti et al.[1] the reference signal is

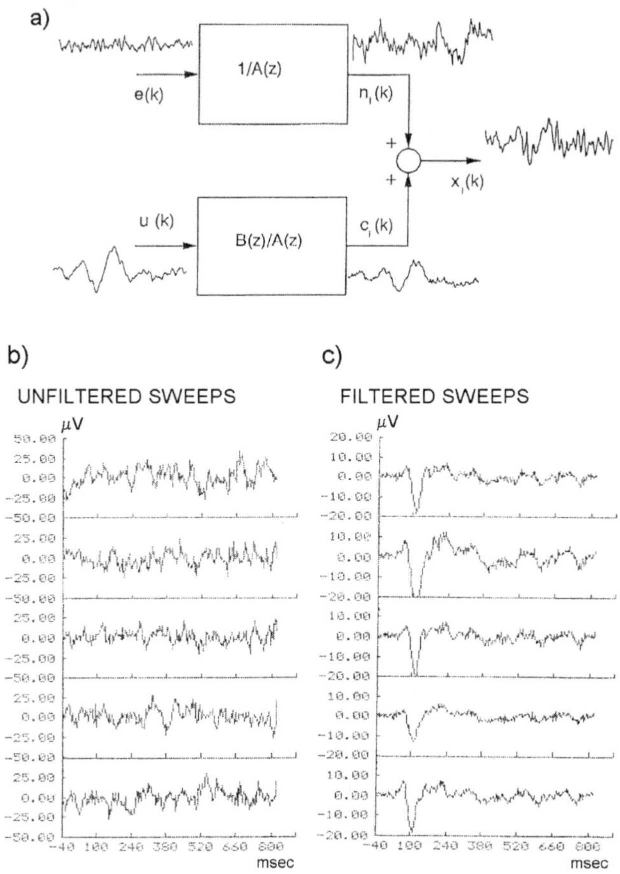

Figure 1. a) Block diagram of the ARX model; b) example of five unfiltered single sweep of a visual evoked potential; c) the filtered single sweeps.

evaluated by averaging the single responses: each single response is *similar* to the reference, the single sweep differences are accounted by the filter.

The block diagram of the ARX filter is shown in Figure 1a, and an example of some single sweeps before and after the filtering procedure is in Figure 1b and c respectively.

According to the model, the recorded signal is described, in the time domain, by the following expression:

$$x(k) = -\sum_{i=1}^{p} a_i y(k-i) + \sum_{j=-d}^{q-d+1} b_j u(k-j) + e(k) \tag{1}$$

where *x(k)* is the recorded signal in correspondence of each stimulation, and *u(k)* is the reference input, *e(k)* represent the stochastic part of the signal, not described by the model, i.e. the *prediction error*; *p* and *q* are the order of the autoregressive and of the moving average section of the model, respectively, and *d* takes into account the possible delays between input and output of the filter.

In the *z-transform* domain the model is represented as:

$$X(z) = \frac{E(z)}{A(z)} + \frac{B(z)}{A(z)} \cdot U(z) \tag{2}$$

where $A(z) = 1 + \sum_{i=1}^{p} a_i z^{-i}$ and $B(z) = \sum_{j=-d}^{q-d+1} b_j z^{-j}$

The a_i and b_j parameters, as well as the variance of the prediction error, are estimated sweep-by-sweep trough a least squares procedure. Though in literature some criteria do exist for the choice of the optimal order,[7,8] the procedure for its evaluation is really time consuming when performed sweep-by-sweep and usually the orders are *a priori* defined on the basis of previous experiences, and are *a posteriori* tested by verifying the whiteness of the prediction error.[9]

The quantitative parameters usually employed for the classification of the evoked potentials (latency and amplitude of some relevant peaks in the signals) can now be evaluated for each sweep and their dynamical evolution can be studied in correspondence of non stationary conditions: photostress recovery test,[2] anesthesia induction,[3] during variable sedation levels.[10]

Dynamic Time Warping (DTW)

The Dynamic Time Warping (DTW) is a technique of nonlinear alignment for locally contracting or dilating the time scale of two signals, in order to reduce distorsions due to normal morphological differences in the waveforms. The method was initially used for automatic recognition of speech and was also applied in the study of EEG[11] and evoked potentials.[12] Problems of sweep alignment may occur during the averaging process producing a jitter in the peaks of clinical relevance. The DTW was proposed by Gupta[13] for an optimal alignment of evoked potentials before the averaging procedure. An example of this application is shown in Figure 2. In panel a) two hypothetical waveforms, *x(i)* and *y(j)*, with similar shape are different for time scales. The procedure is introduced for minimizing latency differences by means of a function *w(k)* called *warping function,* defined as in the following:

$$w(k) = w(i(k), j(k)) \tag{3}$$

Where, in the present example, it assumes the values:

$$\begin{aligned} &w(1) = (1,1), w(2) = (2,1), w(3) = (3,1), w(4) = (4,2), w(5) = (5,3), \\ &w(6) = (6,4), w(7) = (7,5), w(8) = (7,6), w(9) = (7,7) \end{aligned} \tag{4}$$

The *warping function* is mapped in Figure 2.b. as a function of the time indices *i* and *j* of the waveforms *x(.)* and *y(.)*. Figure 2.c puts into evidence the correspondence between sample pairs.

The warping function is obtained through the minimization of the *dissimilarity function D(x,y),* where

$$D(x, y) = \sum d(k) \tag{5}$$

and

$$d(k) = d(w(k)) = d(i, j) = \| x(i) - y(j) \| \tag{6}$$

$\| x(i) - y(j) \|$ is a measure of the distance between the waveforms and can be defined differently. In the current application it is evaluated as:

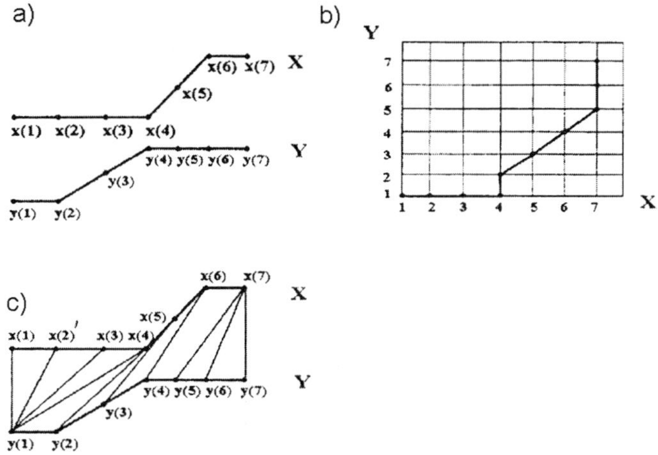

Figure 2. a) Two simulated signals *x(i)* and *y(j)*, with a similar waveform, but with a different time scale; b) the warping function that relates corresponding points in signals *x(i)* and *y(j)*; c) correspondence between sample pairs in signals *x(i)* and *y(j)*.

$$d(i, j) = |x(i) - y(j)| + |\dot{x}(i) - \dot{y}(j)| \tag{7}$$

The warping procedure produces important results and applications: 1) the possibility of a more realistic averaging of the sweeps, by summing samples with a morphological, instead of temporal, correspondence; 2) automatic tracking of fiduciary points on the sweeps; 3) creation of a template of the evoked response.

Time-variant Spectral Estimation

Learning and attentive processes implies changes not only in the event related potentials, but also in the background EEG and in the degree of interrelationship between different cortical areas. The EEG signal is usually characterized and classified on the basis of its frequency content. The traditional spectral analysis, both through parametric and non-parametric techniques, requires the stationarity of the signal, but such a condition is very difficult to obtain in physiological situations. In addition variability and transition between states characterize many cases of interest. For this reason different techniques have been developed in literature for the simultaneous time-frequency representation of a signal. They can be classified as linear decompositions of the signal, which include the short time Fourier transform and the wavelet decomposition; as quadratic distributions, such the Wigner-Ville distribution and the more general Cohen class of distributions;[14-16] and finally as parametric models implemented in time-variant form.

In this chapter time-variant autoregressive models are described.

This approach implies a three step procedure: 1) the definition of a proper model describing the generation mechanism of the signal, in this case an autoregressive model; 2) the identification of the model parameters and its validation; 3) the calculation of the power spectral distribution by inserting the model parameters in a proper expression.[17]

The AR model is made time-variant by evaluating recursively the set of the parameters according to the expression:

$$\mathbf{a}(t) = \mathbf{a}(t-1) + \mathbf{K}(t)e(t) \tag{8}$$

where

$$\mathbf{K}(t) = \frac{\mathbf{P}(t-1)\varphi(t)}{\mu + \varphi(t)^T \mathbf{P}(t-1)\varphi(t)} \tag{9}$$

$$\mathbf{P}(t) = \frac{1}{\mu}\left[\mathbf{P}(t-1) - \frac{\mathbf{P}(t-1)\varphi(t)\varphi(t)^T \mathbf{P}(t-1)}{\mu + \varphi(t)^T \mathbf{P}(t-1)\varphi(t)}\right] \tag{10}$$

where $\phi(t) = [y(t-1), y(t-2), \ldots, y(t-p)]$ is the observation vector and the apex T denotes the transposition operator.[18]

The algorithm is summarized in the block diagram shown in Figure 3.

The equations are obtained by a least squares procedure that minimizes the figure of merit:

$$J = \sum_{i=0}^{t} \mu^{t-i} e(i) \tag{11}$$

The *forgetting factor* μ gives an exponential weighing to the error terms, in such a way that the innovation is mainly based on the most recent terms, while to the oldest ones are progressively discounted. The discount rate is defined by the value of μ. In fact, the time constant of the exponential window is $L = \frac{1}{1-\mu}$.

For each sample in the signal it is thus possible to calculate a new AR model and a new power spectrum distribution (PSD)[19]

$$PSD(f,t) = \frac{\sigma^2 \Delta t}{\left| 1 + \sum_{i=1}^{p} a_i(t) z^{-i} \right|^2_{z=\exp(j2\pi f)}} \tag{12}$$

The AR model can be expressed also in bivariate form, allowing the time-dependent calculation of the coherence and phase relationships between signals at each frequency.[20] From the coefficient matrix the PSD matrix is obtained:

$$\mathbf{PSD}(f,n) = \begin{vmatrix} \mathbf{S}_{11}(f,n) & \mathbf{S}_{12}(f,n) \\ \mathbf{S}_{21}(f,n) & \mathbf{S}_{22}(f,n) \end{vmatrix} \tag{13}$$

From the cross-spectral elements in polar form:

$$\mathbf{S}_{kl}(f,n) = \mathbf{G}_{kl}(f,n)\exp(-j\mathbf{F}_{kl}(f,n)) \tag{14}$$

the squared coherence function can be evaluated:

$$\mathbf{K}^2{}_{kl}(f,n) = \frac{\mathbf{G}_{kl}(f,n)^2}{\left(\mathbf{S}_{kk}(f,n) \cdot \mathbf{S}_{ll}(f,n)\right)} \tag{15}$$

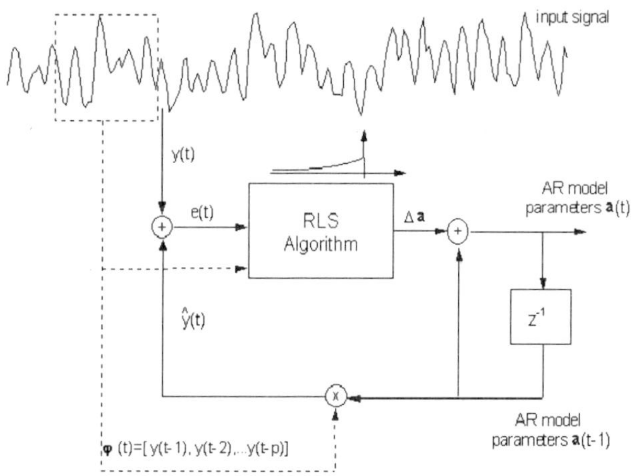

Figure 3. Block diagram for the recursive estimation of the autoregressive models.

Spectral and cross-spectral (coherence and phase relationships at the frequency of interest) can be plotted as function of time.[21]

Results

In this section the methods above presented are applied in the study of the dynamical adaptation of the central nervous system during a learning protocol, thus allowing to put into evidence different dynamical behaviors in correspondence of learning or not learning conditions. In the example here presented the subjects had to execute a right arm movement, shifting a mouse, on a digitizing mouse-pad, towards a visual target shown on a monitor, without looking at the arm. Two different conditions have been considered: i) no information about the performance (control condition); ii) *a posteriori* knowledge of the final position of the mouse at the end of the movement (learning condition). Each performance was scored by two indices: the *mean directional error* and the *directional error evaluated at the acceleration peak.*[22] Subject L modified his movement strategy on the bases of the previous results (learning subject), while NL subject do not modify his movement strategy on the basis of the previous results (not-learning subject). During the task execution, the EEG signal was recorded from 19 electrodes in the 10-20 system at a sample frequency of 128 Hz and stored on a personal computer.

The recorded signals were analyzed from different points of view: the event related potentials were filtered through the ARX model in order to obtain the single sweep responses. They were aligned by the DTW and an automatic tracking of parameters related to the P300 were obtained; the goal of this first step of the procedure was to define the dynamic of the learning process, mainly regarding premotor structures in the programming phase.

The degree of coupling of the EEG signal between premotor and motor cortex, during the programming and moving phase was quantified through the calculation of the coherence function, obtained by a time-variant autoregressive model in the bivariate form.

Figure 4 shows the results obtained from the event related single sweep analysis. After obtaining the evoked responses to each single movement, a new template was evaluated by averaging the sweeps after their alignment through the DTW. In Figure 4a the template related to the electrode F3 is shown. Such an electrode corresponds to the premotor cortex controlateral to the arm that executes the movement. Few characteristic waves are present in the signal: the N140 and the N400 peaks, and in particular the P200-N280-P300 complex, related to cognitive and learning processes. In particular the interlatency between P200 and P300 peaks (A) and the area under the curve (B), seem the parameters that better put into evidence the learning phase. They were evaluated sweep-by-sweep and their values were plotted as function of the sweep number for the two considered subjects and in the two different conditions.

The trend of A parameter, plotted in Figure 4b and in Figure 4c for the L and NL subject, respectively, in light line for the control condition, and in dark line for the learning condition, puts into evidence an increasing during the learning phase, typically after 30-40 movement repetitions, in the L subject. A similar increasing is also evident in the B parameter, plotted in Figure 4d and in Figure 4e for the L and NL subject, respectively, always in the L subject, while in NL subject it is not possible to evidence a precise trend in the above parameters, which can separate the two different conditions.

The dynamical links between different cortical areas in the controlateral side of the moving arm is quantified through the calculation of the squared coherence function on the EEG signal. As the described protocol involves mainly visual, pre-motor and motor cortex, the analysis is restricted to O1, F3 and C3 electrodes. Figure 5 shows the coherence function evaluated sample by sample according to the above described method, between electrode

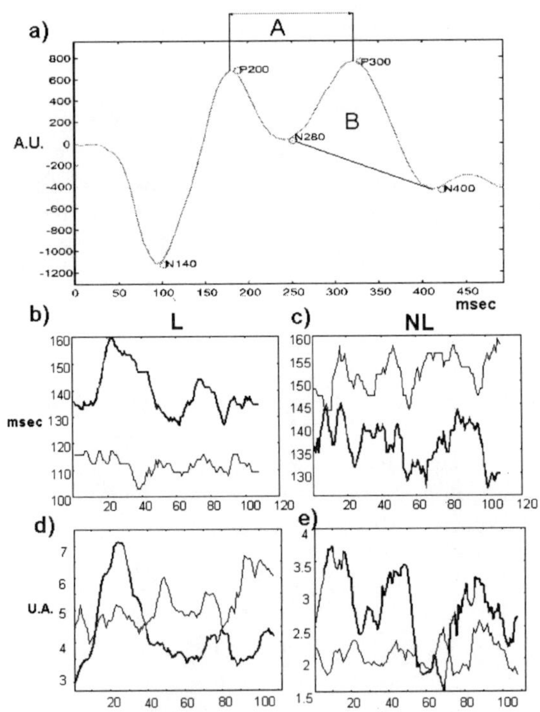

Figure 4. a) Template related to F3 electrode during the learning protocol, in particular the interlatency between P200 and P300 peaks (A) and the area under the curve (B) are put into evidence; b) area under the P300 peak (parameter B) in the learning subject (L) during the learning condition (dark line) and control condition (light line); c) area under the P300 peak (parameter B), in the not-learning subject (NL) during the learning condition (dark line) and control condition (light line); d) latency between P200 and P300 peaks (parameter A) in the learning subject (L) during the learning condition (dark line) and control condition (light line); e) latency between P200 and P300 peaks (parameter A) in the not-learning (NL) subject during the learning condition (dark line) and control condition (light line).

pairs after Laplacian transformation of the raw recorded potentials.[23] The data are referred to the frequency band inside 20.5-26.5 Hz during the learning condition, and the squared coherence is shown as function of the time; *t=0* is referred the appearance of the target on the screen. Usually the movement is started after 400-500 ms and lasts about 700 ms. Thus we may state that after 2 s it is concluded. The signal has been analyzed up to 3 s after the presentation of the target. The main differences between L and NL subjects are in panel a), where the coherence between pre-motor and motor areas is shown. L subject is characterized by a modulation in the links between pre-motor and motor cortex: in particular the coherence decreases before and during the movement, while increases at its end, when the subject knows the result of his/her performance. In NL subject the coherence is not modulated and, in particular in this case, is almost constant at high values. Coherence between C3-O1 and between F3-O1 are shown in panel b) and c) respectively. They do not show a different trend in L and NL subjects. In conclusion the learning process seems to be restricted in the different link between pre-motor and motor cortex along the time, and in particular in its increasing at the knowledge of the results.

FROM STRUCTURE TO FUNCTION

Contribution to a Computational Biology of the Heart

Recent results on HRV signal analysis show that its dynamic behavior involves nonlinear components that also contribute in the signal generation and control.[24]

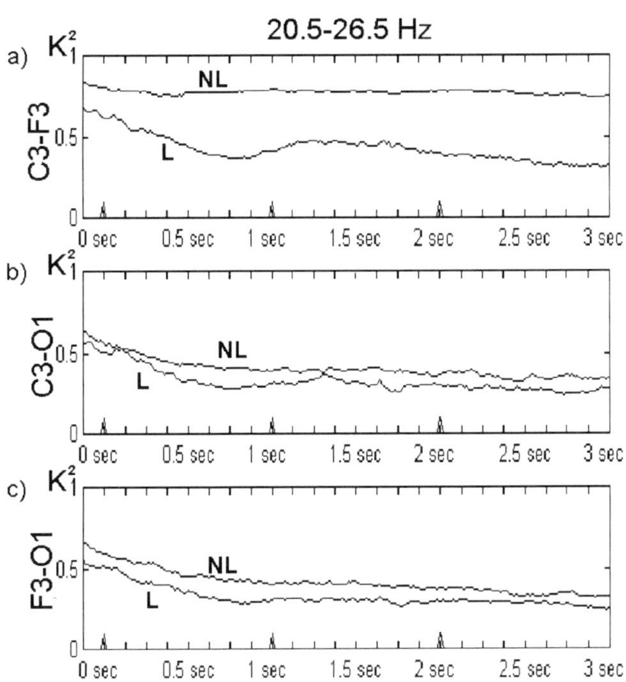

Figure 5. Coherence function in the frequency range 20.5-26.5 Hz, evaluated sample by sample between electrode pairs after Laplacian transformation of the raw recorded potentials. The curve L is referred to the learning subject, while curve NL is referred to not-learning subject, both during the learning condition. The time $t = 0$ is referred to the presentation of the target on the screen. a) coherence between F3 and C3; b) coherence between C3 and O1; c) coherence between F3 and O1.

It is then possible to investigate the structure of the physiological rhythms of the heart involving nonlinear dynamics by verifying how the presence of nonlinear deterministic phenomena could affect the HRV signal both in short and long temporal windows.

Measuring some geometrical properties of the time series, which can be an index of the presence of nonlinear dynamics, fulfills the objective.

A large number of these approaches is based on the reconstruction procedure whose reliability is guaranteed by the Takens theorem.[25] Many others do not presume any a priori hypothesis about the existence of a system attractor with complex chaotic characteristics. They only perform a measurement of some geometrical properties of the time series, which can be an index of the presence of nonlinear dynamics.[26] Among them, we particularly consider methods performing a measurement of *Self-similarity H parameter* in time series.

A time series $x(t)$ that satisfies (after appropriate setting of the initial conditions) the relation

$$x(\lambda t) =_d \lambda^H x(t) \tag{16}$$

is said to be *self-similar*. Figure 6 illustrates the rescaling in time and in amplitude of the time series. The $=_d$ implies that both terms of the relation have the same distribution function. H is a constant ($0 < H < 1$), λ is the time scaling factor. If a signal $x(t)$ is self-similar its distribution remains unchanged even after a rescaling of a factor λ^H.[27]

H characterizes the level of *self-similarity*, providing information on the recurrence rate of similar patterns in time at different scales. $H \cong 1$ implies a strong positive correlation, i.e. an increase of $x(t)$ is likely followed by another increase. $H \cong 0$ means a strong negative correlation: in this case a decrease follow an increase of x(t), and vice-versa. If $H = 0.5$, then the self-similar signal $x(t)$ is completely uncorrelated. The self-similarity parameter is even known as *Hurst exponent, long-memory parameter, long-term correlation parameter* or simply *index H*.

Estimation of the Hurst Exponent

There just exist well-known, accepted methods to estimate the exponent of Hurst in different ways: method of *modulus, variance, periodogram* and *Higuchi*.[28] All these techniques provide a global value that quantifies the self-similarity by the analysis of statistical properties of the sequence.

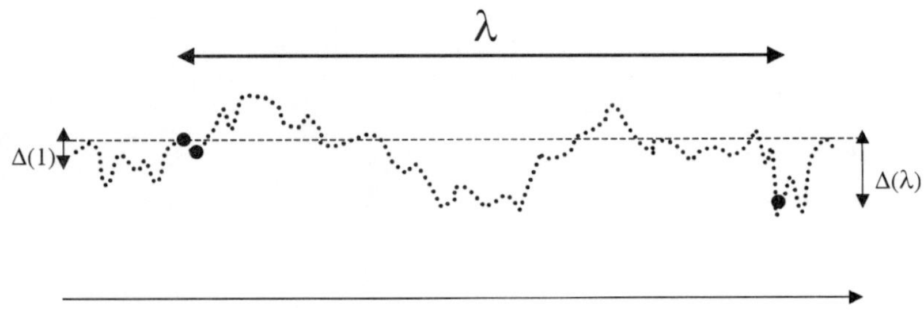

Figure 6. Definition of Self-Similarity: $x(t) \Rightarrow \lambda^{-H} x(\lambda t)$ *(λ is the scaling factor)*; $x(t) =_d \lambda^{-H} x(\lambda t)$. $=_d$ means equality in distributions. H exponent is called self-similarity parameter, long period correlation, Hurst exponent, long period memory. H=0 ⇔ negative correlation; H=0,5 ⇔ uncorrelated signal; H=1 ⇔ positive correlation.

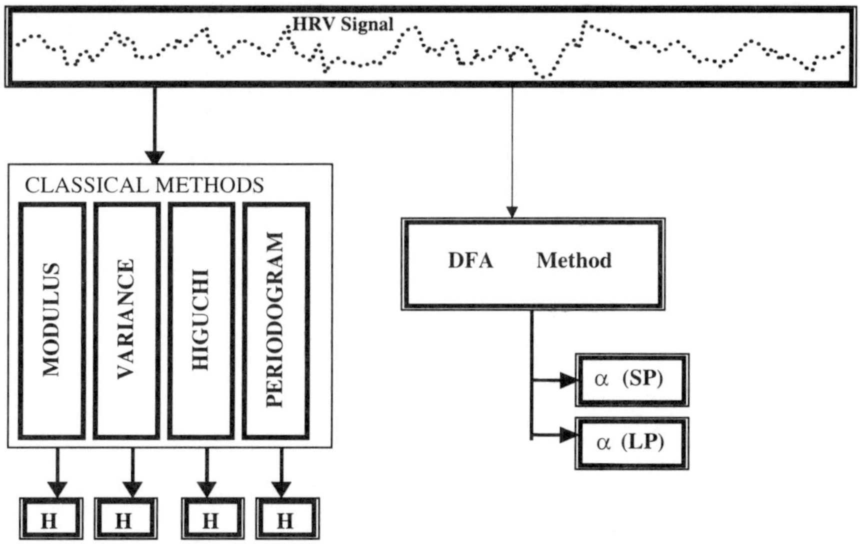

Figure 7. Summary of methods for the H parameter estimation: Absolute values of the aggregated series, Variance of the aggregated series, Higuchi, Periodogram, Detrended Fluctuation Analysis (DFA).

Furthermore, the *Detrended Fluctuations Analysis* (*DFA*) is useful to distinguish the content of fractality in short and long temporal scales. *DFA* method allows extending the definition (16) for different values of the parameter λ.[29] Figure 7 summarizes methods that have been used to estimate self-similar (or self-affine) properties of Heart Rate Variability (HRV) series.

Results

The estimation of exponent H in the HRV sequences by 'classical' methods provides an expected result: Table 1 shows normal subjects have inferior values as regards to the heart diseased ones. In other words, HRVs of normal subjects show patterns that are similar to the 1/f noise. DFA estimates α slope ($H=\alpha-1$) in normal and diseased group, both in short (S: short period, 4,000 beats) and long time scales (L: long period,10,000 beats).

Table 1. H Values Estimated Through Different Methods

H Techniques	*avg±STD* **normal**	*avg±STD* **diseased**	**Test *t***
Variance	0.127±0.034	0.205±0.072	$p<0.05$
Higuchi	0,121±0.022	0.141±0.042	n.s.
Periodogram	0.092±0.054	0.165±0.107	$p<0.05$
Modulus	0.115±0.017	0.153±0.050	$p<0.05$

Figure 8 illustrates an example of DFA for a Normal and for a Heart Failure subject.

DFA evidences two different time scale values in the HRV series. In fact the DFA graph shows two slope for the α values: α_{short} and α_{long} respectively. For normal subjects $\alpha_{long} < \alpha_{short}$ as for Heart Failure population $\alpha_{long} > \alpha_{short}$. Distinction is then possible, in many cases, only by visual inspection of the relevant graphs.

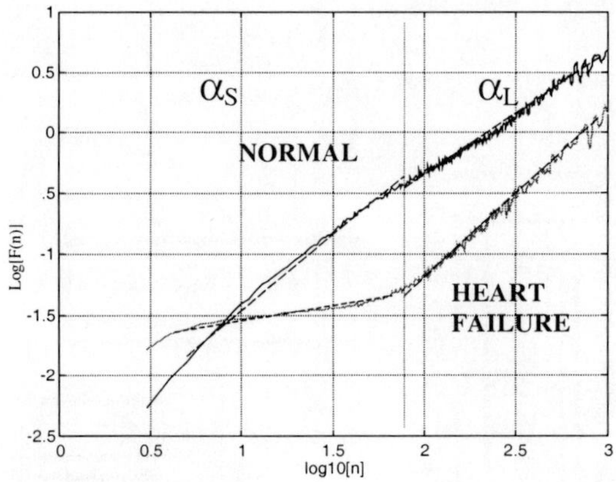

Figure 8. Estimation of fractal properties in HRV signal through the Detrended Fluctuation Analysis, DFA, method. S: short period (4,000 beats); L: long period (10,000 beats). Normal subjects show Random Walk in short period, 1/f noise in long period. HF patients show 1/f noise in both short and long period. Normal subjects are more correlated than HF patients in long period and less correlated in short period.

Normal subjects show Random Walk in short period, 1/f noise in long period. HF patients show 1/f noise in both short and long period. Normal subjects are more correlated than HF patients in long period and less correlated in short period.

Table 2 shows results of discrimination between four categories of parameters (S=short, L=long, N=normal, HF= heart failure).

We have analyzed with the three methods the 24 hours HRV series of normal and heart transplanted subjects.[30] The length N of the RR interval series varies from 70.000 to 130.000 points. Subjects in healthy conditions show a higher value of the slope, which means smaller values in terms of H. We obtained H values (avg.± STD) within the following range: 0.12±0.07 (N) vs. 0.22±0.11 (T) with the *periodogram* (p<0.02 provided by the Student test); 0.17±0.04 (N) vs. 0.31±0.02 (T) with the *absolute value* ($p<2\times10^{-7}$); 0.13±0.04 (N) vs. 0.21±0.03 (T) with *Higuchi's* method (p<0.0002). As the p values demonstrates, *Higuchi's* and *absolute value* method did the best performance.

Results in ICU patients indicated that the considered indexes seem strongly related to the prediction of both positive and negative patient outcome.[31]

Finally, in subjects after myocardial infarction, the H parameter estimation through the periodogram slope calculation was able to separate normal subjects from patients with reduced Ejection Fraction.[32] The analysis always confirms the role of HRV nonlinear dynamics as a global marker of cardiovascular pathologies.

The present results confirm the exponent H is a powerful indicator of the neural control activity on the heart over long time scales. Further decomposition of H value could be used to better understand fractal properties of the heart, which shows nonlinear behavior, thus improving the traditional linear analysis.

Table 2. α Values Estimated Through DFA Method. N=Normal, HF=Heart Failure

DFA	TEST t
$\alpha_{S,N} > \alpha_{S,HF}$	99.9 %
$\alpha_{L,N} < \alpha_{L,HF}$	99.9 %
$\alpha_{S,N} > \alpha_{L,N}$	99.9 %
$\alpha_{S,HF} < \alpha_{L,HF}$	98 %

CONCLUSION

Both neurosensorial and cardiovascular systems are characterized by complex structures, from the cellular level, up to the tissue, the organ, the whole system. Their functions are manifold as they have to respond to different situations, in relation to external and internal stimuli. It is not surprising that the expression of their functions, in the present case the biological signals, are characterized by complex dynamics, that require proper analysis tools able to quantify different aspects and situations.

Modeling of the signal-noise interaction, time-warping procedures, time dependent spectral estimation, multiparametric approaches, the estimation of self-similarity indices, the detrended fluctuation analysis, are only examples of the different methods that are applied at this purpose. Each of them is intended to describe a particular aspect of the system under analysis providing different information to be integrated with anatomical, physiological and biological knowledge.

The learning process is really a complex function, still unknown, in which a relevant role is played by the anatomical connections among different cortical structures and by modulations in their functional relationships. Quantitative indices are here evaluated that may help in highlighting the complex mechanism involved.

Results obtained in the analysis of HRV time series provide an important confirmation that a healthy cardiovascular system is strongly characterized by fractal properties in the time behavior of its variables. When a cardiovascular system illness takes place the system undergoes a reduction of these complex patterns.

In this chapter we have shown a few examples in applying engineering principles and methodologies to study and analyze important physiological and biomedical problems such as the structure to function relationships in the neurosensorial and cardiovascular system organization. In fact only the study of the biological synthesis and organization of the information from genes to the whole system can generate an integrated body of knowledge which will be able to improve the comprehension of the physiological behavior, methods of diagnosis, treatment and prevention of disease.

Not only does biological structure affect the electrical or mechanical function of a system, but electrical and mechanical factors also influence biological function.

The final aim of this integrated Bioengineering is to clarify the role of forces in relating structure to function in biology.

REFERENCES

1. S. Cerutti, G. Baselli, D. Liberati, and G. Pavesi, Single sweep analysis of visual evoked potentials through a model of parametric identification, *Biol Cybern*, pp. 56-111 (1987).
2. R. Magni, S. Giunti, A. M. Bianchi, G. Reni, F. Bandello, A. Durante, S. Cerutti, and R. Brancato, Single sweep analysis using an autoregressive with exogenous input (ARX) model, *Docum. Ophtalm*, pp. 86-95 (1994).
3. E.W. Jansen, M. Nygaard, and S.W. Henneberg, On line analysis of middle latency auditory evoked potentials (MLAEP) for monitoring depth of anaesthesia in laboratory rats, *Med Eng Phys*, pp. 20-722 (1998).
4. S. Cerutti, G. Chiarenza, D. Liberati, P. Mascellani, and G. Pavesi, A parametric method of identification of single-trial event-related potentials in the brain, *IEEE Trans. On Biom. Eng.*, pp. 9-701 (1988).
5. W. Gersch, Spectral analysis of EEG by autoregressive decomposition of the time series, *Math. Biosc.*, pp. 7-205 (1970).
6. S. Cerutti and D. Liberati, Autoregressive modelling and filtering of EEG signal generating mechanism in patients under surgical interventions, *Sign. Proc.*, pp. 9-25 (1985).
7. H. Akaike, Statistical predictor identification, *Ann. Inst. Statist. Math.*, pp. 22-203 (1970).
8. G.E.P. Box, Time series analysis, forecasting and control, *Holden-Day* S. Francisco, (1976).
9. D. Liberati, S. Cerutti, E. Di Ponzio, V. Ventimiglia, and L. Zaninelli, The implementation of an autoregressive model with exogenous input in a single sweep visual evoked potential analysis, *J. Biomed. Eng*, pp. 11-285 (1989).

10. LT Mainardi, J Kupila, K Nieminen, I Korhonen, AM. Bianchi, L Pattini, J. Takala, J Karhu, and S Cerutti, Single sweep analysis of event related auditory potentials for the monitoring of sedation in cardiac surgery patients, *Comp.Meth Prog Biomed*, in press.
11. H. C. Huang and B. H. Jansen, Eeg waveform analysis by means of dynamic time warping, *Int. J. Biomed. Comput.*, pp. 17-135 (1985).
12. K. Roberts, P. Lawrence, A. Eisen, and M. Hoirch, Enhancement and dynamic time warping of somatosensory evoked potential components applied to patients with multiple sclerosis, *IEEE Trans on Biiomed. Eng.*, pp. 34-397 (1987).
13. L. Gupta, D. L. Molfese, R. Tammana, and P.G. Simos, Nonlinear alignment and averaging for estimating the evoked potential, *IEEE Trans. On Bimed. Eng*,. pp. 43-348.
14. F. Hlawatsch and G.F. Boudreaux-bartels, Linear and Quadratic Time-.frequency Signal representations, *IEEE Signal Processing magazine*, pp. 4-21 (1992).
15. O. Rioul and M. Vetterli, Wavelets and signal processing, *IEEE SP Mag.*, pp. 10-14 (1991).
16. L. Cohen, Time-frequency distribution, A review *Proc. IEEE*, pp. 7-941 (1989).
17. S. M. Kay and S.L. Marple, Spectrum analysis: a modern perspective, *Proc. IEEE*, pp. 69-1380 (1981).
18. L. Liung and T. Soderstrom, Theory and Practice of Recursive Estimation, *Cambridge, MA: MIT Press*, (1983).
19. M. Bianchi, L. T. Mainardi, E. Petrucci, M. G. Signorini, M. Mainardi, and S. Cerutti, Time-variant power spectrum analysis for the detection of transient episodes in HRV signal, *IEEE Trans. On Biom Eng.*, pp. 40-136 (1993).
20. L. T. Mainardi, A. M. Bianchi, R. Furlan, S. Piazza, R. Barbieri, V. di Virgilio, A. Malliani, and S. Cerutti, Multivariate time- variant identification of cardiovasdcular variability signals: a beat-to-beat spectral parameter estimation in vaso-vagal syncope, *IEEE Trans. on Biomed. Eng.*, pp. 44-978 (1997).
21. M. Bianchi, T. Locatelli, L. T. Mainardi, M. Cursi, G. Comi, and S. Cerutti, Event related brain potentials: laplacian transformation for multichanel time-frequency analysis, *Meth. Of Inf. In Med.*, pp. 39-160 (2000).
22. J. Gordon, M. F. Ghilardi, and C. Ghez, Impairmnets of reacing movements in patients without propioception, I. Spatial errors, *J. Neurophys.*, pp. 73-347 (1995).
23. Bianchi,M., T. Locatelli, L. Mainardi,T., M. Cursi, G. Comi, and S. Cerutti, Event-related brain potentials: laplacian transformation for multichanel time-frequency analysis, *Methods of information in medicine*, pp. 39-160 (2000).
24. L Glass et al. eds., Theory of the heart. Biomechanics, Biophysics and Non Linear Dynamics of Cardiac Functions, Springer Verlag (1989).
25. H.D.I Abarbanel, R. Brown, J.J. Sidorowich, and L.S. Tsimiring, The analysis of observed chaotic data in physical systems, *Rev. Mod. Phys,* pp. 65-1331 (1993).
26. B.B. Mandelbrot, Self-Affine Fractals and Fractal Dimension, *Physica Scripta*, pp. 32-257 (1985).
27. J. Beran, *Statistics for Long-Memory Processes*, Chapman & Hall, New York, (1994).
28. M.S. Taqqu, V. Teverovsky, W. Willinger, Estimators for long-range dependence: an empirical study, *Fractals*, pp. 4-785 (1996).
29. C.K. Peng, S. Havlin, J.M. Hausdorff, J.E. Mietus, H.E. Stanley, A.L. Goldberger, Fractal mechanisms and heart rate dynamics, long-range correlations and their breakdown with disease, *Journal of Electrocardiology* 28 Suppl:59 (1995).
30. M.G. Signorini, Evaluation of self-similarity and regularity patterns in heart rate variability time series. "Chaos and Noise in Biology and in Medicine", S. Chillemi and M. Barbi eds., World Scientific Publishing, Singapore, pp. 7-249 (1998).
31. L.T. Mainardi, A. Yli-Hankala, I. Korhonen, M.G. Signorini et al., Monitoring the Autonomic Nervous System in the ICU Through Cardiovascular Variability Signals, *IEEE-EMB Mag.*, pp. 16-64 (1997).
32. F. Lombardi, G. Sandrone, A. Mortara, D. Torzillo, M.T. La Rovere, M.G. Signorini, S. Cerutti, and A. Malliani, Linear and nonlinear dynamics of heart rate variability after acute myocardial infarction with normal and reduced left ventricular ejection fraction, *Am. J. Cardiol.*, pp. 77-1283 (1996).

ARCHITECTURES FOR ARTIFICIAL MENTATION AND BEHAVIOR

Igor Aleksander

Intelligent and Interactive Systems Group
Department of Electrical and Electronic Engineering
Imperial College of Science, Technology and Medicine
London SW7 2AZ
e-mail: i.aleksander@ic.ac.uk

INTRODUCTION

This work supports the hypothesis that behavior is the result of rich but mediated representations of the real world and the organism within it. 'Rich' means containing all the detail necessary for appropriate action taking and 'mediated' means that representations of 'self' are at work which include decisions on whether and how to act. The central issue is an understanding of the emergent depictive properties of brain-inspired complex architectures of which the components are neural network modules each with their own emergent properties. This work contrasts markedly with Artificial Intelligence techniques, a point which is made early in the chapter.

The consideration of artificial systems that think, act and decide, generally evokes traditional techniques in artificial intelligence where an artificial organism, faced with a complex sensory scene attempts to process signals to assign object-related symbols to the scene and generating actions from a rule-base which also receives data as symbols from a user that indicates the user's commands. On the other hand, primate brains recreate the world by rich depiction in neural firing patterns in areas distributed throughout the brain. It is from these depictions that actions are driven in a contemplative manner, that is inner exploration of action-result scenarios take place before action is unleashed.

In this chapter we indicate some of the complexity of these architectures and draw attention to the fact that achievement of thought and action depends heavily on the design of brain-inspired architectures that have the depictive property (this will be defined). This leads to the creation of artificial neural organisms in which depiction, evolution and emergence are the forces that lead to success rather than recognition or symbolic representation which is characteristic of more classical approaches in artificial intelligence and neural networks. The chapter contains two examples, the first of which simply defines the depictive property in the context of a planning problem, and the second is a model of a visual system that has the depictive property and, on the basis of a moving visual transducer, an interface with a simple virtual world and the use of a single-word verbal interface.

Human and Machine Perception 3: Thinking, Deciding, and Acting, edited by Cantoni *et al.*
Kluwer Academic/Plenum Publishers, New York, 2001.

REVERSE ENGINEERING THE BRAIN = REVERSE ENGINEERING THE MIND?

Designing brain-inspired architectures is tantamount to asking whether the complexity and architectural properties of the brain are relevant to the emergence of mind. By '*mind*' is meant the capacity for thought and hence the sum total of all the possible thought generated by the architecture. *Thought* (or, mentation) is defined in classical ways: perception, memory, and planning. Normal human thinking makes heavy use of internalized models of the sensory world as the prerequisite to action. We shall formalize such models later by referring to them as *depictions.* Such models represent the world with sufficient accuracy to favor survival in an evolutionary sense. So the strong assumption in the methodology used in this chapter is that, indeed, reverse engineering the brain by the creation of a virtual brain is relevant to the study of both real and artificial systems in which purposeful mentation precedes action. That is, it is argued that a virtual brain creates a virtual mind, promoting a study of the relationship between the two.

It is vitally important to distinguish this activity from the traditional aims and methods of artificial intelligence, and the next two sections are designed to do this.

ARTIFICIAL INTELLIGENCE AS THE ART OF SEARCHING

It is now fifty years since the first program of artificial intelligence was proposed. This piece of innovation was interestingly initiated by the same person who pioneered the use of Boolean Algebra in digital system design, and Information Theory: Claude Shannon.[1] This is important in the arguments of this chapter as Shannon's chess playing algorithm focused on look-ahead searching with appropriate curtailment (the minimax and alpha-beta methods of avoiding exhaustive searching). Searching became the salient tool even in problem solving (thought-action) problems. Typical is the classical block re-stacking problem. For four blocks it turns out that a typical search space has 85 block positions which may have to be traversed before the solution is found.

We contrast this with the mental mode in humans where the solver is not aware of more than the few moves that lie between the starting and finishing states. It is not advocated that the reason for this is that the whole search is somehow 'unconscious', rather, it will be argued below, the induced start and end states act as controls on the state trajectory of a knowledge-holding dynamic neural system. It is not implied that Artificial Intelligence did not come up with some very smart programs (particularly in Chess Playing), but it is implied that notice should be taken of the way that our inner sensation in solving a problem tells us that the search methodology should have an alternative.

Artificial intelligence has never really got to grips with the operating principles of the brain. Of course, an airplane does not have to flap its wings, but this does not mean that the operating principles of bird flight are not of interest to the aeronautical engineer.

It is also important to look hard at acts that involve intelligence in living creatures that AI has not been able to reach. These lie largely in areas where interactions with real worlds are involved (Robot visual cognition) or areas where natural language interactions are to be supported.

EMERGENCE, EVOLUTION, AND DEPICTION

The above three concepts are the key differentiators between the operating principles of the brain and those that are commonly used in AI. It is best to deal with them in turn.

Emergence

The word emergence has been used since the late 1960s to *mean a surprising behavior of a cellular system which is not easily derived from a knowledge of the function of the cells and their interconnections*. The word has, indeed, been used in some celebrated work on neural nets[2] in the context of an emergent stability resulting from the training of a fully connected network. Emergence is also a controversial word – it has overtones that prompt both mysticism and incredulity. There are those who argue that given a large enough cellular system, mind will emerge unaided, while others suggest that emergence points to properties of cellular structures that will never be understood. The computing skeptic simply sees surprises as bugs in the program.

In this chapter the view is taken that emergent stability is a property of some of the modules. This is relatively well understood not only through the Hopfield energy-based analysis which applies to neurons with differentiable transfer functions but also automata-theoretic arguments which apply to the digital neurons used in systems reported a few years ago.[3] But more important, is the fact that achievement of an intelligent task, is seen as the result of behavior emerging from the interaction of neural modules, each of which might have emergent stability. This is to be contrasted with the achievement of intelligence in AI where reliance is placed on search through a large base of rules and facts and, indeed the emergence of stability in single neural modules.

Far from emergence being something that can never be understood, it is vital in brain-inspired systems that, due to the central role that this force plays in performance, that theory should be deployed to understand it.

Evolution

The reverse engineering of the brain has to cope with a design that has evolved over millennia. This has two major implications. First, it is not clear how the pattern of interconnections between the 100 or so distinct modules (of just the cerebral cortex) contributes to the emergence of mind. Second, it suggests that evolution should be at work, or, at least, be understood as a mechanism for optimizing the structure of the artificial system.

In previously reported work,[4] we showed how evolution has surprising effects on the optimality of a visual system which sees, names and imagines (virtual awareness). The task was to name and imagined a combination of features not present in the training set. Two modes were considered, evolution through improvements made by a designer to remove erroneous performance and an automatic evolution based on random variations in the design. Some of the results of this work are reproduced in Figure 1. These show how the system behaved when unevolved, when evolved by design and when evolved automatically. The 'features' column refers to performance obtained when the training set includes all combinations of features.

In this particular case, the fitness function was the accuracy of the internal visualization of the external world. This points to one of the difficulties of this work: the fitness function of a real brain is the very quality of conscious experience, and this begs the question of what conscious experience actually is.

Depiction

As depiction is the focus of this chapter, it is fully treated in the next section.

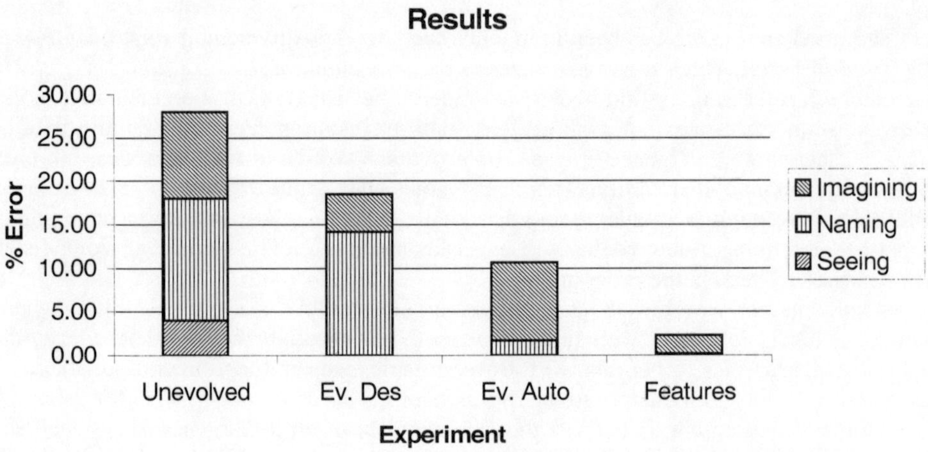

Figure 1. The effect of evolution on a virtual awareness system.

DEPICTION: THE BASIS OF BRAIN-INSPIRED ARCHITECTURES

In addition to search, a hallmark of the AI paradigm is the symbolic representation of problems and environments. In contrast, the living brain only uses symbols as means of linguistic communication and not means of processing. It is also notable that it is unlikely that species other than humans use symbols. However, it is clear that survival of living organisms depends on inner states that accurately represent environments and allow the organism to imagine the consequences of its own actions. Such inner states are assumed to be due to neural firing patterns and they are described as *depictions* to distinguish them from representations which have symbolic overtones.

In our work, we have assumed that depictions have a direct relationship to world events (rather than being transformations of world events) for the simple reason that this resolves the 'conscious experience' problem. Sensation is of a real and not a transformed world, so were depiction transformed this leads to some mystical theories about the reverse transformation. Depictions in the brain are distributed in the sense that they occur in different parts of the brain for the same elemental event. This gives rise to discussions about binding or even of how sensations arise without binding.[5]

Depiction allows the definition of 'sensation' even in an artificial system. In the brain, the question of which neural cells contribute to sensation and which do not needs to be faced. This has been hypothesized in Aleksander and Dunmall[5] and Aleksander[6] as being the cells which have the necessary support for depiction. This support is both innate (as for musculature) or learned, as in the case of language.

VIRTUAL NEURONS AND VIRTUAL BRAINS

The word 'virtual' is used rather than the word 'simulated', as the virtual object needs to be manipulated as if it were the real thing. Therefore the neuromodelling starts with the 'weightless' neuron (see Aleksander and Morton[3] and Aleksander et al.[4] for a full description). Weightless neurons are lookup tables, using one-shot learning with nearest-neighbor generalization. Color is encoded as the state of three binary lines, that is 8 colors. The neurons are probabilistic in the sense that if input is outside a 'nearest neighbor' radius, the neuron outputs an arbitrary signal. That is, the system is more exploratory for unfamiliar inputs than familiar ones. The computational advantage inherent in the proposed methodology may be illustrated as follows. In recent work, we were able to build a model of (adjectival phrase) visual imagination and verbal interpretation of simple scenes involving 36 modules, a total of 0.5 million neurons and a total of about 40 million connections. Each processing step (training or computation) takes 0.4 seconds on a Pentium 500 MHz machine with 128 Megabytes of RAM. As a rough assessment, most continuous models (error BP or radial basis systems) would require at least an order of magnitude longer to train and would need an order of magnitude more memory to study the same architectural effects.

The methodology is facilitated by the existence of the Neural Representation Modeller (NRM) which may be accessed on the internet.[5] Advanced versions of this contain not only neural models that achieve a function through learning, but also neurons with fixed functions such as Mexican hat filters of their receptive fields. NRM is a kind of Meccano for building brain inspired systems in order to study their emergent properties.

DEMONSTRATION OF A DEPICTIVE SYSTEM FOR SOLVING A RESTACKING PROBLEM

In the classical problem-solving area, there is a set of block re-stacking problems that were highly prominent in the 1960s and '70s. This consists of having been given a stack of colored blocks G,B,R,Y say stacked as {GBRY} where left-right corresponds to top to bottom with the rightmost element assumed to be standing on a surface. The search space is developed by a single rule that says that any topmost element may be placed as topmost

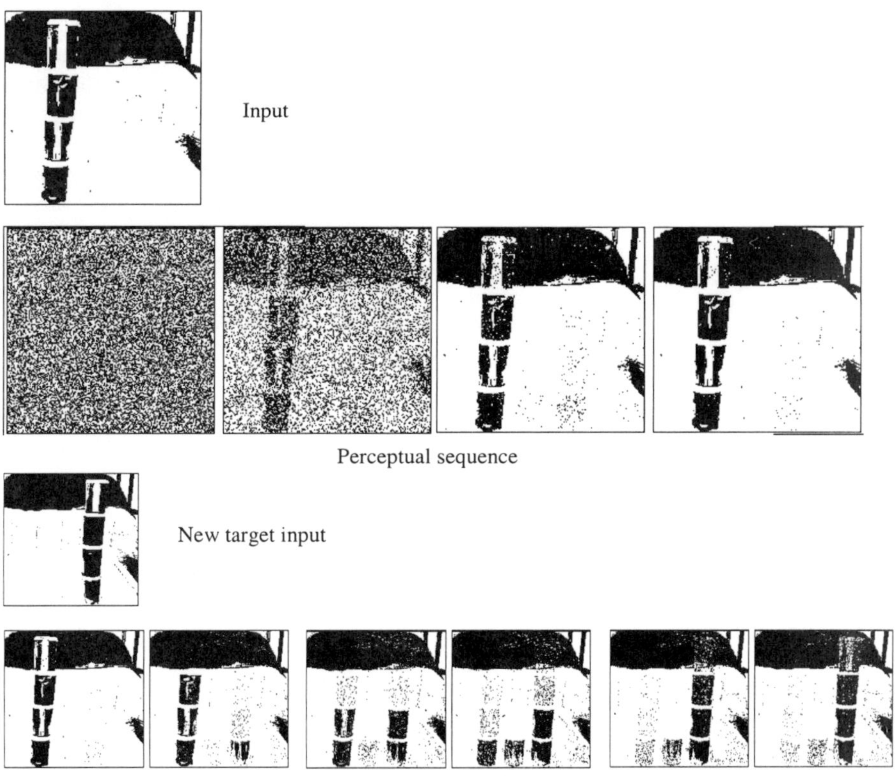

Figure 2. Depiction and emergence in a single neural state machine.

element on any existing stack including an empty one. Say then that the target is {RYGB}, the classical system performs a search (with whatever shortcuts are available) until the desired re-stack is found. The state space contains 85 possibilities. Of course there are programmer induced rules which say something like 'unstack until all the blocks are on the table then build any stack'. This requires a total of 6 moves (or 2(n-1)) for n blocks.

In neural methodology two things that distinguish the methodology from the classical one happen. First, the current state of the problem is the state of a state machine, with the desired state as controlling input. In this way legitimate moves may be learned and the search space becomes a state space. Second, in visual models, the state represents of the entire visual scene including the perception of the current and desired state of the solution.

An example is shown in Figure 2. Here it can be seen that the initial state of the problem can be perceived by a simple recursive net with input. The subsequent trajectory represents experience of rearranging the objects according to a new desired input.

It is interesting to note that depiction in this instance treats the movement of objects as a learned relationship between subsets of neurons. This gives the inner 'sensation' the character of movements in the external world – a primary property of depiction. It should be stressed that this is not a neuromodel of the brain just an illustration of the way that depiction and emergence can work together in a single neural system. The system discussed in the next section *is* a model of brain architecture.

A DEPICTIVE MODEL OF VISUAL AWARENESS

The major questions about the neural underpinnings of visual awareness, including imagined images relate to the reconstructional abilities of a depiction which is compensated for effects such as eye and head movement. In Aleksander and Dunmall[5] we have suggested a hypothesis of how a muscular referent can avoid the need for binding. Here is described the model on which this hypothesis is based. It is known that in the primate brain there is an area crucial to the visual system: the superior colliculus. This receives retinotopic projections and causes the foveal region of the eye to move to high-information areas of the visual scene. At the same time information of eye position must be fed to perceptual areas (e.g. extraxtriate modules for depicting color, motion and shape). Such areas are known to have gaze correction which is controlled by the eye position feedback. The system is

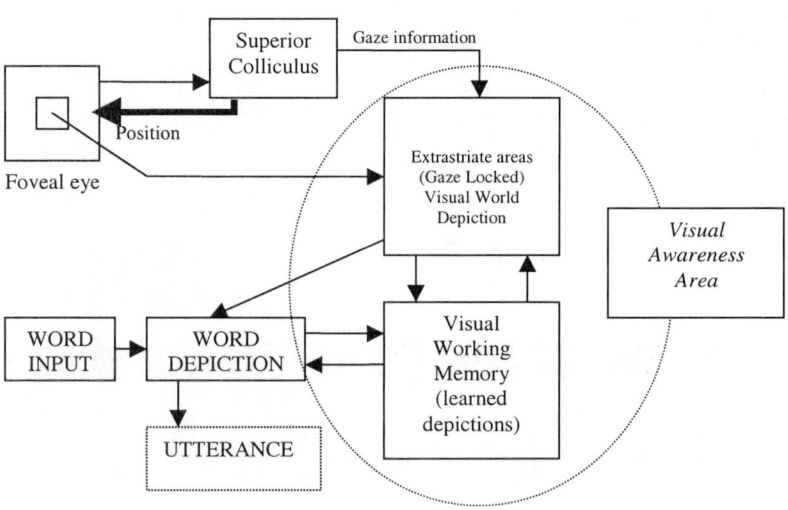

Figure 3. The generation of visual awareness.

Figure 4. Gaze performance of the superior colliculus model

extended to include visual working memory driven by single-word verbal interfaces. That is, it must be able to imagine through depiction the input word and utter the correct word as the input is being investigated. The arrangement as run in NRM is shown in Figure 3.

In this particular model low resolution perifoveal images are fed to the superior colliculus. This moves the gaze to areas that contrast with white, recognizing localized regions of such contrast and 'clearing them out'. The resulting gaze exploration on two different types of images is shown in Figure 4.

One cannot help remarking that such plots are similar in kind to that which is obtained in measuring the gaze of primates.

The rest of the model uses this gaze information to reconstruct the world image which upon reconstruction can acquire a word label and at the same time create a reconstructed image in working memory. The result of this is that on seeing a known object the awareness areas anticipate this object, as is known to occur in living systems. This effect is shown in Figure 5.

The memory mechanism operates by triggering the associated word as reconstruction is in progress, which, in turn, causes the working memory module to enter the appropriate state. As argued in Aleksander and Dunmall,[5] the awareness of these two areas coincides giving the result in Figure 5. Similarly, the system can have its inner awareness stimulated by purely by an input word.

FUTURE PROSPECTS

Referring to the use of Emergence, Evolution and Depiction (EED) as the basis of a new paradigm for achieving mental prowess in artificial systems, it becomes evident that

Without memory With memory

Figure 5. Visual awareness during the reconstruction process with and without memory.

this is a very young paradigm within which many important questions still lie in the future. The first question is how far can one progress through the use of virtuality alone. Or, in other words, how important is it to use this methodology in conjunction with situated or embodied machines. It appears to us that there is a great deal of work which needs to be done even in the virtual domain. The focus of such work is in the domain of natural language. Our current program may be summarized as having the following phases: positional language, action language and planning language

Positional Language

A kitchen scenario is defined with objects such as crockery and cutlery.

Examples of phrases: The cup is on the saucer.
The spoon is in the glass.
The knife is to the right of the plate.
The plate is behind a cup.

The key to this work lies in extending the visual attentional mechanisms which have been described above. A moving input window (eye) has both a foveal (high resolution color sensitive) center and a perifoveal (low resolution content) which is moved over the scene by a model of the superior colliculus in the brain to areas of high spatial frequency in the scene. What is now researched under this heading is the addition of higher level attentional neural mechanisms that help to disambiguate possible descriptions as these are being constructed in the depictive areas. Both naming of scenes on which the system has *not* been trained and the depiction of phrases *not* seen during training, but using words that were, is being investigated.

Action Phrases

Here phrases are being used referring to movable objects (represented by the activity of the virtual world) and involve the moving input window (with foveal/perifoveal attributes as described above).

Examples of important phrases: Follow the moving ball
Find a green ball moving to the left
Move towards the cup.

Note how the last of these implies the generation by the neuromodel of actions that are input to the virtual world. Attentional mechanisms are again of prime concern in this work, and both description and imagination is being developed and tested.

Planning and Control

This is a major study using scenes that represent the classical four-block stacking problem as outlined earlier. This continues to be done with images of real blocks as states of the virtual world. Specifically, the virtual world will represent the entire search space while the neuromodel will plan moves internally (by depiction) and externalize the solution by verbal statements.

CONCLUSIONS

In broad terms this chapter has continued offering neuromodelling as a means of achieving cognitive performance in an artificial system which goes beyond that which can be achieved by preprogrammed AI methods alone. It has been argued that the principles involved in this are depiction rather than symbolic representation, evolution that optimizes the system according to its own subjective tests rather than predesign and emergence which also has evolved to maintain a functional stability within the system rather than the unconstrained growth of. Work is being developed at the moment on developing planning behavior within such systems, using them to throw light on mental deficits (Parkinsonism and the like), applying them to mobile and torso robots, the latter being applied to surgery.

On a wider front, this methodology not only contrasts with classical AI, but, despite being neural, points neural network research in a new direction towards studying the properties of interacting modules and evolutionary techniques that give rise to such architectures.

In greater detail, some philosophical points need to be made. The classical approaches of artificial intelligence go back in history to times when computer memory was scarce and the world within a system had to be represented with great parsimony. That is, any world concept had to be represented by the fewest number of bits, namely, a *symbol*. This naturally brought computer reasoning into the domain of solving logical equations based on expressions of these symbols. The category mistake came in when it began to be said that these computer-based methods reflect the operating principles of the brain, that is, the human mind and its collection of thoughts. The key thrust of this chapter has been to argue that through the concept of depiction, the internal handling of concepts can retain some of their 'aboutness' in the world. But the operating principles of intelligence within the brain have evolved to transform raw depictions into actions and language in a controlled way. That is, contemplation about action but without action is the evolutionary base and purpose of brain process. This is what we call 'thought' and we refer to 'mind' as a signifier of the capacity to think or the sum total of thinking done by the organism.

The experiments described in this chapter are very early examples of an alternative form of computing which seeks to make virtual these complex brain processes. This demonstrates that the requirement of symbolic parsimony is no longer necessary in computation and that brain architectures provide the clues as to how these virtual machines may be built.

But this does not mean that 50 years of symbolic work with artificial intelligence systems has been wasted. In a practical sense the symbolic and the depictive are complementary. It is very hard to see, and a lot of effort, for a depictive system to learn how to do logical symbolic processing. This does happen in brains, but may leave the possibility of hybrid practical systems that use depiction and logic open. These would not be the classical neural/logic hybrid systems, but would be systems in which the interfaces between the depictive and the symbolic still have to be worked out.

REFERENCES

1. C.E. Shannon, Programming a computer for playing chess, *Philosophical Magazine*, pp 256-275, (1950).
2. J. Hopfield, Neural networks and physical systems with emergent collective properties, *Proc. Nat. Acad. Sci. USA,* 79, pp. 2554 – 2558, (1982).
3. I. Aleksander and H. Morton, *Introduction to neural computing*, London, Thompson Press, (1995).
4. I. Aleksander, B.E. Dunmall and V. Del Frate, Neurocomputational models of visualisation: a preliminary report, in *Foundations and tools for neural modelling*, Mira and Sanchez-Andres eds., Heidelberg, Springer Verlag, (1999).
5. I. Aleksander and B.E. Dunmall, An extention to the hypothesis of the asynchrony of visual consciousness, *Proc Royal Society, London B*, 270, pp. 197-200, (2000).
6. I. Aleksander, *How to build a mind*, London, Weidenfeld and Nicholson, (2000).
7. www.sonnet.co.uk/nts

DEVELOPING RATIONAL AGENTS

Marco Colombetti and Pier Luca Lanzi

Department of Electronics and Information
Politecnico di Milano
Piazza Leonardo da Vinci, 32
20133 Milano, Italy
e-mail: {colombet,lanzi}@elet.polimi.it

An agent is a system that interacts with an environment continually and without human assistance in order to carry out a predefined task. We are interested in developing artificial agents that act rationally, in the sense that they are able to maximize a suitable utility function. In this chapter, we describe the main problems underlying the realization of rational agents and present commonly adopted mathematical models. In particular, we consider the case in which the environment can be modeled as a finite state stochastic process and address the problem of developing agents that can learn to act rationally through their own experience.

INTRODUCTION

In recent years, agents have become a central concern in various fields of computer science. The term 'agent' is used to denote different kinds of artificial systems, from autonomous robots to programs that search the Web or take part in electronic auctions (see for example Wooldridge[1]). In this chapter we shall be concerned with robotic agents, but much of what will be said also applies to "soft agents", that is, to systems that do not possess anything similar to an artificial body.

Intuitively, an *agent* is an artificial system interacting with an *environment* continually and without human assistance in order to carry out a *task*. An agent's environment is understood as the part of the world that can affect, or be affected by, the agent; for example, the environment of a mobile robot is the part of the physical world that is perceived by the robot's sensors and can be affected by the robot's effectors. A task is what the agent is designed for; for example, the task of a surveillance robot is to discover possible intruders, signal their presence and describe their positions.

In the relevant literature, agents are often dubbed as "autonomous", "intelligent", or "rational". These terms express important aspects of agent design but are difficult to define rigorously. In this chapter we want to analyze the concept of *rational agent*. Rationality will be interpreted from an admittedly narrow utilitarian standpoint: an agent will be said to be rational if it acts so that it maximizes some measure of expected utility. We shall discuss a number of models that have been used to develop agents of this kind, starting from the

Human and Machine Perception 3: Thinking, Deciding, and Acting, edited by Cantoni *et al.*
Kluwer Academic/Plenum Publishers, New York, 2001.

simplest models, then analyzing their limitations and moving to more complex ones. In particular, we shall devote some attention to the problem of developing artificial agents that learn to act rationally from their own experience.

The chapter is structured as follows. Firstly, we give an informal definition of the concept of agent. Then, we introduce the mathematical apparatus we shall use to model agents and their environments. Later we describe reactive agents, that is, agents that base their actions on their current perceptions only; in particular, we briefly sketch how a reactive agent may learn a rational action policy from experience. Then we introduce cognitive agents, that is, agents that maintain an internal representation of their environment. Finally we consider the role of motivations. We conclude by sketching some directions for future research.

AGENTS AND THEIR ENVIRONMENTS

At the highest level of abstraction, the relationship between an agent and its environment is described by the scheme of Figure 1. The agent performs *actions*, a, that affect the *environment's state*, x; in turn, the environment's state affects the agent and, eventually, the actions it performs.

Both the agent and the environment are physical systems. For the moment, we make no restrictive hypothesis of the environment: we assume only that it can be modeled as a dynamic system. On the contrary, it is coherent with the scope of this chapter to suppose that the agent is a *digital system* (i.e., a digital computer running a program). As such, the agent has an operation cycle (that we shall call the *agent cycle*), which induces a discrete structure on time: at a generic cycle, t, the agent samples the *environment's state*, $x(t)$, and executes an action, $a(t)$. The delay between the sampling of the environment and the execution of an action is treated as negligible with respect to the time interval between two consecutive samplings.

Example 1

Consider a simple mobile robot in a closed territory, schematized as a two-dimensional grid like the one presented in Figure 2. At each cycle the robot, represented by the diamond, can move one cell apart in the four directions shown by the arrows ('Up', 'Down', 'Left', and 'Right'). The robot cannot cross walls, depicted as thick borders.

In a situation like this, it is very important to understand what counts as the environment and what counts as the agent. Intuitively, our world consists of two entities: a grid-shaped territory and a mobile robot. In turn, the robot consists of a "body" and a "mind", the latter being understood as the computer program controlling the robot's behavior. The agent coincides with the robot's mind, while the territory *and* the robot's body count as the environment. More precisely, the position of the robot in the territory will be regarded as the state of the environment. In the example of Figure 2 there are therefore 16

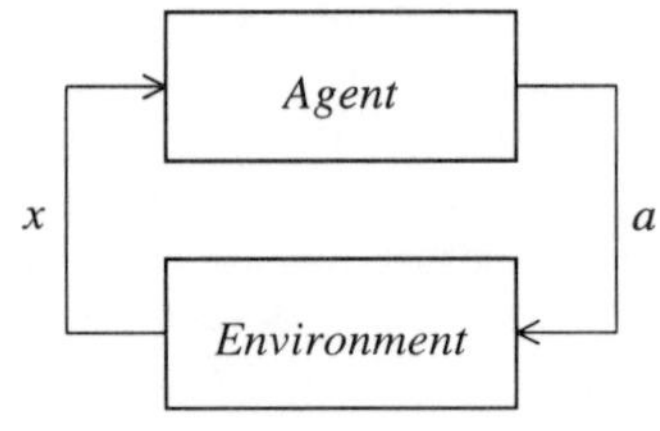

Figure 1. An agent and its environment.

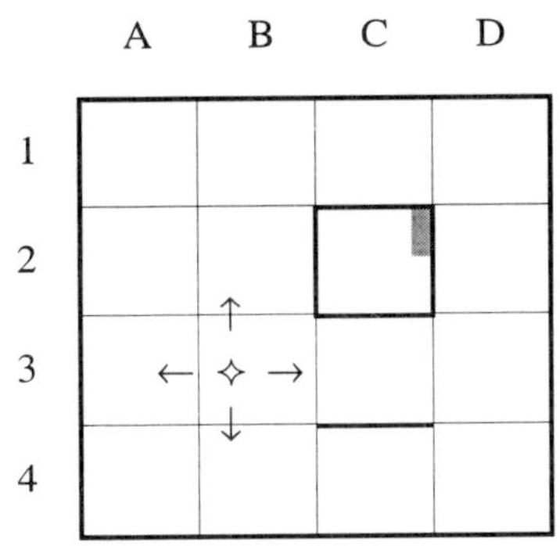

Figure 2. A mobile robot in a grid environment.

different states, from A1 to D4. At every cycle, the agent may decide to move. As we have already seen, there are 4 possible actions, to which we may add the *null action* ('Stay Still'). If the action succeeds, the robot's body will change its position accordingly, thus affecting the state of the environment.

Let us now push our analysis one step further. We can model the agent as the sequence of two processes: *perception* and *action* (Fig. 3). Perception and action interface the agent to the environment through input transducers (i.e., *sensors*) and output transducers (i.e., *effectors*). The output of the perception module is fed into the action module.

Example 2

In a grid world like the one of Figure 2, perception can be modeled in different ways. One possibility is that the robot accurately perceives its absolute position on the grid. In such a case, there will be 16 different perceptions, in a one-to-one correspondence with the 16 states of the environment. In the situation of Figure 2, for example, the robot may correctly perceive its position as B3. In such a case, we say that perception is *faithful*. A more realistic model will allow for *perceptual noise*. Again in the situation of Figure 2, for example, the robot may incorrectly perceive its position as A3.

Example 3

Another possibility is to assume that the robot has telemetric sensors, allowing it to measure the distances from walls in the four possible directions. If we list such measures clockwise (in the order: up, right, down, and left), the robot of Figure 2 will thus have the

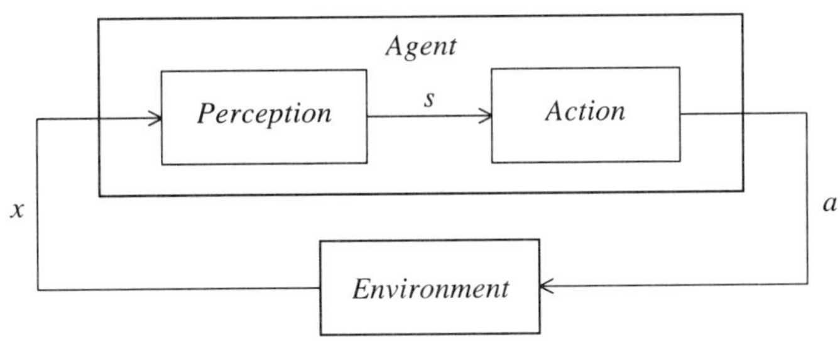

Figure 3. The basic processes of an agent.

perception 2211. It is interesting to remark that the robot may have the *same* perception in *different* environment's states. For example, the robot's perception would be 0102 both in C1 and in C4. In such circumstances, we say that the robot's perception is *aliased* (which means it is no longer in a one-to-one correspondence with the states of the environment). Also in this case there might be perceptual noise. For example, the robot of Figure 2 may incorrectly perceive its position as 2111.

The complexity of an agent's behavior depends crucially on three factors: the internal architecture of the agent, and the structure of the environment, and the structure of the task. In the rest of the chapter, we shall introduce a number of models, from the simplest to the more complex, in order to account for real agents of different complexity. In particular, we shall discuss agents of three categories:

- *reactive agents*, whose actions are a function of current perceptions only;
- *cognitive agents*, which entertain an internal representation of their environment; and
- *conative agents*, endowed with an internal motivational system.

MODELING AGENTS

We shall now see how the environment and the processes of perception and action can be modeled mathematically.

The Environment

We shall model the environment in terms of the *state transitions* induced by actions. In general, we suppose that $x(t+1)$, the state of the environment at time $t+1$, is stochastically determined by the previous state, $x(t)$, and by the action performed at t, $a(t)$, in a way that does not change with time. If X is the set of all possible environment's states, A is the set of all possible actions, and both sets are finite, this amounts to assuming that

$$Pr(\, x(t+1)=x' \mid x(t)=x \text{ and } a(t)=a\,) = T(x,a,x') \tag{1}$$

where T is a function

$$T\colon X \times A \times X \rightarrow [0,1] \tag{2}$$

with the requirement that

$$\sum_{x' \in X} T(x,a,x') = 1, \quad \text{for all } x \in X \text{ and } a \in A \tag{3}$$

It is customary to express (2) and (3) with the more concise notation

$$T\colon X \times A \rightarrow \Pi\,(X) \tag{4}$$

which means that T is a function from elements of $X \times A$ to probability distributions on X.

Technically speaking, we have modeled the environment as a stationary finite-state Markov process. Therefore, we shall call an environment that can be adequately modeled as above a *Markov environment*.

Perception

Being based on measurements carried out by sensors on the environment, the perception at time t, $s(t)$, is related to the actual state of the environment at time t, $x(t)$. As we have already remarked, it is often unrealistic to assume that this relationship is one-to-one or deterministic: different environment's states can be perceived in the same way (due to perceptual *alias*), and the same environment's state may be perceived differently on different occasions (due to perceptual *noise*). However, we shall at least assume that the probability of perceiving a given environment's state in a certain way is stationary (i.e., it does not change with time).

With such an assumption, we can model the relationship between environmental states and perceptions as follows. Let X be the set of all environment's states, and S the set of all possible perceptions. Both X and S are assumed to be finite sets. The conditional probability of having perception s, given that the environment's state is x, is defined as

$$Pr(\ s(t)=s \mid x(t)=x\ ) = \sigma(x,s) \tag{5}$$

where

$$\sigma: X \rightarrow \Pi\ (S) \tag{6}$$

Noiseless perception can be treated as a special case, in which perception is modeled as an ordinary function

$$\sigma: X \rightarrow S \tag{7}$$

When perception is *faithful* (i.e., neither noisy nor aliased), function σ is a bijection (i.e., a one-to-one map) from X onto S. In such a case, we can safely assume that $S = X$, and model the agent as if it had direct access to the environment's state.

Before moving on, a final comment may be useful. In our treatment, we equate perceptions with environmental states in the special case of faithful perception, but keep S and X distinct in the general case. The reader might wonder why one has to distinguish between perceptions and objective environmental states anyway, given that only perceptions can influence action (see Fig. 3). The answer is that the model we are developing does not represent the point of view *of the agent*, but rather the point of view *of the designer*. In general, an agent's designer needs to reason not only on the agent's perception, but also on the objective state of the environment, and on the relationship between the two.

Action and Utility

In actual applications, the complexity of the action module may vary a lot, depending on the structure of the environment and on the agent's task. It is in fact customary to classify agent according to the architecture of this module. As we have anticipated, we shall distinguish three types of agents: *reactive*, *cognitive*, and *conative*, which will in turn be subdivided into subtypes and described from next session on. It must be noted, however, that these terms are not used consistently throughout the relevant literature.

The action process is crucial for the success of an artificial agent. In particular, it is relative to this process that an agent can be regarded as *rational*. Being rational means to act according to reasons. In the case of human beings, reasons for action can be based on a variety of factors, like economical utility, emotional states, external laws and regulations, moral principles, religious beliefs, and superstitions. In the case of artificial agents, it is up to designers to develop systems whose actions are justified by suitable reasons. In many applications, designers will be able to base their choices on some objective utility function.

In such cases, the only reason for an agent to perform an action is that it contributes to maximizing utility.

In general, utility is regarded as a function of the effects of actions. Given that the effect of action $a(t)$, performed at time t, can be viewed as the state transition caused by $a(t)$ on the environment, we can model utility as a numeric function of environmental state transitions:

$$r: X \times X \rightarrow \Re \qquad (8)$$

The value of the utility function is usually called the *immediate reward* of the environment's state transition. In any concrete application, the definition of r is crucial. In fact, the immediate reward function can be regarded as an abstract specification of the agent's task and, if properly defined, it is a source of information that the agent can exploit in order to learn how to act rationally.

In order to evaluate the degree of rationality of an agent's behavior, it is important to regard utility as a function of the action performed by the agent, rather than as a function of an objective state transition in the environment. To do so, given function r we compute the expected value, $R(x,a)$, of immediate reward associated to an action, a, performed when the environment is in state x:

$$R: X \times A \rightarrow \Re \qquad (9)$$

$$R(x,a) = \sum_{x' \in X} T(x, a, x') \cdot r(x, x') \qquad (10)$$

$R(x,a)$ is called the *expected immediate reward* associated to action a in state x.

In general, immediate rewards are not sufficient to characterize rational behavior. Suppose for example that, at instant $t = 0$, the environment is in state x_1 and the agent has two alternatives. As shown by the *transition diagram* of Figure 4, the agent can:

- perform action a_1, with an immediate reward of 10 units, bringing the environment to state x_2, from which no further reward can be gained; or
- perform action a_2, with an immediate reward of 5 units, leaving the environment in state x_1.

If we suppose that the agent goes on acting forever, it appears to be more rewarding, and therefore rational, to perform action a_2, even if the immediate reward associated to a_1 is higher. This shows that it may be essential to rational behavior not to be greedy and to "delay satisfaction".

On the other hand, an agent might value rewards to be gained in a distant future less than those that can be gained immediately or in the near future, and this makes the matter more complex to judge. To model this fact, it is customary to introduce a *discount factor* γ, such that $0 \leq \gamma \leq 1$, which multiplies future rewards. More precisely, if we compute values at time 0, an immediate reward r_0 is valued

$$\gamma^0 \cdot r_0 = r_0 \qquad (11)$$

a reward r_1 to be gained at time 1 is valued

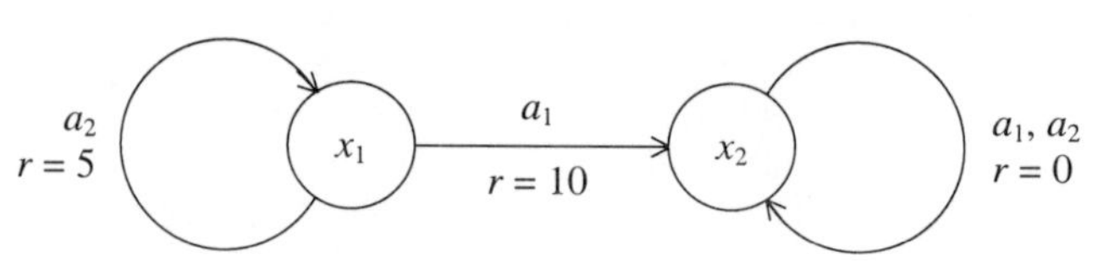

Figure 4. The importance of delaying satisfaction.

$$\gamma^1 \cdot r_1 = \gamma \cdot r_1 \tag{12}$$

and, in general, a reward r_t to be gained at time t is valued

$$\gamma^t \cdot r_t \tag{13}$$

If we assume that the agent can go on acting forever, the *total discounted reward* to be gained is given by

$$\sum_{t=0}^{t=\infty} \gamma^t \cdot r_t \tag{14}$$

To give an example, let us go back to the transition diagram of Figure 4, taking $\gamma = .9$. In such a case the total discounted reward to be gained starting from state x_1 and performing action a_2 forever is:

$$\sum_{t=0}^{t=\infty} (.9^t \cdot 5) = (\sum_{t=0}^{t=\infty} .9^t) \cdot 5 = \frac{1}{1-.9} \cdot 5 = 50 \tag{15}$$

On the contrary, the total discounted reward to be gained by first performing action a_1 in state x_1 and then doing either action forever is:

$$10 + \sum_{t=1}^{t=\infty} (.9^t \cdot 0) = 10 \tag{16}$$

Therefore, with a .9 discount factor it is rational for the agent to do a_2 when in state x_1. The situation may change, however, for different values of γ. Take for example $\gamma = .5$. In this case the total discounted reward to be gained starting from state x_1 and performing action a_2 forever is:

$$\sum_{t=0}^{t=\infty} (.5^t \cdot 5) = \frac{1}{1-.5} \cdot 5 = 10 \tag{17}$$

and therefore equals the total discounted reward to be gained by first performing action a_1 in state x_1 and then doing either action forever. And the latter option becomes more rewarding than the former when $\gamma \leq .5$.

The discount factor allows one to characterize the "personality" of an agent as more or less greedy. In particular with $\gamma=0$ an agent is extremely greedy, because only the immediate reward has a value for it, given that

$$\sum_{t=0}^{t=\infty} \gamma^t \cdot r_t = r_0 \quad (\gamma = 0) \tag{18}$$

On the other hand, when $\gamma=1$ discount has no effect, because the agent gives the same value to rewards whether they are close or far in the future. In such a case we have:

$$\sum_{t=0}^{t=\infty} \gamma^t \cdot r_t = \sum_{t=0}^{t=\infty} r_t \quad (\gamma = 1) \tag{19}$$

However, under the assumption that the agent will go on acting forever, this sum will in general be infinite, which prevents a comparison of different courses of action. For this reason, γ is always taken to be strictly less than 1.

Let us now turn to the simplest class of agents.

REACTIVE AGENTS

An agent is *reactive* if it acts on the basis of a *reactive policy*, that is, of a direct mapping from perceptions to actions. A reactive policy can be viewed as a function

$$\pi: S \rightarrow A \tag{20}$$

This simply means that, at every time t, the action a made by the agent can be computed as:

$$a = \pi(s) \tag{21}$$

In some situations, as we shall see later, it can be useful to adopt a reactive policy that is stochastic, and therefore defined as a function

$$\pi: S \rightarrow \Pi(A) \tag{22}$$

from perceptions to probability distributions on actions. In such a case, at every time t the conditional probability of doing action a given perception s will be:

$$Pr(\ a(t)=a \mid s(t)=s\ ) = \pi(s,a) \tag{23}$$

The action module of a reactive agent is extremely simple: the agent always knows what to do, given its current perception. We shall now consider a special subclass of reactive agents.

Markov Agents

A *Markov agent* is a reactive agent with faithful perception of a Markov environment. As already remarked, faithful perception allows us to assume that the agent directly reacts to the states of the environment, so that we can simplify our model accordingly (Fig. 5).

The model of a Markov environment is made up by four elements:

- the finite set X of all possible environment's states;
- the finite set A of all possible actions;
- the state transition distribution T: $X \times A \rightarrow \Pi(X)$; and
- the immediate reward function r: $X \times X \rightarrow \Re$.

This model is known in the literature as a *Markov Decision Process* (MDP).[2-4]

Given a Markov environment (X,A,T,r), to define a Markov agent we only need to specify a reactive policy π: $X \rightarrow A$. Now suppose that, at time $t = 0$, the environment is in state x. Given a discount factor γ, it is possible to compute the *value* $v_{\pi,\gamma}(x)$ of state x, defined as the expected value of total discounted reward to be gained starting from state x, and acting forever according to policy π. Such a value can be computed as follows.

For every $x \in X$, $v_{\pi,\gamma}(x)$ can be written as

$$v_{\pi,\gamma}(x) = R(x, \pi(x)) + \gamma \sum_{x' \in X} T(\ x, \pi(x), x') \cdot v_{\pi,\gamma}(x') \tag{24}$$

In this way one gets a linear system of equations (one equation for each environment's state $x \in X$) that can be solved for the unknowns $v_{\pi,\gamma}(x)$.

It is a remarkable property of Markov agents that, for each choice of γ, there is always at least one *optimal policy* π^*, that is, a policy that attributes to every state a value greater or equal to the value attributed by any other policy. Moreover, it can be proved that no stochastic policy could do better than a deterministic policy.

Optimal policies can be computed from T and r by Dynamic Programming methods.[2,5] Therefore, when T and r are known to the agent designer, the problem of determining a rational course of action has a general solution: it is sufficient to compute an optimal policy π^* and implement a reactive agent that follows it.

In many applications, however, it is not possible to compute an optimal policy in advance, because T and r are not known at design time. In such cases, one has to rely on the agent's ability to learn from experience. By "experience" we mean the collection of events the agent goes through. At a given instant, the environment will be in state x, and the agent will perform action a. As a result, the environment will switch to state x', and the agent will get reward r (possibly equal to 0). Experience is then made up by the collection of all quadruples of the form (x,a,x',r) collected during an operative session. Learning to behave optimally from experience is known as *reinforcement learning*.[6,7]

There are basically two ways of exploiting experience to learn to behave optimally. The first is to use experience to build a representation of the environment (in terms of T and r), from which the agent can then compute an optimal policy by Dynamic Programming methods. This approach is possible in principle but very inefficient, because it requires a huge amount of computational resources. The second option is to exploit experience in order to compute an optimal policy directly, without building an explicit representation of the environment. This approach, usually called *model free*, is more efficient and has been successfully used in many applications. In the next subsection we shall sketch one of the most widely applied methods of this kind, known as *Q-learning*.[8]

Reinforcement Learning for Markov Agents

The idea underlying model-free reinforcement learning is that an agent can use its own experience to search the space of action policies. In the case of Markov agents the search is guaranteed to converge to an optimal policy under fairly general assumptions. These usually require that the agent visits every possible environmental state during its life and performs all the available actions for every state. In addition, it is required that as the agent's search for the optimal policy proceeds recent experience becomes less and less relevant as time passes. These simple assumptions are easy to justify. The assumption requiring that the agent visits every possible state is needed to prevent the agent from getting stuck in local policies. The assumption requiring that the agent tries every possible action is needed to

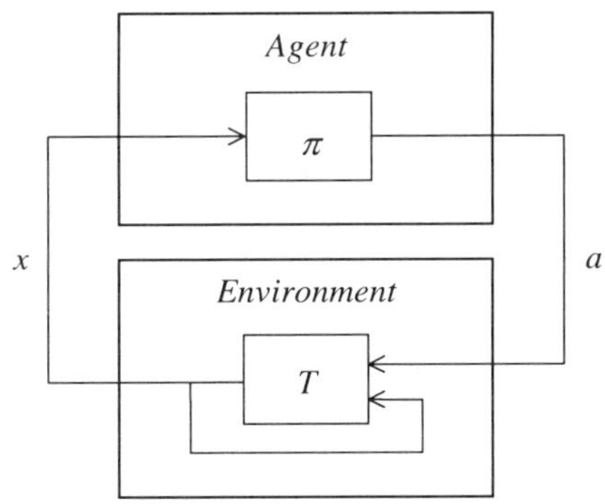

Figure 5. A Markov agent.

guarantee that learning asymptotically approaches the optimal policy even if the environment is noisy.

Watkin's *Q*-learning (1989) is probably the best-known reinforcement learning algorithm. The algorithm computes, by successive approximations, the table of all $Q(x,a)$ values, named *Q-table*. A $Q(x,a)$-value, shortly *Q*-value, is defined as the reward predicted under the hypothesis that the agent performs action a in state x, and then carries on always selecting an action with the highest *Q*-value (i.e., predicting the highest reward).

Q-learning works as follows. Initially, for each state-action pair, the *Q*-table is initialized with random values. At each time step $t > 0$, when action $a(t)$ is performed in state $x(t)$ the *Q*-table is updated according to the formula:

$$Q(x(t-1),a(t-1)) \leftarrow (\alpha - 1)Q(x(t-1),a(t-1)) + \alpha(r(t) + \gamma \max_{a \in A} Q(x(t),a)) \tag{25}$$

where $r(t)$ is the actual reward gained at time t, and γ is the *discount factor* introduced in the previous section. The term α ($0 \leq \alpha \leq 1$) is called *learning rate*, and determines how much the most recent experience (i.e., the value $r(t)$) is taken into account. A value of α close to zero means that recent experiences have little influence on *Q*-values. A value of α close to one means that recent experiences have strong influence on the *Q*-values.

It has been proved[9] that the *Q*-learning algorithm approaches, with probability one, the *optimal Q-table* Q^*, which maps optimal state-action pairs onto the associated expected reward. Convergence is guaranteed under the assumption that every action is tried an unlimited number of times in each state, and that the learning rate asymptotically approaches zero with order between $1/t$ and $1/t^2$.

Given the optimal *Q*-table Q^*, the optimal policy π^* is then defined as follows:

$$\pi^*(x) = \arg\max_{a \in A} Q^*(x,a). \tag{26}$$

Q-learning is simple and easy to implement but it is infeasible for problems of realistic extent, because the size of the *Q*-table (which is $|X|\cdot|A|$) grows exponentially with the problem dimension. This is a major drawback in practical applications since with bigger *Q*-tables, (i) more experience is required to converge to a good estimate of the optimal policy, and (ii) more memory is required to store the table.[7] To cope with the complexity of the tabular representation, the agent must be able to *generalize* over its experiences, that is, to produce a good approximation of the *Q*-table from a limited number of experiences, using small amount of storage. In reinforcement learning, generalization is usually achieved by using function approximation techniques such as neural networks, gradient-descent methods, coarse coding, and tile coding (see Sutton & Barto,[7] for a review).

Non-Markov Reactive Agents

So far we have assumed that:

- the environment can be modeled as a finite-state Markov process;
- the agent has faithful perception of the environment's states.

While the first hypothesis is often realistic, the second cannot be adopted in many interesting applications. Let us now move to a slightly more general model, by assuming that perception is not necessarily faithful. We can model this by describing perception through a stationary conditional probability distribution:

$$\sigma: X \rightarrow \Pi(S) \tag{27}$$

where $\sigma(x,s)$ is the probability of having perception s when the environment is in state x (Fig. 6).

Our problem is determined by six elements:

- the finite set X of all possible environment's states;
- the finite set S of possible perceptions;
- the finite set A of all possible actions;
- the perception distribution σ:: $X \to \Pi(S)$;
- the state transition distribution T: $X \times A \to \Pi(X)$; and
- the immediate reward function r: $X \times X \to \Re$.

This model is known in the literature as a Partially Observable Markov Decision Process.[10]

The problem now is whether the methods described in the previous subsection can still be applied. The idea is that we may be able to do so if we can, so to speak, move perception from the agent to the environment. More technically, we start from our model (X,S,A,σ,T,r) and try to define a new model (S,A,T_σ,r), so that the function

$$T_\sigma: S \times A \to \Pi(S) \tag{28}$$

defines a Markov Process on the set of *perceptions* (and no longer on the set of environment's states). Simple probabilistic considerations show that function T_σ should be defined as

$$T_\sigma(s,a,s') = \sum_{x,x' \in X} \sigma'(s,x) \cdot T(x,a,x') \cdot \sigma(x',s') \tag{29}$$

where $\sigma'(s,x)$ is the probability that the environment is in state x when the agent has perception s:

$$\sigma'(s,x) = Pr(\ x(t)=x \mid s(t)=s\ ) \tag{30}$$

Now, for (S,A,T_σ,r) to be a Markov Process, the probability distribution defined by T_σ must be stationary. In view of equation (29), given that both T and σ have been assumed to be stationary, we need to prove that σ' is stationary. Unfortunately, it isn't. In fact, by Bayes's theorem we have:

$$\sigma'(s,x) = \frac{\sigma(x,s) \cdot Pr(x)}{Pr(s)} \tag{31}$$

where $Pr(x)$ is the absolute probability that the environment is in state x, and $Pr(s)$ is the

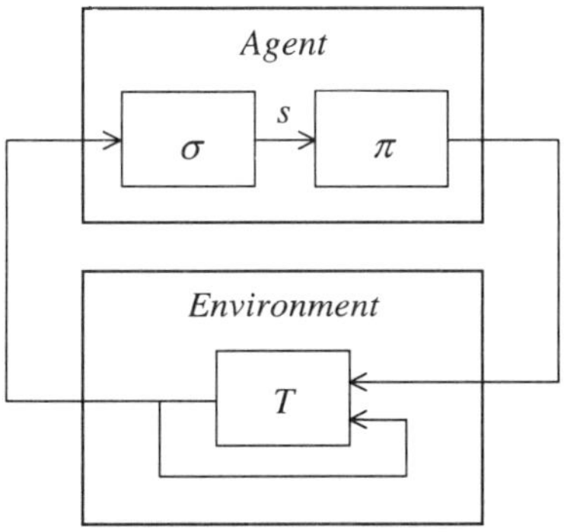

Figure 6. A non-Markov reactive agent.

absolute probability that the agent has perception s, with

$$Pr(s) = \sum_{x \in X} Pr(x) \cdot \sigma(x, s) \tag{32}$$

This shows that σ' is stationary if the absolute distribution of environmental states, $Pr(x)$, is itself stationary. Now, from the theory of Markov Processes we know that $Pr(x)$ asymptotically approaches a stationary distribution if the agent's reactive policy is kept constant. In other words, if an agent follows a fixed policy, the environment will tend to be perceived as a stationary Markov Process as times goes by. But, if the agent changes its policy, the environment will then be perceived as a *different process*. This fact is an obstacle to learning: obviously, learning involves changing one's policy, and we see that such changes will impact on how the environment is perceived. It can therefore be expected that, in general, algorithms that learn in an MDP setting will not work for a POMDP. A further interesting result is that in the case of non-Markov reactive agents, a stochastic policy can do better than a deterministic one.[11]

We conclude that to act optimally in the case unfaithful perception, it is necessary to go beyond reactive behavior. This matter is addressed in the next section.

COGNITIVE AGENTS

When perception does not give sufficient information, an agent has to rely on some form of internal representation of the environment. Intuitively, a representation is a state *of the agent* that estimates the state *of the environment*. When the agent needs information about the environment that it cannot collect through perception, it can rely on its representation. A representation of this kind will be called a *belief*, and an agent possessing beliefs about the environment will be called a *cognitive agent*.

Bayesian Agents

A Bayesian agent is a cognitive agent that maintains an internal representation of the probability distribution of environment's states. We can model a Bayesian agent as follows. As before, X will be the finite set of possible environment's states, S the finite set of possible perception, and $\sigma: X \to \Pi(S)$ the probability distribution of perceptions conditioned by environment's states. Moreover, we assume that the agent implements a representation of an absolute probability distribution of environment's states:

$$b \in \Pi(X) \tag{33}$$

that we call *belief*. The intuitive interpretation of b is that at any time instant, $b(x)$ represents the probability that the environment is in state x at that very instant. If σ and T are known, at every cycle distribution b_t can be updated to b_{t+1} using the following formula (derived from Bayes's theorem; see Kaelbling et al.,[12] for a similar treatment):

$$b_{t+1}(x) \quad = Pr(\, x(t+1)=x \mid s(t+1)=s \text{ and } a(t)=a\,) = \frac{\sigma(x,s) \cdot \sum_{x' \in X} T(x',a,x) \cdot b_t(x')}{Pr(s(t+1)=s)} \tag{34}$$

Therefore, a Bayesian agent computes its current beliefs from its current perception, its previous beliefs, and the action previously performed (Fig. 7).

A first difficulty with this method is that the Bayesian estimation of the environmental state can be computed only if σ and T are known (in other words, this method is not viable within a model-free approach). But even in such a case, in general policy π cannot be reactive if the agent is to act optimally. In some simple cases, it has been shown that an optimal policy can be implemented as a finite-state automaton.[12] But in more complex cases, an optimal policy may have to be an algorithm of arbitrary complexity.

The conclusion is that the problem of acting optimally in a Markov environment may be extremely complex to solve for agents with unfaithful perception. Many researchers have therefore concentrated on efficient methods that find a suboptimal but still acceptable action policy.

Agents with Memory

Agents with memory are cognitive agents that maintain a representation of the environment much simpler than Bayesian agents. Let us start with the hypothesis that an agent has an aliased but noiseless perception

$$\sigma: X \to S \tag{35}$$

and that the environment can be viewed as a deterministic system

$$T: X \times A \to X \tag{36}$$

The only problem in this case is that there are distinct environment's states that cannot be distinguished through perception alone. Suppose that x_0 and x_1 are the only indistinguishable states, and that

$$\sigma(x_0) = \sigma(x_1) = s \tag{37}$$

It is possible to disambiguate x_1 and x_2 using a 1-bit memory register m, for example by stipulating that

$$\begin{aligned} x(t) = x_0 \quad &\text{iff} \quad s(t) = s \text{ and } m(t) = 0 \\ x(t) = x_1 \quad &\text{iff} \quad s(t) = s \text{ and } m(t) = 1 \end{aligned} \tag{38}$$

The memory register can be set to 0 or 1 by internal actions (a_{mem}), to be kept distinct from ordinary actions performed on the environment (a_{env}; Fig. 8). For instance, consider again the agent of Example 3. The robot's perception is 0102 both in C1 and C4, and thus the two states are indistinguishable. However, the robot can enter such a state only from

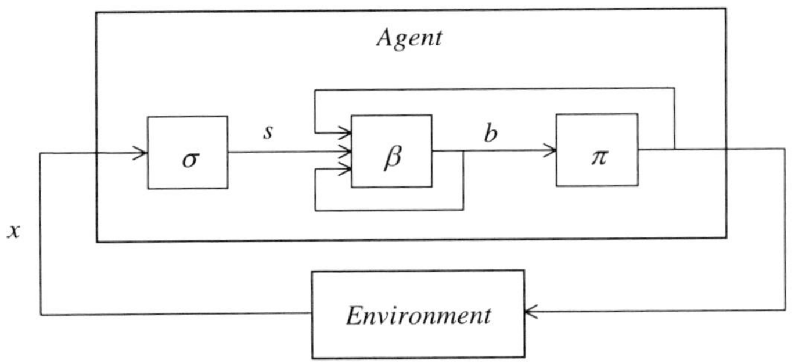

Figure 7. A Bayesian agent.

adjacent positions, which on the contrary are perceived differently (Fig. 9).

Now assume that C1 and C4 are represented as follows:

$$\begin{aligned} x(t) = \text{C1} \quad &\text{iff} \quad s(t) = 0102 \text{ and } m(t) = 0 \\ x(t) = \text{C4} \quad &\text{iff} \quad s(t) = 0102 \text{ and } m(t) = 1 \end{aligned} \tag{39}$$

To guarantee a correct representation it is sufficient to define a_{mem} coherently. For example, to represent the action policy

$$\text{B1} \longrightarrow \text{Right} \qquad\qquad \text{C1} \longrightarrow \text{Right}$$

we can define:

$$\begin{aligned} \pi_{env}(0231,-) &= \text{Right} \qquad\qquad & \pi_{env}(0102,0) &= \text{Right} \\ \pi_{mem}(0231,-) &= 0 & \pi_{mem}(0102,0) &= - \end{aligned}$$

where the symbol '–' means that the corresponding value is irrelevant.

As it turns out, for an agent with memory it is not trivial to learn an optimal action policy from experience. In particular, it is not easy to learn the π_{mem} component of an optimal policy (for recent work on this problem see Lanzi,[13] and Lanzi & Wilson[14]).

THE ROLE OF MOTIVATIONS

So far, we have assumed that expected immediate reward is stationary (i.e., it does not change with time). This assumption is often unrealistic.

In particular, consider an agent with multiple subtasks. For example, it may be an autonomous robot that has to carry objects from a place to another as its main task, but also has to reach home and recharge batteries when the current supply is low. In such a case, it is reasonable to assume that the immediate reward of battery recharging depends on the present charge level (eating is much more rewarding when one is hungry, isn't it?). In principle, it is possible to treat charge level as part of the state of the *internal environment*, where 'internal' should be understood as 'internal to the robot's body.' But this approach is most often practically unfeasible, and we shall now try to explain why.

Let us discuss the example of battery recharging. In principle, the level of current

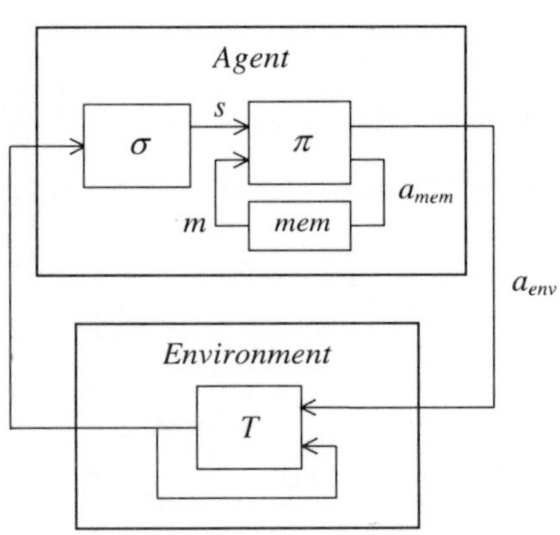

Figure 8. An agent with memory.

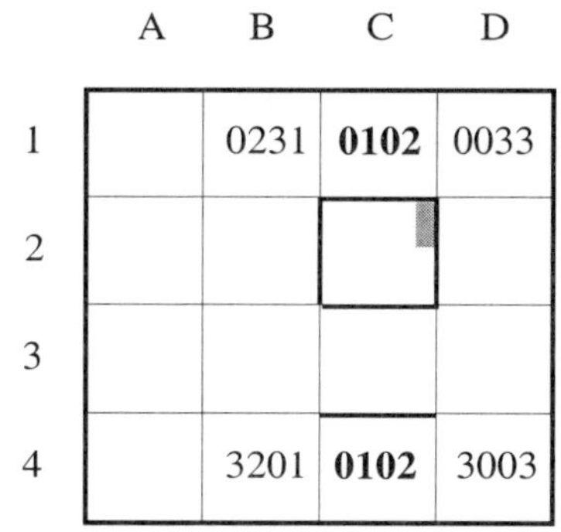

Figure 9. Dealing with undistinguishable states.

supply is a continuous variable. To turn it into a discrete variable, we could define a threshold, θ, and identify two discrete levels for the robot's internal state, x_{int}, as follows:

$$
\begin{aligned}
x_{int} &= low &\quad \text{iff } current < \theta \\
x_{int} &= high &\quad \text{iff } current \geq \theta
\end{aligned}
\tag{40}
$$

The problem, however, is that in general the effect of an action on the internal state will not follow a stationary probability distribution. The environment (including both external and internal states) will not be Markovian, and all the models we have presented so far will be no longer applicable.

One solution, which has been tested successfully,[15] is to treat internal states not as part of the environment's states but as *motivations*. For example, one could stipulate that the estimated utility of an action is to be multiplied by a factor proportional to the "need of charge" (e.g., maximum charge level minus current charge level). In such a way, an action leading to recharging becomes more and more attractive when charge decreases. As can be expected, agents with motivations can produce highly flexible and adaptive behavior. However, only empirical work has been carried out so far in this direction. No general mathematical models have been developed, at least to our knowledge, and therefore no results are available on optimality, learnability, and so on.

CONCLUSIONS

As we have tried to show in the chapter, the need for a principled approach to the design of rational agents has stimulated the development, analysis and computer implementation of interesting mathematical models, more or less related to the methods of Dynamic Programming.

Researchers who want to proceed further along the lines we have described are now facing problems of various kinds. A first type of difficulty concerns the expressive power of models based on the notion of finite state Markov process. It is clear that modeling the environment as a finite state system, even if a stochastic one, is a strong limitation. However, it is not easy to overcome such a limit without losing the mathematical properties that make Markov processes so useful.

A second difficulty has to do with the computational efficiency of learning algorithms. Even with finite state processes, learning becomes unmanageable as the number of states becomes large. So far, this limitation has made it difficult to apply the methods described in this chapter to problems of realistic size.

Finally, it is clear that to account for flexible and adaptive behavior one needs to treat in a uniform way aspects related to the reactive, cognitive, and conative components of an agent. However, whether it is possible to do so within a unified mathematical approach is still to be proved.

REFERENCES

1. M. Wooldridge, *Reasoning about rational agents*, MIT Press, Cambridge, MA (2000).
2. R. Bellman, *Dynamic programming*, Princeton University Press, Princeton, NJ (1957).
3. A.R. Howard, *Dynamic programming and Markov processes*, The Technology Press of MIT, Cambridge, MA, and John Wiley & Sons, New York (1960).
4. D.P. Bertsekas, *Dynamic programming and optimal control*, Athena Scientific, Belmont, MA (1995).
5. M. Putermann, *Markov decision processes: Discrete dynamic stochastic programming*, John Wiley, New York (1994).
6. L.P. Kaelbling, L.M. Littman, and A.W. Moore, Reinforcement learning: A survey. *Journal of Artificial Intelligence Research*, 4, 237–285 (1996).
7. R.S. Sutton and A.G. Barto, *Reinforcement leaning: An introduction*, MIT Press, Cambridge, MA (1998).
8. C.J.C.H. Watkins, Learning with delayed rewards, Ph.D. dissertation, University of Cambridge, Cambridge, UK (1989).
9. C.J.C.H. Watkins and P. Dayan, Technical note: *Q*-learning, *Machine Learning*, 8, 279–292 (1992).
10. F.J. Sondik, The optimal control of partially observable Markov processes over the infinite horizon: discounted case, *Operations Research*, 26, 282–304 (1978).
11. S.P. Singh, T. Jaakkola and M.L. Jordan, Learning without state-estimation in partially observable Markovian decision processes, in W.W. Cohen and H. Hirsch eds., *Proc. 11th Int. Conf. on Machine Learning*, Morgan Kaufman, San Francisco, CA, 284–292 (1994).
12. L.P. Kaelbling, L.M. Littman, and A. Cassandra, Planning and acting in partially observable stochastic domains, *Artificial Intelligence*, 101, 99–134 (1998).
13. P.L. Lanzi, Adaptive agents with reinforcement learning and internal memory, *Proc. 6th International Conference on the Simulation of Adaptive Behavior* (*SAB 2000*), Paris, F(2000).
14. P.L. Lanzi and S.W. Wilson, Toward optimal classifier system performance in non-Markov environments, *Evolutionary Computation*, 8 (4), 393–418 (2000).
15. A. Bonarini and M. Matteucci, Learning context motivation in coordinated behaviors, *Proc. 6th Intelligent Autonomous Systems Conference* (*IAS-6*), IOS Press, Amsterdam, NL, 519–526 (2000).

OVERVIEW ON DECISION MAKING

Giuseppe Mosconi and Laura Macchi

Department of Psychology,
University of Milano-Bicocca,
Piazza dell'Ateneo Nuovo 1
20126 Milano – Italy
e-mail: laura.macchi@unimib.it

INTRODUCTION

The systematic empirical study of judgement and decision making did not begin to emerge as a discipline in its own right until the 1960s, when there was an upsurge of interest in the broader and more general field of cognitive psychology that includes memory, thinking, problem solving, mental imagery and language.

At that time, the study of probabilistic reasoning and decision-making processes reflected the traditional trend of research into reasoning in general, which was considerably influenced by the logistic theorists who represented the human mind as being guided by rules substantially corresponding to those of formal logic. The initial results offered a vision of the judgement of statistically naive subjects that was in line with the principles of the inferential and decision-making models of the theory of probability and, until the mid-1970s, the studies of intuitive probabilistic judgement led to the conclusion that most people make judgements on the basis of rational methods that are in line with the main rules of statistics and probability theory.[1-3]

However, two factors subsequently modified the line of research: psychological researchers identified the presence of systematic errors in contrast to what was assumed by logistic theories and, in the economic field, Allais demonstrated the existence of a paradox of normative theory that leads to its failure as a descriptive model of decision making. Better descriptive models would lead to better normative models and a revision of their implied concept of rationality.

In the field of psychology, the search for descriptive models of human behavior that are better than those offered by normative theories led to studies of the way in which ordinary people reason and evaluate the probability of an event, and the identification of **systematic errors** in carrying out judgement tasks led to the relationship between norms and thought being rethought by studying reasoning methods at the level of intuitive statistics.[4-7]

The emphasis on errors has already been overcome in other areas of reasoning, but it is only recently that the same has been done in the field judgement and decision making. Formal logic was and still is used as a yardstick for the study of deductive reasoning but, in

Human and Machine Perception 3: Thinking, Deciding, and Acting, edited by Cantoni *et al.*
Kluwer Academic/Plenum Publishers, New York, 2001.

addition to the fact that it accounts for only a small part of the psychology of thinking, this approach has been increasingly criticised. The development of pragmatics has led to a shift from the **formal** to the **content and contextual** characteristics of thinking (see, for instance, the development of the studies on selection tasks and syllogisms), and the use of formal logic now plays only a minor role or has been abandoned altogether.

On the other hand, even in the research of descriptive models, normative models have been the most important research tools in the psychology of judgement and decision making ever since its beginnings. The most well-known models of judgement are probably Bayes' theorem and the multi-attribute utility models; the normative model of decision making is the Subjective Expected Utility (SEU) model.

The most extensive and influential work in this context has been that of Tversky and Kahneman concerning probabilistic thinking. According to these authors, probability judgements are often severely biased because people rely on heuristics in many situations, and this leads to systematic errors.

HEURISTICS AND BIASES PROGRAMME

According to this perspective, reasoning is guided by heuristic strategies of thought that simplify judgement but often (or in the majority of cases) lead to error. The errors or biases that may derive from the activation of one or more heuristics represent the difference between actual and normatively correct performances.[5-7,8-10] Some examples of biases and heuristics are given below.

Table 1. Examples of Principal Biases and Heuristics

Biases	Heuristics
gambler fallacy	representativeness
base rate fallacy	representativeness specificity causality
conjunction fallacy	representativeness availability causality
overestimate of the sample size	availability
over or underestimate of probability	anchoring
subadditivity (support theory)	availability
. . .	. . .

Availability

There are situations in which people assess the frequency of a class or the probability of an event by the ease with which instances can be brought to mind. Instances of large classes are usually recalled better and faster then instances of less frequent classes. However, since ease of retrieval is also influenced by factors other than frequency or probability (e.g., recent occurrence) the reliance on availability leads to systematic judgmental errors.[8]

The causes of death that are more easily recoverable by the memory (because their spectacular or violent nature means that they are more emphasised by the mass media) are overestimated in comparison with their real frequency: i.e. air crashes are overestimated, whereas deaths due to diabetes or asthma are underestimated.[11]

In an elementary demonstration of this effect, subjects heard a list of well-known personalities of both sexes and were subsequently asked to judge whether the list contained more names of men than women. Different lists were presented to different groups of subjects. In some of the lists, the men were relatively more famous than the women; in others, the women were relatively more famous than the men. In all cases, the subjects erroneously judged that the class (sex) that had the more famous personalities was the more numerous.[8]

Representativeness

People often judge the probability of an event on the basis of the *representativeness* it has for the underlying population or the generating process. However, "this approach to the judgement of probability leads to serious errors, because similarity or representativeness is not influenced by several factors that should affect judgements of probability". Such factors are, for instance, base rates, sample size and the reliability of the information.

Gambler Fallacy. The law of large numbers states that only very large samples are highly representative of the population from which they are drawn: for example, only in the case of a very large number of lottery draws is it reasonable to expect (and therefore bet on) the extraction of a particular number. When we find ourselves expecting an almost regular alternation of black and red numbers to come up at the roulette wheel, we are committing the error of believing that our small sample of cases follows the same laws as those governing the population to which they belong: i.e. the tendentially infinite set of spins of the wheel itself. We believe that the alternation of the elements of a random and limited sequence is more probable than it actually is, and so we underestimate the degree of repetition of the elements themselves (it is considered very unlikely that the same number will come up twice in succession); we expect a compensatory alternation of events which, in the case of roulette, means an equal number of reds and blacks. Tversky and Kahneman[12] defined the common intuition that the law of large numbers also applies to small samples as the law of small numbers.

Base Rate Fallacy. Consists of evaluating the probability of an event by considering only the specific information (likelihood) and neglecting the prior probability (base rate): i.e. that relating to the entire population from which the event has been drawn.[13]

One crucial experimental task consists of two types of information: the base rate (in the following example, the incidence of the illness) and likelihoods (the hit rate and false alarms based on the proportion of sick and healthy people with a positive mammography).

The probability of breast cancer is 1% for women at age forty who participate in routine screening.

If a woman has breast cancer, the probability is 80% that she will get a positive mammography.

If a woman does not have breast cancer, the probability is 10% that she will also get a positive mammography.

A woman in this age group had a positive mammography in a routine screening. What is the probability that she actually has breast cancer? ___%

To assess the requested probability, it is first necessary to consider 80% (the probability that someone who has breast cancer will have a positive mammography) of 1% (the probability of having breast cancer) (=0.8%). It is then possible to determine the probability of having a positive mammography in the case of ill (= 0.8%) or healthy people (10% of 99%= 9.9%), and consider the proportion of cases of illness and positive mammography (0.8%) among all of the cases of positive mammography (0.8% + 9.9%= 10.7%, then 0.8/10.7= 0.07). In problems of this type, people typically ignore the incidence of the disease and concentrate on the likelihood of cancer given a positive test result, thus giving rise to the well-known phenomenon of base-rate neglect. In this kind of task, the implicated heuristic principle is said to be specificity (i.e. more importance is given to specific information) and/or causality (i.e. the base-rate has no causal relevance).[14-16,12]

1000
sample

10
ill people

of whom

8 (80%) ill with positive mammography 2

990
healthy people

of whom

99 (10%) healthy with positive mammography 891

Bayes'Theorem

$$p(I/+) = \frac{p(I \& +)}{p(I \& +) + p(H \& +)} = \frac{8}{8 + 99} = \frac{8}{107} = 0.07$$

Figure 1. Diagram of the medical diagnosis problem.

Conjunction Fallacy. A great deal of research concerning probabilistic reasoning has highlighted what is called an error in the evaluation of the probability of the conjunction of two events.[17] According to the third basic axiom of probability theory [P(A & B)= P(A/B) P(B)= P(B/A) P(A)], the conjunction of two events (A and B) cannot be considered as being any more probable than each of the single constituent events (A) and (B).

In a series of studies, Tversky and Kahneman showed that people tend to consider the conjunction as more probable than the constituent events. In their well known Linda Problem, Tversky and Kahneman[17] constructed a brief personality sketch:

Linda is 31 years old, single, outspoken, and very bright. She majored in philosophy. As a student, she was deeply concerned with issues of discrimination and social justice, and also participated in anti-nuclear demonstrations.

The personality sketch was followed by eight possible outcomes, including the following three, which were defined by the authors as being representative, unrepresentative, and a conjunction of the two:

Linda is active in the feminist movement. (F)
Linda is a bank teller. (T)
Linda is a bank teller and is active in the feminist movement.(T and F)

The subjects were asked to rank the statements of outcome in the order of their probability, starting from the most probable. According to the authors, of the two characteristics in the final item ("T and F"), one is representative and the other unrepresentative of Linda. Faced with this problem, 85% of the subjects tend to consider "T and F" as being more probable than "T" ("Linda is a bank teller"), thus falling into what Tversky and Kahneman call the "conjunction fallacy" which, according to them, is an effect of the *representativeness* heuristic.

Other studies with the same experimental design have confirmed the general nature of the effect, which is always described as an error in probabilistic reasoning.[18-23]

CRITICAL REVIEW

There are some problems left unsolved by the theory of heuristics:

• If the cause of the bias is a given heuristic, it should not occur in the absence of the heuristic itself. However, there are cases in which the presence of a certain bias (for instance, the base-rate fallacy as in the lawyers/engineers problem in Tversky & Kahneman)[4] still occurs in the absence of the heuristic principle (i.e. the representativeness) considered as its explanation (see the case of the non-diagnostic -non representative-description). The proponents of the theory conclude that the bias is very "strong", but do not see the explanatory incongruence of their theory.

• The same heuristic conditions can lead to different and opposite effects. One example of this is the *cabs* vs. *books* problem:[13] neither have a heuristic (causality), but the cab problem leads the majority of subjects into committing the error (the base-rate fallacy) and the book problem does not.

• There are doubts concerning the authenticity of the errors raised by the frequentist approach[24] and, from another perspective, by the pragmatic approach.[25-27]

FREQUENTIST APPROACH

The traditional research programme supporting the heuristic thesis has been opposed by the "frequentist" approach of Gigerenzer,[28,29] for whom the traditionally encountered errors are due to the fact that probability (%) is applied to single events or relative frequencies (e.g. "the probability that Mr. X has the disease is 1%") rather than to sequences of observable events (e.g. "1 of the 100 have the disease"), which is considered to be much closer to the intuitive method of reasoning under conditions of uncertainty. Difficulties with base rate problems are said to be caused by the probabilistic formulation of single-case oriented information.[30,31] By giving classical problems a frequentist format, the authors obtained a reduction in the number of classical errors (for similar studies involving the conjunction fallacy see Fiedler,[32] Hertwig and Gigerenzer,[33] Jones, Taylor and Frisch,[34] Reeves and Lockhart[35]).

The frequentist hypothesis has also been used with reference to the conjunction fallacy, and has recently attributed this presumed error to the non-extensional nature of the term "probability" and the fact that the problem used (the Linda problem) refers to a single case rather than a succession of events (which would be more natural according to frequentist theory[28]).

Experiments in this direction already began with Tversky and Kahneman,[17] who compared the probabilistic and frequentist versions of similar problems [also see Fiedler;[32] Jones, Taylor and Frisch;[34] Reeves and Lockhart[35]]. However, the results of these studies vary a great deal. In their experiment, Tversky and Kahneman found 42% of correct inclusion judgements using the probabilistic version and 75% using the frequentist version which, instead of referring to a single case, referred to a sample of 100 individuals. On the other hand, Fiedler found different results depending on the type of problem, even if there was a general tendency to reason following the rule of inclusion in the case of the frequentist version. Unlike these authors, Reeves and Lockhart did not compare different (frequentist vs. single case) versions of the same problem, but also varied the format of the versions by changing their content, thus not controlling the albeit statistically non-significant influence of the latter on the results. Finally, Jones et al. obtained 5% of inclusive answers using the probabilistic version of the Linda problem vs. 15% when the frequentist version was used.

These differences in results reflect the substantial differences in the type of frequentist version adopted. Tversky and Kahneman did not describe a particular individual but only gave a reference class, whereas Fiedler[32] also described an individual (for example, Linda) and provided variable reference classes (100 in some cases, 170 in the case of Linda). Reeves and Lockhart[35] asked for a probabilistic (not frequentist) judgement, but provided the data in frequentist form and also asked for an evalution of the disjunction (C or F). Jones, Taylor and Frisch[34] used the Linda problem by asking the subjects to imagine 100 women with similar characteristics and evaluate the number (out of 100) of "tellers" and "feminist tellers". This type of text is identical to that used by Fiedler, although he used a reference class other than 100 and provided additional items to those used by Jones, Taylor and Frisch [*ibid.*]. The same also emerges from the work of Hertwig and Gigerenzer[36] who, using variant versions of the Linda problem, reproduced the results obtained by Fiedler by similarly providing a reference class other than 100 (200 in this case).

The effects obtained by these authors have been explained in different ways: Tversky and Kahneman considered that the presence of extensional cues, such as the reference to the size of the class, facilitated reasoning following the inclusion rule; Fiedler, Hertwig and Gigerenzer thought it was the polysemia of the word "probability", which usually recalls only non-extensional meanings; Reeves and Lockhart spoke of differences in the ways of reasoning between frequentist (the "distributional" approach) and single cases (the "single" approach).

The differences in the results encourage us to offer the cautious consideration of the presumed generality of the frequentist effect, which seems to depend more on other aspects of the experimental procedure.

PRAGMATIC APPROACH

People interpret messages by following the hypotheses they make concerning the meaning the speaker intends to communicate. A failure to consider the role that conversational rules play in governing inferential processes may lead to rational answers being classified as erroneous and erroneously attributed to cognitive shortcuts.[25-27,37-40] The main conversational rules are the principle of cooperativeness[41] and the maximisation of the informative content of a message.[39,42-45]

The pragmatic approach revealed the importance of context – the source of information (point of view, state of knowledge, intention of the speaker) and the (syntactic, semantic) structure of the discourse – when interpreting a task that cannot be interpreted by applying an exclusively logical analysis to the experimental material and the replies of the subjects. Consideration of the differences between what is "said" and what is "communicated" (formal vs natural language) indicates the representation of the task, which is no longer assumed to be equivalent to that which logic would require.

For example, in the Linda problem the question formulated by the researchers violates a conversational rule, and that leads to an interpretation of the task by the subjects that is different from that required by logic. Then, the comparison becomes:

Linda is a bank teller (and *not* active in a feminist movement)
vs.

Linda is a bank teller and active in a feminist movement

The comparison of two utterances:
"Linda is a teller " (A)
"Linda is a teller and a feminist" (A & B)
is not pragmatically acceptable because of its obviousness. The response is already stipulated contextually. In order to make the question relevant, the subjects produce an implicit structure on the basis of which the utterance:
"Linda is a teller" (A)
is interpreted as
"Linda is a teller and not a feminist" (A & not B)
Furthermore, if the request is not interpreted in these terms, the description would be totally irrelevant for the purposes of evaluating the inclusion of A & B in A.

This intuition is supported from both the theoretical and experimental points of view:

Grice:[41] "maxim of quantity". You must not give or ask for too much or too little information. To make the comparison meaningful/not banal, an implicit structure is generated in which A vs. A and B becomes A and not B vs. A and B.

Kasher:[46] "principle of rationality". Everyone thinks that the speaker is or tends to be rational. The person asked the question excludes the possibility that the speaker actually means such a banal comparsion, and therefore generates the implicit structure.

Sperber & Wilson:[45] "principle of pertinence". Linda-type problems are misleading because they induce the false idea in the receiver that the context given by the description of Linda is pertinent and should therefore be used, whereas it is totally irrelevant in terms of indicating whether the included or the including class is the more probable/larger.

Mosconi & Macchi[27] have shown that the answers provided by subjects cannot be appropriately interpreted as violating a logical rule, because they obey discursive criteria that take into account requirements that logic legitimately ignores. They have also shown[27,47] that comparisons of the kind required by the "Linda problem" are unacceptable to common speakers by giving subjects a text in which a court witness, who had been asked to indicate the more probable of two possibilities, applied a logical criterion and indicated the including event as the more probable (the "correct" answer in terms of classical studies of the conjunction fallacy). However, when the subjects were asked to evaluate this answer as "truthful", "false" or "reticent", all of them considered it false or reticent. If these results are applied to the "Linda" problem, it seems clear that common speakers do not consider "bank teller" more probable because otherwise they would violate pragmatic rules.

With the aim of demonstrating the pragmatic nature of this type of bias, after making a pragmatic analysis of the problem and the indication of the way to block or manifest the implicature (not F), Politzer & Noveck[25] carried out an experiment based on the hypothesis that the implicature (from "T" to "T not F") is made when one item is compared with another that is exactly the same except for the addition of a qualifier. The compared items in the "Linda" problem are "T" and "T+F", and the explicit absence of "F" from the first induces the implicature of the qualifier "not F" in the event of "T" alone; according to the authors, the bias could be reduced by eliminating the use of "and" and lexically highlighting the inclusion relationship between the items. This is exactly what happened during an experiment in which, after having been told that Daniel was a brilliant High School student and very determined, the subjects were asked to rank the probability (from the most to the least probable) of the following three possibilities:

1) Daniel entered Medical School
2) Daniel dropped out of Medical School for lack of interest
3) Daniel graduated from Medical School

Only 31% committed the fallacy using this text, as against 53% using the control text containing the more qualifying "and" form ("Daniel entered Medical School and graduated" in place of option 3 above).

According to Dulany & Hilton,[37] Tversky and Kahneman's experiment with the "Linda" problem involved a *model* condition (i.e. with a description) in which the wealth of detail in the description of Linda may justify the inference that the researcher knows a lot about Linda, and so the subjects may assume that the omission of her involvement in a feminist movement in the statement "Linda is a bank teller" is due to the fact that the researcher knows the contrary to be true. However, in a *model-free* condition in which subjects only receive information concerning Linda's age, they may think that the researcher has omitted to say that Linda is a feminist because he does not know whether it is true or not. Consequently, when subjects are asked to judge whether "T+F" is more probable than "T", it is more likely that they will interpret "T" as "T and not F" in the model than in the model-free condition, a hypothesis that the authors subsequently confirmed.

DECISION MAKING

The studies of decision making have led to the construction of a descriptive theory that gives more weight to the contextual and representational factors that enrich and reveal the fact that the decision making process is more complex than that suggested by the normative parameter.

The normative approach adopted by economists and mathematicians indicated how someone should make a rational decision.[48] The descriptive approach used by psychologists of thought and decision making has led to the construction of models for describing and predicting the decision-making process underlying the produced choices. Within the framework of studies of decision making, we have the structuring of a descriptive theory (Prospect Theory[49]) that addresses the consideration of contextual and representational factors which, in comparison with the normative parameter, have the intent to enrich and reveal the greater complexity of decision making.

Prospect Theory

Prospect theory explains how and why our choices deviate from the normative model of expected-utility theory. As a modification of expected-utility theory, it has two main parts: one concerning probabilities and the other concerning utilities.

The theory retains the basic idea that we make decisions as though we multiplied something like a subjective probability by something like a utility: the more probable a consequence is, the more heavily we weigh its utility in our decision. According to the theory, however, we distort probabilities, and we think about utilities as changes from a reference point. The reference point is easily affected by irrelevant factors, and this leads us to make different decisions depending on how it is presented to us.

Probability. Let us first look at the probability part. In essence, prospect theory begins with the premise that we do not treat probabilities as they are stated, but distort them according to a particular mathematical function (the "pi function").

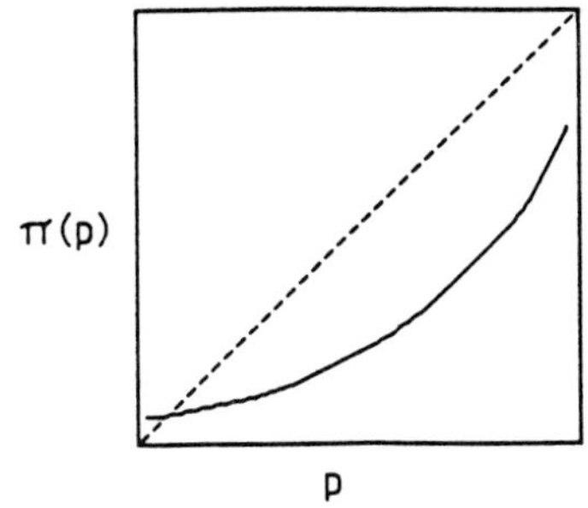

Figure 2. The "pi function".

The Certainty Effect. Subjects appear to be attracted by the absolute certainty of \$30. The pi function explains this effect because the certainty is overweighted. For example, in a gamble some people prefer (\$30) to (\$45, .80), but they prefer (\$45, .20) to (\$30, .25). Here (\$30) means \$30 for shure ; (\$45, .80) means \$45 with probability .80 and \$0 with probability .20, and so forth.

- \$30 for sure
- \$45 , .80

- \$45 , .20
- \$30 , .25

If you prefer (\$30) to (\$45, .80), this would imply – using the expected-utility formula – that

u(\$30) > .80 . u(\$45)

If you prefer (\$45, .20) to (\$30, .25), this implies that

.25 . u(\$30) < .20 . u(\$45)

both inequalities cannot be true. Each side of the second inequality can be derived from the corresponding side of the first by multiplying by .25.

.25 . u(\$30) > (.25 . .80) . u(\$45)

This pattern of choices is called the *certainty effect*, because subjects appear to be attracted by the absolute certainty of the \$30.

Utility: the Value Function and Framing Effects. Let us now look at the part of prospect theory concerning utility. According to the theory, individuals evaluate outcomes as changes from a reference level, which is usually the individual's status quo. Because we take different conditions as the reference point, depending on how a decision is described to us, we can make different and inconsistent decisions for the same situation depending on how it is described. An individual views monetary consequences in terms of changes from a reference level, which is usually the individual's status quo. The value of the outcomes for both positive and negative consequences of the choice then has the characteristic of diminishing returns. The resulting value function is steeper for losses than for gains: risk aversion for gains, and risk seeking for losses.

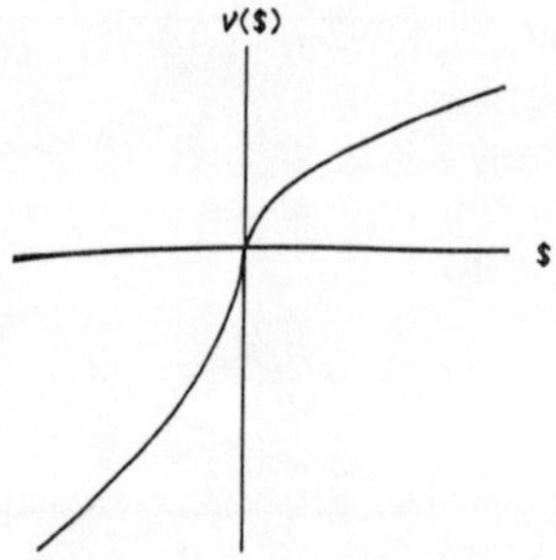

Figure 3. The value function.

1. The value function shows that we treat losses more seriously than equivalent gains. We consider the loss in value of losing $10 is greater than the gain in value of gaining $10.

2. The value function is convex for losses (an increasing slope as we move to the right, as shown in the lower left of the figure) and concave for gains (a decreasing slope). This means that subjects avoid risks in the domain of gains and seek risks in the domain of losses, when gains and losses are defined in terms of expected monetary change from their reference point.

Framing Effect. Imagine that the U.S. is preparing for an out break of an unusual Asian disease that is expected to kill 600 people. Two alternative programmes to combat the disease have been proposed. Scientific estimates of the consequences of the programmes are as follows:

-If programme A is adopted, 200 people will be saved.
-If programme B is adopted, there is 1/3 probability that 600 people be saved and 2/3 probability that no people will be saved.
Which programme would you favour?

Most subjects choose programme A, presumably because they do not want to take the substantial risk (probability .67) that nobody would be saved at all. However, when the same problem is presented in a different way:

-If programme C is adopted, 400 people will die.
-If programme D is adopted, there is 1/3 probability that nobody will die and 2/3 probability that 600 will die .

most subjects choose programme D, presumably because 400 deaths seemed almost as bad as 600 and because it seemed worthwhile taking a chance that nobody would die at all.

The two identical versions are described differently, with the first emphasising the saving of life (a gain) and the second emphasising death (a loss). This constitutes a violation of the invariance principle, because the choice made is dependent on how the situation is presented or "framed", whereas the invariance principle asserts that choices ought to depend on the situation itself and not on the way it is described.

Mental Account. Mental accounts affect the effort we make to save money when we buy things, which seems to depend on the subjective value of the money saved rather than on the amount. Many of us are willing to spend as much effort to save $1 on a $10 purchase as we would to save $100 on a $1,000 purchase. Here is a good example:

Imagine that you are about to purchase a jacket for $125 and a calculator for $15. The calculator salesman informs you that the calculator you wish to buy is on sale for $10 at the other branch of the store located 20 minutes drive away. Would you make the trip to the other store?

Most subjects asked were willing to make the trip to save the $5 but, in another version of the problem, very few were willing to make the trip to save $5 on the jacket. In both cases, the question is whether you would be willing to drive 20 minutes for $5. If the saving is part of your mental account for the calculator, the saving is subjectively greater than it is if is part of your account for the jacket because of the convexity of the curve of the value function for losses.[50,51]

CRITICAL PERSPECTIVES

The area of decision making is also coming under critical review as a result of the reintroduction of previously unconsidered elements that make the choices of the subjects more reasonable.

According to the meta-rationality argument,[52-54] decision behavior that violates principles of rationality can be described as perfectly rational if the cognitive costs of being rational are taken into account. For example, when somebody wants to buy a book at a train station because he is going on a long trip, he will probably not check all of the available books in order to find the best of all but will look at a few and then buy the first that he finds reasonably attractive. He can be described as working on the satisfying principle (and could be seen as irrational from the perspective of the SEU model, but he would probably be intuitively considered non-rational if he wanted to check all of the books in the store. With finite time and resources available, it is not rational to spend infinite effort on the exploration of all potential consequences of all options. In this model it is assumed that people make meta-decisions between the various decision strategies they have in their repertoire, taking into account the characteristics of the situation (e.g. familiarity with the problem, time pressure) and the subject (e.g. knowledge, ability).

Hogarth's continuity argument[55] represents judgements and decisions as moments in a continuous process. They may sometimes appear biased or deficient if treated as discrete events, but may in fact be very functional when considered as moments in a continuous and changing environment. In an experiment by Ronen,[56] the subjects had to choose between two options with identical (positive) payoffs and probabilities of success that depended on the results of two sequential events: in one option, there was a higher probability of success at the first step and a lower probability at the second step; in the alternative option, it was the other way around. Since the SEUs of the two options were equal, rationality would imply indifference; but the subjects preferred the option with a higher probability at the first step. Hogarth[55] provides the following explanation: if people are used to a changing environment, they may give less weight to the probability of the second step because the situation might have changed when this phase is reached. More generally, many previous experiments have excluded the possibility of feedback and redundancy, but this is characteristic of the environment, which may be assumed to be either stable or changing. Many of the biases illustrated by the discrete approach can be interpreted as indicators of cognitive mechanisms that are actually very functional in a continuous environment (see Hogarth[55] on the use of availability). Other discussed assumptions are the stability of preferences and goals, and the degree of ambiguity of preferences. In general, this approach attributes deficiencies in judgement and decision making to the inadequate assumptions of the SEU model, such as fixed time horizons or the very low ecological validity of the experimental situations.

Another argument against studies considering deviations from the SEU model as irrational behavior comes from Berkeley and Humphreys[57] and Phillips.[58] The claim is that demonstrations of cognitive deficiency are questionable because they fail to take into account the subjects' internal structural representation of the problem, which is considered to be the same as that of the experimenter.

Finally, another modification of expected-utility theory proposes that the utility of an outcome depends not only on the outcome itself, but also on the other outcomes that would occur in different parts of the utility table: in particular, the emotions that we experience when we compare an outcome that does occur with other outcomes that could have occurred. Emotions are the real consequences of our choice: if the same monetary reward is associated with two different emotions, these imply different consequences. Whenever we make a decision, we anticipate these emotional factors and take them into account – see the Regret Theory of Loomes & Sudgen[59] and Bell,[60] the Disappointment Theory of Loomes & Sudgen[61] and Bell,[62-63] and the Decision Affect Theory of Mellers, Schwartz and Ritov.[64]

CONCLUSIONS

Research into judgement and decision making has been mainly structured around the idea of formulating a descriptive theory that considers non-rational aspects: heuristics and contextual factors.

However, the explanatory principles that have been used to explain discrepancies in relation to the normative apparatus leave the latter in the unchanged position of judge and indicator as to how people should reason in order to make rational choices, but – paradoxically – had also the a-problematical assumption that the internal representation of a task coincides with its logical and normative representation.

It seems possible to say that, along the lines of what happened in the case of research into reasoning and problem solving, the pragmatic aspects and emotional components of decision-making processes have:

- reduced the generality of the phenomena identified (the base-rate and the conjunction fallacies) when the same representation of the problem is shared by the researchers and the experimental subjects (respecting the rules of pragmatics);
- and broadened the concept of rationality (which in fact coincides with the assumptions underlying the maximisation of expected utility – seen mainly, if not exclusively, in terms of economic advantage) (see studies of emotional or ethical aspects, individual differences, dynamic vs. static decisions).

We think that the consideration of pragmatic and emotional aspects has contributed towards a greater understanding of the representation of a task by ordinary people, and that perspective seems to be crucial for studies of judgement and decision making that have been constructed around the idea of being a descriptive theory of our behavior.

REFERENCES

1. W. Edwards, The theory of decision making, *Psychological Bulletin*, 51, pp. 380-417, (1954).
2. W. Edwards, Conservatism in human information processing, in B. Kleinmuntz ed., *Formal representation of human judgment*, Wiley, New York (1968).
3. C.R. Peterson and L.R. Beach, Man as an intuitive statistician, *Psychological Bulletin*, 68, 1, pp. 29-46, (1967).
4. A. Tversky and D. Kahneman, Judgment under uncertainty: heuristics and biases, *Science*, 185, pp. 1124-1131, (1974).
5. P. Slovic, Information processing, situation specificity, and the generality of risk-taking behavior, *Journal of Personality and Social Psychology*, 22, pp. 128-134 (1972).
6. I.L. Janis and L. Mann, *Decision making: a psychological analysis of conflict, choice, and commitment*, New York: Free Press (1977).
7. R.E. Nisbett and Ross, *Human inference: strategies and shortcomings of social judgment*, Englewood

Cliffs, NJ: Pertinence-Hall (1980).

8. D. Kahneman and A. Tversky, On the psychology of prediction, *Psychological Review*, 80, pp. 237-251 (1973).
9. D. Kahneman and A. Tversky, On the reality of cognitive illusions: a reply to Gigerenzer's critique, *Psychological Review* (1996).
10. D. Kahneman, P. Slovic, and A. Tversky eds., *Judgment Under Uncertainty: Heuristics and Biases*, Cambridge, Cambridge University Press, (1982).
11. S. Lichtenstein, P. Slovic, B. Fischhoff, M. Layman, and B. Combs, Judged frequency of lethal events, *Journal of Experimental Psychology: Human Learning and Memory*, 4, pp. 551-578, (1978).
12. D. Kahneman, P. Slovic, and A. Tversky eds., *Judgment under uncertainty: heuristics and biases*, Cambridge, Cambridge University Press (1982).
13. A. Tversky and D. Kahneman, Causal schemas in judgments under uncertainty, in M. Fishbein ed., *Progress in Social Psychology*, Vol. 1, Hillsdale, NJ: Erlbaum, pp. 49-72 (1980).
14. M. Bar-Hillel, Back to base rates, in R.M. Hogarth ed., *Insights in Decision Making*, Chicago, University of Chicago Press (1990).
15. W. Casscells, A. Schoenberger, and T.B. Graboys, Interpretation by physicians of clinical laboratory results, *New England Journal Of Medicine*, 229, pp. 999-1001, (1978).
16. D.M. Eddy, Probabilistic reasoning in clinical medicine: problems and opportunities, in D. Kahneman, P. Slovic, and A. Tversky eds., *Judgment Under Uncertainty: Heuristics and Biases*, Cambridge, Cambridge University Press (1982).
17. A. Tversky and D. Kahneman, Extensional versus intuitive reasoning: the conjunction fallacy in probability judgment, *Psychological Review*, 90, pp. 293-315, (1983).
18. J. Leddo, R.P. Abelson, and P.H. Gross, Conjunctive explanations: when two reasons are better than one, *Journal of Personality and Social Psychology*, 47, pp. 933-943, (1984).
19. A. Locksley, and C. Stangor, Why versus often: causal reasoning and the incidence of judgmental bis, *Journal of Experimental Social Psychology*, 20, pp. 470-483, (1984).
20. G.L. Wells, The conjunction error and the representativeness heuristic, *Social Cognition*, 3, pp. 266-279, (1985).
21. E.B. Shafir, E.E. Smith, and D.N. Osherson, Typicality and reasoning fallacies, *Memory and Cognition*, 18, pp. 229-239, (1990).
22. M. Bar-Hillel and E. Neter, *How alike is it versus how likely is it: a disjunction fallacy in probability judgments*, Communication presented at the SPUDM, Friburg (1991).
23. B.W. Carlson and J.F. Yates, Disjunction errors in qualitative judgment, *Organizational Behavior and Human Processes*, 44, pp. 368-379, (1989).
24. G. Gigerenzer and D.J. Murray, Thinking: from insight to intuitive statistics, in *Cognition as Intuitive Statistics*, Hillsdale, New Jersey, Laurence Erlbaum Associates (1987).
25. G. Politzer and I.A. Noveck, Are conjunction rule violations the result of conversational rule violations? *Journal of Psycholinguistic Research*, 15, pp. 47-92, (1991).
26. L. Macchi, Pragmatics aspects of the base-rate fallacy, *The Quarterly Journal of Experimental Psychology*, 48A(1), pp. 188-207, (1995).
27. G. Mosconi and L. Macchi, The role of pragmatic rules in the conjunction fallacy, *Communication presented at the 3rd International Conference on Thinking, British Psychological Society Cognitive Science*, 29-31 August 1996.
28. G. Gigerenzer, On cognitive illusion and rationality, in E. Eells & T. Maruszewski ed., *Reasoning and rationality. Essays in honour of L. J. Cohen*, Amsterdam, Rodopi (1991).
29. G. Gigerenzer, Why the distinction between single-event probabilities and frequencies is important for psychology (and vice versa), in G. Wright and P. Ayton, *Subjective Probability*, Chichester: John Wiley and Sons (1994).
30. L. Cosmides and J. Tooby, Are humans good intuitive statisticians after all? Rethinking some conclusions from the literature on judgment under uncertainty, *Cognition*, 58, pp. 1-73, (1996).
31. G. Gigerenzer and U. Hoffrage, How to improve bayesian reasoning without instruction: frequency formats, *Psychological Review*, 102, pp. 684-704, (1995).
32. K. Fiedler, The dependence of the conjunction fallacy on subtle linguistic factors, *Psychological Research*, 50, pp. 123-129, (1988).
33. R. Hertwig and G. Gigerenzer, The "conjunction fallacy" revisited: how intelligent inferences look like reasoning errors, *Journal of Behavioral Decision Making* (1999).
34. S.K. Jones, K. Taylor, and D. Frisch, Biases of probability assessment: A comparison of frequency and single-case judgments, *Organizational Behavior and Human Decision Processes*, 61, pp. 109-122, (1995).
35. T. Reeves and R.S. Lockhart, Distributional versus singular approaches to probability and errors in probabilistic reasoning, *Journal of Experimental Psychology: General*, 122, pp. 207-226, (1993).
36. R. Hertwig and G. Gigerenzer, The "conjunction fallacy" revisited: How intelligent inferences look like reasoning errors, Unpublished manuscript (1997).

37. D.E. Dunlay and D.J. Hilton, Conversational implicature, conscious representation, and the conjunction fallacy, *Social Cognition*, 9(1), pp. 85-110, (1991).
38. L. Macchi, Partitive formulation of information in probabilistic problems: beyond heuristics and frequency format explanations, *Organizational Behavior and Human Decision Processes*, 82, 2, pp. 217-236, (2000).
39. G. Mosconi, *Discorso e Pensiero*, Il Mulino, Bologna (1990).
40. Z. Ginossar and Y. Trope, Problem solving in judgment under uncertainty, *Journal of Personality and Social Psychology*, 52, pp. 464-474, (1987).
41. H.P. Grice, Logic and conversation, in P. Cole and J.L. Morgan eds., *Syntax and Semantics*, Vol. 3: Speech acts, New York, Academic Press (1975).
42. G. Mosconi and V. D'Urso, *Il farsi e il disfarsi del problema*, Firenze, Giunti-Barbera (1974).
43. G. Politzer, Laws of language use and formal logic, *Journal of Psycholinguistic Research*, 15, pp. 47-92, (1986).
44. G. Politzer, Immediate deduction between quantified sentences, in K.J. Gilhooly, M.T.G. Keane, R.H. Logie, and G. Erdos eds., *Lines of thinking. Reflections on the psychology of thought*, London, Wiley (1990).
45. D. Sperber and D. Wilson, *Relevance: Communication and Cognition.*, Oxford: Basil Blackwell (1986).
46. A. Kasher, Conversational maxims and rationality, in A. Kasher ed., *Language in focus: foundations, methods and systems*, Dordrecht, Holland, Reidel (1976).
47. G. Mosconi and L. Macchi, The role of pragmatic rules in the conjunction fallacy, *Mind & Society*, in press.
48. J. von Neumann and O. Morgenstern, *Theory of games and economic behavior* (2nd ed.), Princeton: Princeton University Press (1947).
49. D. Kahneman and A. Tversky, Prospect theory: an analysis of decision under risk, *Econometrica*, 47, pp. 263-291, (1979).
50. A. Tversky and D. Kahneman, The framing of decisions and the rationality of choice, *Science* , 221, pp. 453-458, (1981).
51. R. Thaler, Mental accounting and consumer choice, *Marketing Science*, 4, pp. 199-214, (1985).
52. L.R. Beach and T.R. Mitchell, *A contingency model for the selection of decision strategies*, Academy of Managment Review, 3, pp. 439-449, (1978).
53. H.J. Einhorn and R.M. Hogarth, Behavioral decision theory: processes of judgment and choice, *Annual Review of Psychology*, 32, pp. 53-88, (1981).
54. J.W. Payne, Contingent decision behavior, *Psychological Bulletin*, 92, pp. 382-402, (1982).
55. R.M. Hogarth, Bayond discrete biases: functional and dysfunctional aspects of judgmental heuristics, *Psychological Bulletin*, 90, pp. 197-217, (1981).
56. J. Ronen, Effects of some probability displays on choices, *Organizational Behavior and Human Performance*, 9, pp. 1-15, (1973).
57. D. Berkeley and P. Humphreys, Structuring decision problems and the 'bias heuristic', *Acta Psychologica*, 50, pp. 201-252, (1982).
58. L.D. Phillips, A theoretical perspective on heuristics and biases in probabilistic thinking, in P.C. Humphreys, O. Svenson, and A. Vari eds., *Analysing and aiding decision processes*, Amsterdam: North Holland (1983).
59. G. Loomes and R. Sudgen, Regret theory: an alternative of rational choice under uncertainty, *Economic Journal*, 92, pp. 805-824, (1982).
60. D.E. Bell, Regret in decision making under uncertainty, *Operations Research*, 30, pp. 961-981, (1982).
61. G. Loomes and R. Sudgen, Disappointment and dynamic consistency in choice under uncertainty, *Review of Economic Studies*, 53, pp. 271-282, (1986).
62. D.E. Bell, Regret in decision making under uncertainty, *Operations Research*, 30, pp. 961-981, (1985).
63. D.E. Bell, Disappointment in decision making under uncertainty, *Operations Research*, 33, pp. 1-27, (1985).
64. B. Mellers, A. Schwartz, and I. Ritov, Emotion-based choice, *Journal of Experimental Psychology: General*, 128, 3, pp. 332-345, (1999).

PANEL SUMMARY: **THINK ABOUT THIS: IS THINKING NECESSARY TO DO ACTIONS?**

Jörg D. Becker,[1] Mariagrazia Semenza,[2] and Alessandra Setti[3]

[1]Universität der Bundeswehr München
ET / Physik
85577 Neubiberg, Germany
e-mail: Joerg.Becker@unibw-muenchen.de
[2,3]Computer Vision Laboratory
Computer and Systems Engineering Department
University of Pavia
27100 Pavia, Italy
e-mail: {lella,ale}@vision.unipv.it

Emitte Spiritum tuum et creabuntur, Et renovabis faciem terrae.

INTRODUCTION *(J.D. Becker)*

When Virginio Cantoni, Vito Di Gesù, and one of the authors (JDB), in a very creative meeting in Palermo, decided about the title of this panel discussion, the original formulation simply was: Is thinking necessary?

Jean Beaudrillard has argued that in the future thinking will be a useless luxury. Josef Lechner has stated that there is a complementarity between thinking and happiness. Niko Tinbergen has conjectured that intelligence in man has developed for the ability to predict the actions of other people to a higher degree (so here we already have a connection between thinking and action), and that thinking about general relativity or artificial intelligence is an abuse of the brain. All these statements are not very encouraging.

On the other hand, the computing power and the number of computers in the world grows faster than the population of India, so we may easily predict that in some not too distant future the supply of Indians will no longer suffice to meet the demand for programming. As far back as in 1987, Hans Moravec has demonstrated that in the year 2020 the computing power of the human brain can be purchased at the cost of a PC, and since then we have little evidence that we should abandon his arguments. This means that we have to aim at ASP (Automatic Software Production) if we want to avoid that (at least in the future) the hardware is willing but the software is weak.

One of our students once said that since we cannot decide whether God exists or not we may add his existence to our set of axioms. We can do the same with thinking: we may take as a working hypothesis that thinking is necessary, and discuss the consequences of this

Human and Machine Perception 3: Thinking, Deciding, and Acting, **edited by Cantoni *et al.***
Kluwer Academic/Plenum Publishers, New York, 2001.

axiom. But we may also say that, if we take necessity of thinking as a thesis and superfluousness of thinking as an antithesis, freedom of thinking is the synthesis.

THINKING IS CONSTRUCTION

Most of traditional occidental philosophy has got two serious limitations: it is primarily analytic, and it is primarily language oriented. Here we shall postulate that thinking in general is a constructive process, and that the constructs of thinking may belong to very different spaces. So we may think not only in words but also in shapes, in pictures, in motions and muscular sensations (like Einstein), in tunes, in actions, in scenes and emotions (like Fellini), and so on.

We call our approach "Constructive Constructivism" to make a distinction to Radical Constructivism (as developed by Glaserfeld, von Foerster, Maturana and others). Radical Constructivism claims that reality is only constructed in the brain. We do not want to discuss this approach here even if it is at least useful to assume that the tax office is a reality which exists also outside of our brains. Independently from the question of existence and reality we may ask how the mechanisms of construction work; and that is what we mean by Constructive Constructivism.

The basis of our approach is the general model of an autonomous learning system. In this model there are three persons:

- the *robot* who has (given or found) *building blocks* and *construction rules* at his disposition;
- the *theologian* who sets the *norm* to the robot - and we assume that there is a *vision* behind this norm;
- the *psychologist* who reflects the *sanity or madness* of the theologian.

In addition we have:

- the *environment* with its *resources* and *constraints*.

The three persons can also be realized in one being; thus, a scientist is a robot, a theologian, and a psychologist at the same time. (This scheme gives us a first hint for understanding the Holy Trinity.) Originally this model has been designed for learning systems, where the construction rules include the learning rules (for example, those for neural networks or those for evolutionary systems). However, this scheme may also be used as a model of constructive thinking, because thinking is a playground for reality. Complex constructs (thoughts) are composed from basic building blocks using the construction rules. A given set of basic building blocks and construction rules may be called a *style*. The thinking process is controlled by a norm or vision.

We note that this scheme is quite general; it includes digital and analog representations, computer programming and arts. We also note that it can be formalized, but we do not want to discuss this issue here. It is a framework within which we can discuss phenomena like creativity (producing many constructs which make sense in a given environment) or innovation (finding the golden middle between stagnation and permanent revolution).

Thinking seen as a constructive process also means changing ourselves because we gain information.[1] Any thinking process "reflects" back to us, is an ipsoflexive process. We can even study self-consciousness in this approach. Thomas Aquinus says about the Holy Trinity that the Father perceives Himself in His Son through the mediation of the Holy Spirit. This reminds us of Piaget's model of self-consciousness: a child observes a second child having a model of a third one, and then (by identifying the persons) observes that it has a model of itself. Source, goal, and mediation are of the same nature, and, like in the Japanese Ba theatre, players, spectators, and the play emerge in the process of interaction.

PROTOCOL OF THE DISCUSSION

Aleksander Igor. Considering individuals with physical damage to the brain, you could assert that thinking is definitively necessary. It is important to distinguish about artificial domain and human/natural domain. For example, bees have a contemplative representation of the world (flowers, ...) and do actions. Bees are autonomous, robots are not. So thinking is a necessary requirement to be autonomous; it is a constructive process within a limited domain.

Richard Walker. Robots can do sophisticated actions without thinking. Let us try to change the question: when do thoughts become necessary? What are the problems which require thoughts? (Compare the artificial world and the experiment of Dario Floriano - Losanna).

Jörg D. Becker. You said that we have robots that do sophisticated actions without thinking, but before the robot works there is a great investment in thinking for constructing it.

Aleksander Igor. What is thinking necessary for? The difficulty is again with the words.

Settimo Termini. It seems we begin doing actions before we are aware of doing actions. The mechanism is not clear. Is thinking necessary? No?! Why should we start the discussion with the working hyphotesis yes? What is it that characterizes thinking?

Thinking is a constructive process, but it is not the only one. Building a program requires thinking, but the working of a computer program is not thinking. A program is a constructive process.

Jörg D. Becker. If a computer program only performs a given task I would not call that thinking. But if the program acquires new knowledge, if it produces unforeseen constructs which increase pragmatic information[2] (as it happens in evolutionary systems), I would call that thinking.

Bertrand Zavidovique. I do not have more difficulty to define thinking than to define acting. There are programs that create free variables: maybe such a program started to think? Is thinking really necessary to survive? There is a 6 neurons crab that has no predators and lives since millions of years but it does not think.

Thinking does transform a representation into another (or one instance of a representation into another), in a similar manner as acting transforms an object (or an instance of it, or a state of it) into another. This is by no means a definition of thinking, but thinking does that and it is a good starting point for machines.

A sufficient property for an activity to be called thinking could better be in the ability to create and manipulate free variables (a free variable is one to be automatically instanciated at execution time of a procedure) on some regular basis (once is not enough!). So there is no reason to ask for a more precise definition since, from there, thinking would be the process by which we discuss ill-defined objects!! And along the same recursive line acting is not better defined than thinking: for instance "mimicking" would not be acting more than obeying some program (plan, strategy, ...), designed by some other entity, is thinking. This is probably a reason why it is commonly accepted that a significant difference lies between adaptivity and intelligence.

Another clue, very close to the latter, deals with precision: as Lotfi Zadeh said one matter of and possible reason for thinking is mastering (adapting) precision.

I would rather say that thinking aims at putting a human or a system in a situation that makes them independent of precision contingencies. Thinking then would be a process by which a system becomes precision invariant (more generally it makes the link between observation and action much less tight). In that respect "thinking" certainly helps "acting" for the same reason why prospective geometry is a more abstract (efficient) representation of our world, but a more difficult one as well.

Edoardo Boncinelli. There are bacteria: they exist and they certainly do not think!

Jörg D. Becker. Is thinking necessarily restricted to neurons? Also a biochemical process can be constructive and information-increasing; amebas may be able to "think" in this way. Actually, evolution itself is a thinking process, and if a crab with six neurons can survive for millions of years without thinking evolution has thought before and thus produced the crab.

Mariagrazia Semenza. Richard Feynman[3] has reported an observation of a paramecium in a droplet of water under a microscope. In the textbooks you read that the paramecium has an extremely simple behavior, but what Feynman observed was something quite different which did not at all resemble a "machine" driven by a simple program. Even a one-celled animal can do relatively complex actions: it can evaluate a situation somehow, it can choose an action, moreover it can decide to change its "decision" and change its behavior. It looks like a simple kind of thinking.

Antonio Restivo. Is it an action to walk during sleep? When a child learns to ride a bike it has to think how to do that, but when it has really learnt, it does not think any more about how to do it. There is an inner model of what we are doing (drive a car, ...), we are not consciently thinking about that. Thinking is to be consciously aware.

Lotfi Zadeh. There is a tendency to be categorical. It is not possible to be categorical on the definition of thinking. We do not need a high degree of precision, there are different degrees in nature (not only crisp answers have to be considered). The real challenge is automation of things that require high level processes.

Jörg D. Becker. I agree completely with you; this is what our work is about.

Sergio Cerutti. The fundamental definition of thinking is that to reproduce the thinking mechanisms in a machine is now impossible. Sophisticated machines that adapt very well are "intelligent" but they do not think. Thinking is a consciousness of itself: a function of man, that requires plasticity of the brain, adaptation to the external world, answers to stimula, experience. It is a long process. Do you think of thinking as a superior activity of the mind - not the brain?

Thinking is the result of very complex transformations inside the human brain or - perhaps - the highly evolutionary primates.

Thinking involves self-consciousness and is strongly correlated to a highly complex nervous system. Probably it is born out of the evolution of living systems, starting from genetic predisposition and involving adaptation and plasticity mechanisms with strong interaction with the environment of the external world.

Due to the extremely complex system involved in thinking and the variety of stimulations to obtain it, given the actual knowledge in neurophysiology, it seems unlikely to provide some "thinking" elements in artificial intelligence applications.

Salvatore Gaglio. In our introspection there is a lot of mental activity. Thinking is to be aware, not something to be put into machines; we can implement logic. Is thinking something more than self awareness?

Jörg D. Becker. When I talk about thinking, I think about creative processes, for example about a composer who is writing a symphony.

Aleksander Igor. A crab lives since millions of years and has few neurons; a rock can survive billions of years and it does not think at all. Where does thinking start in the evolution of more complex organisms? Thinking becomes an interference with automatic actions (cf. University of Copenhagen: "Neurobiological bases of thoughts").

Thesis: Thinking is the internal activity which is necessary to predict the result of alternative actions to a given stimulus and choosing one of these. Therefore thinking is necessary when survival depends on the choice of the best of possible actions from a possible set.

Antithesis: Organisms exist with excellent survival properties- amebas, bacteria, some crabs. They do not think but they survive.

Synthesis: Survival is a bad indicator of the necessity of thinking. Given that survival supports a wide range of complexities in organisms it is possible to partition the set into those that think and those that do not. The synthesis states that this partition maps exactly into those organisms that need to select actions to survive and those that do not. Machines: when designing machines, the first step is to identify to which group they should belong.

Virginio Cantoni. The goal of this workshop is the comparison between artificial and natural systems. Thinking is a goal oriented system.

The problem of the phylogenetic evolution from reacting or reflection abilities towards higher levels of tactical or strategic capabilities has a correspondent problem when you are designing autonomous artificial systems. In fact, broadly speaking at least two hierarchical levels of control can be envisaged:

1. a low level control loop, usually the "periphery" (that is "closed" to the sensory environment), which acts immediately and guarantees the survival of the system in cases of extreme danger (e.g. a robot close to a crash or to a ravine);
2. a high level loop, usually goal-oriented, with planning and strategic capabilities.

The two levels have situation-dependent priority: the reactive system should work on-line, with a very short time for intervention, meanwhile the planning system must take the control as soon as possible for not loosing the possibility of reaching the final goal. The integration of the two loops is one of the basic problems in the sophisticated human vision system. The framework between time (for perception) and reasoning (for planning) is implemented with at least three hierarchical control levels (e.g. consider something arriving against you):

1. fixation with head and body movement toward the target (requiring seconds);
2. saccades for fast eyes movements requiring around 250 msec;
3. the "searchlight" of attention without mechanical movements, the cover attention, requiring around 10-25 msec.

Richard Walker. The mathematical characterization in a thinking machine is a terribly long path (dynamic paths with lots of possibilities).

Gianni Rigamonti. Consider the point of view of Chomsky about language: the creative aspect of language; the generative grammar is similar to thinking as a constructive process.

Jörg D. Becker. This is one example - others are fuzzy sets, pictures, biology, ... - life itself is a kind of thinking process. The question which building blocks we should use is a fundamental one. Could life develop in a world without carbon? You may replace water by some other electrolyte (life in alcohol would probably be possible), but for many reasons it would be extremely hard to replace carbon by anything else. Carbon compounds also emerge into various intermediate levels of structure; I wander whether one finds something similar in artificial systems.

Aleksander Igor. As a system evolves in response to emergence that happened, you definitively see intermediate levels.

Edoardo Boncinelli. I want to point out that even in the presence of carbon, life has to be considered an accident, multicellularity another accident and human mind an additional accident.

Marco Colombetti. Do we need the Holy spirit in the construction of artificial intelligent systems?

Jörg D. Becker. Father Busa of the Gallarate monastery near Milan claims that nowadays the Holy Spirit is active in our computers anyway.

Vito Di Gesù. We are in some level of the evolutionary state (homo sapiens, ...). Maybe ants do think and they suppose that human beings do not think: there is no communication.

Conclusions of the Discussion

In the discussion we can identify two basic types of arguments, the first one dealing with various aspects of construction, the other one dealing with consciousness. Of constructivism we have discussed its nature, various problem classes, its relationship with logics, the integration between high level and low level constructs, and a wider range of applications. Some speakers have pointed out the relationship between thinking and consciousness, but it remains an open question whether consciousness is a prerequisite of thinking, and whether consciousness can be achieved in artificial beings.

CONSTRUCTIVISM IN PHILOSOPHY (*J.D. Becker*)

One of the participants has also asked the question whether Hegel is connected to constructivism. We think that he is, and actually his principle of creating something new out of the pair thesis - antithesis has been used several times in the discussion. However, we want to have a closer look at three other philosophers who have contributed to constructivism: Leibniz, Locke, and Whitehead.

Leibniz has proposed that monads are the basic building blocks of everything. Each monad reflects the world from a different perspective. It is eternal, unique, and a center of metaphysical force. There is no interaction between monads, but they act in perfect "prestabilized harmony". Complex objects are aggregates of monades. Time and space are consequences of their existence. Identity of the undistinguishable is a basic assumption. All distinctions are qualitative. Everything is organized in a way that from a minimum of principles a maximum of variety emerges.

Locke claims that ideas and consciousness arise only from (internal and external) experiences. An idea is the immediate content of a perception. Complex ideas arise from linking, comparing, and relating simpler ones. Abstraction means leaving aside the concrete.

Whitehead suggests to use events, or *actual occasions*, as universal building blocks (cf. Luhmann who sees decisions as building blocks of social organizations). Every actual occasion has a degree of freedom - it must add something to its causes. It perceives, or *prehends*, data from other actual occasions. Becoming means *concrescence* of actual occasions. Durable objects are formed by societies (*nexus*) of actual occasions. Stability rests on continuous flows of action.

CONSTRUCTIVISM IN MATHEMATICS

In a way, all of mathematics is construction. Frequently mathematicians have conjectured that in logics we find the "genes", the basic building blocks, of mathematics. But, first of all, logics can be traced back to a more basic structure, and secondly (as far as we know) it is not yet clear whether the rest of mathematics can be based on logics alone.

Spencer Brown[4] has noticed that the most basic object of thinking is a distinction. His axioms of distinctions (which he calls "laws of form") are:

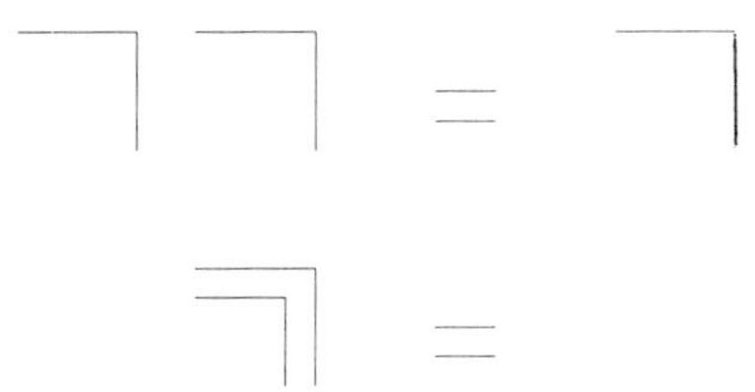

The "corner" is the symbol for a distinction. The right-hand side of the second equation (no symbol) refers to the undivided space ("universe") without distinction. This reminds us of the ring with two elements, called 0 and 1, with the axioms 1x1=1 and 1+1=0. Introducing the symbols D for distinction and U for undivided space we may rewrite Spencer Brown's axioms in the form DxD=D (confirmation of a distinction) and D+D=U (removal of a distinction). This is not a ring, however, because the resulting structure is not distributive. The first equation reflects the similarity between distinction and projection; for every projector P we have PxP=P. After having translated Spencer Brown's equations may be into more familiar symbols it is wise to return to his style because it is much more transparent.

Simple as it looks, this scheme is extremely powerful. Logics and arithmetics appear as consequences, the four-color map problem has been solved with it, and one may even introduce imaginary truth values which correspond to oscillations in time. To give an illustration of the idea behind it, let's have a look at the barber's paradox. A village barber shaves all men in the village who don't shave themselves. Leaving aside the question of political correctness (there are men who don't shave at all), the question is whether the barber should shave himself or not. However, in order to decide whom he should shave he first has to observe. Let's say on the first day he observes that he doesn't shave himself; this is why on the next day he will do so etc. The barber thus creates time by oscillations in his cycle of observing, deciding, and acting (all three are irreversible processes).

Looking for other possible building blocks in mathematics, the Bourbaki group tries to construct everything from three basic structures ("mother structures"): algebra (connections), order, and topology (neighborhood).

We may argue, of course, that for practical purposes it is not necessary to start from the most basic structures. A molecular biologist would not start from quarks. We mention two possible starting points which may be more promising. The first one is a vector space (every vector can be composed from a set of basis vectors); also shapes, pictures, etc. can be seen as belonging to vector spaces. The other one is graph grammars: you have a start graph and a set of rules how to construct more complex graphs from simple ones.

OUTLOOK

The question which building blocks and which construction rules to start with to arrive at intelligent and complex behavior has not been answered yet. However, the limits to most systems lie in the systems themselves. If we remove the artificial fences which surround our playgrounds we may be astonished to see our toys live, grow, and flourish.

REFERENCES

1. Jörg D. Becker, *Learning as gain of information*, in G. Dalenoort ed., The paradigm of self-organization II, Newark, Gordon & Breach (1994).
2. Ioannis Antoniou and Jörg D. Becker, *Pragmatic Information and the Misra Entropy Operator*, in preparation.
3. Richard P. Feynman, *Surely you're joking, Mr. Feynman!* Adventures of a curious character, London, Vintage (1992).
4. George Spencer Brown, *Laws of Form*, London, George Allen and Unwin (1971).
5. Jörg D. Becker and Ioannis Antoniou, *Bergsonian Time and System Maturation*, in Rosolino Buccheri, Vito Di Gesù, Metod Saniga eds., Studies on the Structure of Time - From Physics to Psycho(patho)logy, Dordrecht, Kluwer (2000).

LEARNING FROM MISTAKES

Attilio Giordana and Alessandro Serra

Dipartimento di Scienze e Tecnologie Avanzate
Università del Piemonte Orientale
C.so Borsalino 54, 15100 Alessandria – Italy
e-mail: attilio@al.unipmn.it

INTRODUCTION

Learning from mistakes is probably the most familiar form of learning for every human being. Every time we make a mistake, i.e., we suffer from an inconvenient due to our behavior, we try to modify our knowledge about the world in order to avoid suffering again in the future for the same reason. However, human learning is something of very complex and still poorly understood, so that it is hard to say which are the mechanisms triggered by the detection of a mistake. Even more difficult is to relate this form of learning to other human typical forms of learning, such as learning from textbooks.

For a long time interest in learning processes was substantially aimed at education, in order to make more effective teaching in school classes. Many attempts to understand the knowledge organization and the learning processes in human beings can be found in Philosophy, Psychology, and Pedagogy, starting from the past century.

Recently, the fast growing of computer power made it possible to conceive software agents emulating human performances in many fields. Then, designing and implementing learning algorithms operating on computing machines became a must. In fact, an agent can be said *intelligent* only when provided with adequate learning capabilities.

The research efforts for developing learning programs begun about two decades ago,[1] and kept a continuous growing until today. With respect to psychological investigation about human learning, computer algorithms yield the advantage of being clearly defined and easy to analyze and to evaluate. Even if some inspirations have been taken from cognitive psychology, learning algorithms have been built up with a machine oriented perspective, and, then, cannot be considered as a model of human learning. Even so, the different kinds of learning scenarios have been preserved moving from human learning to machine learning.

In a machine learning context, what we call *learning from mistakes* is probably the most widely investigated form of learning, even if usually it is not called in this way. According to the common meaning, a *mistake* is a *wrong action*, caused by a *bug* (i.e. an *error*) in the agent's knowledge, which prevent it from reaching its goal, or leads it to state

Human and Machine Perception 3: Thinking, Deciding, and Acting, edited by Cantoni *et al.*
Kluwer Academic/Plenum Publishers, New York, 2001.

wrong assertions. Therefore, in many cases the word *error* and *failure* are used instead of mistake.

In the following we will discuss the learning tasks, which are set in the learning from mistake framework. Goal of the chapter is to provide an insight of the different approaches developed in this learning scenario, and to provide a brief review of the most significant literature that should be known by a novice entering this field.

The chapter is organized as in the following. The next section characterizes the scenario in which learning from mistakes takes place, and briefly relates the different learning paradigms to it. The sections that follow review the work done in the area. Specifically, the third section deals with supervised concept learning, giving particular emphasis to incremental algorithms. The fourth section discusses the extension of the basic concept learning task to *supervised learning of classification theories*. In this context, a wide literature is surveyed. Then, the fifth section deals with the connectionist paradigm. A distinction is made between Open Field Networks (OFN) and Locally Receptive Field Networks (LRFN). LRFNs bear properties interesting in order to support incremental learning from mistakes. In the sixth section we will introduce a weaker form of learning, called *reinforcement learning*, which can be considered as a natural generalization of the learning from mistake scenario.

Finally, the seventh section briefly reviews some complexity bounds for an algorithm that learns from mistakes, obtained following the theoretical approach to machine learning. The last section draws some conclusions.

LEARNING FROM MISTAKES: THE SCENARIO

The typical scenario, where an intelligent agent learns from mistakes, is represented in Figure 1. The agent is provided with a knowledge base used to decide its behavior and interacts with the external world by executing actions or by making assertions related to a specific task it is accomplishing.

Periodically the agent receives a feed-back from the world, which allows mistakes be identified. When a mistake is detected, the agent revises its knowledge base in order to avoid the repetition of the same mistake in the future. We will call this cyclic process *learning from mistake*.

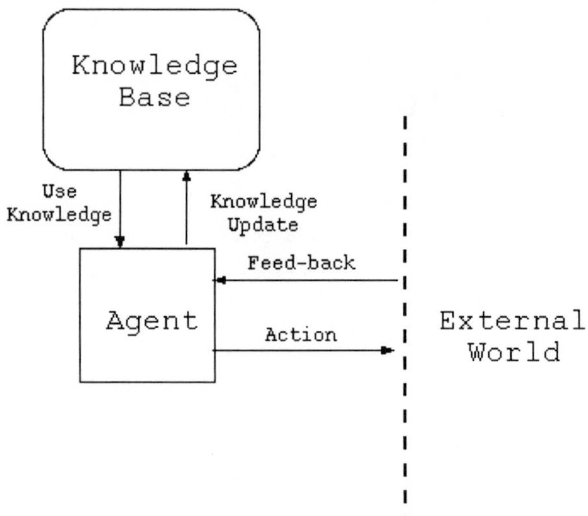

Figure 1. Architecture of a learning agent. When a failure is reported, the mistake is identified, and then, the knowledge base is revised.

In this scenario, being active is a necessary condition for the agent in order to learn. However, the way the learning process may be approached depends primarily on two factors:

- the kind of information fed-back from the external world;
- the method used for encoding the agent's knowledge.

Combining these factors, a variety of more specific learning scenarios can be obtained.

Feed-back from the World

Let us introduce the discussion about this point through an example. Suppose we are designing an agent for playing chess. Suppose, moreover, that the rule of the game, i.e., the possible moves, are given, and we want the agent to learn a game strategy by playing against some opponent. The knowledge source for learning is the information fed-back from the opponent.

The first case we will consider is when the opponent behaves like a teacher aimed at training the agent: every time the agent chooses a move the opponent immediately reports if it is a good move or a bad move. In machine learning this scenario is called supervised concept learning and what is learned is the concept of "good move". Notice that in this case, there is a considerable flow of information from the teacher to the student (the agent in our case).

An alternative case we will consider is when the opponent is a real player, and the agent receives a feed-back only at the end of the game, reporting whether the game is won or lost*. Again, this is a case of supervised learning, but the information flow from the teacher to the agent is much tinier than in the previous case, and so we expect a much greater difficulty. More specifically, in this case, the taught concept is the one of "successful strategy" and so the agent must discover by itself which rule set represents a successful strategy.

The last case is when the teacher does not evaluate the agent's performance on the basis of win or loss, but assigns a score depending on whether he/she liked the match or not. For instance, the teacher can account for the strength of the opponent. Then a loss against a world champion can still be an excellent performance if well played, whereas a victory against a beginner may not be significant for the quality of the strategy used by the agent. In this case, the task of the agent is to modify the strategy in order to obtain good grades from the teacher, instead of simply looking at the result obtained against the opponent.

In the first case, the feed-back from the teacher is the immediate report of a mistake, when a wrong rule is chosen. In the second case, the mistake has to be searched in the sequence of moves that has been selected. It is clear that at least one of them is a wrong rule.

Finally, in the third case, it is not always easy to identify what a mistake is. Suppose the agent won the game, but the score is poor because both opponents played without any skill. How is it possible to identify a mistake? Of course, the sequence of moves will probably contain many mistakes, but it is not possible to identify them because no one can be indicted as responsible of a failure. We will see in the sixth section how it is possible to deal with this case, which requires a framework more general than the one of learning from mistakes.

Knowledge Representation

We can identify three basic paradigms for representing the knowledge of an agent, deriving from three different fields: Artificial Intelligence, Connectionism, and Pattern Recognition.

Artificial Intelligence contributed the symbolic paradigm, which offers a variety of Logic Languages. In Machine Learning the attention has been restricted to propositional, and first order sub-languages of classical logic. In a logical framework, the agent's

---

*In this case the feed-back is obtained automatically from the rule of chess. However, this is not always the case, and in general we can assume to have a referee declaring the succes or the failure.

knowledge is a theory, and a mistake is a false conclusion that follows from a sequence of deduction steps through the theory. Then, learning means to correct the theory in order to eliminate the inconsistency. Notice that the complexity of this class of problems is exponential in the case of propositional logic, whereas checking for theory consistency in first order logic is a semi-decidable problem. For this reason, full first order logic languages have never been used, in practice, but only decidable subsets.

The connectionist approach contributed a set of tools, which, quite improperly, go under the name of *neural networks*. Broadly speaking, a neural network is a function approximator structured as a graph of elementary functions, which can be tuned by adjusting a set of numeric parameters associated to the edges, in order to approximate whatever function with arbitrary precision. Boolean functions may be encoded using a neural network, as a special case*.

In the general case, when a neural network is used to approximate a continuous function, the notion of mistake does not hold any longer. On the one hand an approximation is never perfect and, in average, we will always have an error. On the other hand, the cause of the observed error is distributed among all the constituents of the network.

Instead, the notion of mistake makes sense again when a neural network encodes a boolean function. In this case, every wrong output is a mistake. Nevertheless, the subsequent correction is very different with respect to the case of a logic theory. No single point in the network can be blamed for the mistake, and all parameters of the network are to be modified in some measure. Moreover, whereas in a logic theory the correction problem is combinatorial in nature, in the case of a neural network learning algorithms exploiting the continuous structure of the function approximator are available, such as the error back-propagation algorithm.[2]

The third paradigm is the *instance based* one, and appeared for the first time in Pattern Recognition.[3] In the basic formulation, the agent's knowledge is encoded as a set of reference cases. When the agent must take a decision, or select an action, it compares the current status of the world I(t) with the cases recorded in its library; then, it selects the one more closely resembling to I(t) and behaves consequently.

The most well known algorithm based on this paradigm, is *k*-NN,[3,4] where *k* indicates the number of cases (most similar to the current input), which are retrieved from the library in order to take a decision. The output may be chosen as the most frequent output generated in the retrieved cases, or an interpolation of them.

Considering again our example, a case library can be a set of check-board configurations having associated a recommended move to be done when the check-board has that specific setting. When the agent makes a mistake, it revises the check-board library, possibly adding new configurations or modifying the association between a check-board and the corresponding move.

Being this paradigm easy to apply, and effective in practice, it has been widely investigated in the literature, and many variants have been proposed extending the basic model under many respects. In the fifth section, we will briefly review the fundamental variants of this paradigm, which are related to the neural network paradigm. Other extensions in the direction of Artificial Intelligence, namely the *case base reasoning* in a symbolic framework, are outside the scope of this chapter.

SUPERVISED LEARNING IN A LOGICAL FRAMEWORK

In the previous section we introduced supervised learning and we made the distinction between learning elementary concepts such as the one of "good move", and complex

*Considering our example of the playing chess agent, the choice of the move may be encoded as a boolean function with multiple output: one output for each move.

concepts such as the one of "good strategy". In this section we will be concerned with methods for learning elementary concepts encoded in symbolic languages.

Supervised concept learning, using a symbolic representation, is the most investigated task in the Artificial Intelligence approach to Machine Learning.[1] Even if several different symbolic representation languages have been used in the literature, such as decision trees,[5] or attribute vectors,[6] all of them can be immediately translated into an equivalent logic (propositional or first order) language. Then, we will use the logic framework for our discussion.

A concept is a mapping *h* from a set *D* to the set {*True, False*}. We will denote D^+ the subset of *D* where *h* takes the value *True* and D^- the subset of *D* where *h* takes the value *False*.

A definition of *h* in a logic language L is a formula φ belonging to L, such that:

$$\frac{given: \quad d_i, \quad \varphi}{follows: \quad h(d_i) = True}, \qquad \forall d \in D^+ \tag{1}$$

and

$$\frac{given: \quad d_i, \quad \varphi}{follows: \quad h(d_i) = False}, \qquad \forall d_i \in D^- \tag{2}$$

In the case *D* is known together with the classification $h(d_i)$ of every $d_i \in D$, the concept is known and there is nothing to learn. However, it could still be interesting searching for a definition φ of *h* having a small syntactical complexity. This is a different kind of task and is identified as *information compression*.

Nevertheless, *D* may be so large that, in most practical cases, only subsets of it are known. In this case, learning is defined as the task of generating a hypothetical description φ', by induction from a known subset D_L of *D* and the correct value of *h* for all elements of D_L, that is possibly correct also for the elements of *D* not belonging to D_L. It is clear that φ' can only be an approximate definition of concept *h* because a data *d* not belonging to the D_L may always contain information that has not yet been accounted for and leads to doing a mistake.[7]

On the one hand, it is likely to expect a decreasing rate of mistakes as long as D_L grows. On the other hand, given D_L two different hypotheses φ'_1, φ'_2 may have very different performances on unknown data even if both commit no mistakes on D_L. Most literature in Machine Learning deals with methods for selecting inductive hypotheses.

Generalization versus Specialization

From the above considerations it turns out that a learning algorithm cannot simply record the data D_L together with their classification h_{D_L}, because this form of enumerative description will not classify any data, which is not yet in D_L. On the contrary, a good algorithm should try to *generalize* the data in D_L in order to capture the characteristics typical of the concept and can be found in all instances, also in the ones not yet seen.

Algorithms for supervised concept learning fall into three major categories:

- **Specific-to-general**. The algorithm starts with a hypothesis, which is the description of D_L and keeps relaxing constraints generating more and more general hypotheses. The generalization process stops when the classification error on D_L begins to increase. Examples of this strategy are available in many algorithms, for instance in SHARADE.[8]
- **General-to-specific**. The algorithm starts with a formula Ψ set to the maximally general hypothesis *True*, which is verified by all instances in D_L, and keeps adding

constraints to Ψ until a hypothesis φ', possibly consistent with D_L, is obtained. A classical example of this strategy is represented by the system ID3.[5]

• **Mixed**. Several algorithms perform a bi-directional search by alternating specific-to-general and general-to-specific strategies. Examples in this category can be found in Induce[7] and in RIGEL.[9]

Notice that, independently from the learning strategy, the obtained concept description is always a generalization of D_L. Moreover, all strategies are bound by the error rate on D_L. If the mapping h_{D_L} does not contain errors (i.e., it is noise free), it is possible to obtain a φ' that does not commit mistakes on D_L. Otherwise, one may be obliged to accept inductive hypotheses, which have a non-null error rate even on D_L. Moreover, when D_L is small with respect to D, hypotheses having a higher error rate on D_L may be more accurate on the entire D.

For such reasons, most induction strategies do not limit themselves to generate the maximally general hypothesis compatible with D_L but make use of some heuristics to guide the selection of the inductive hypothesis. Statistics and Information Theory are the baseline for most widely used heuristics.

An excellent introduction to concept learning and to the problems connect to this task can be found in a recent book by Mitchell.[10]

Incremental Concept Learning

Considering the scenario in which we defined "learning from mistake", the learning set D_L and the classification h_{D_L} are obtained incrementally while the agent accomplishes its activity. Then, also the corresponding learning process has to be incremental, reacting to mistakes notified by the external world, as long as they come.

Several learning algorithms for incremental concept learning have been proposed in the literature. The best known algorithm in this category is probably Mitchell's Version Space,[6] because of the methodological issue it addresses. More specifically, Mitchell[6] shows how pure incremental learning is feasible only when the learning set is error-free and when certain restrictions of language L are posed. Being real data not error-free, in practice, the Version Space is almost never applicable.

Nevertheless, other incremental algorithms exist, which are able to deal with the problem of noise (error) in the learning set. A first attempt is represented by STAGGER.[11] An algorithm more advanced than STAGGER is ID5,[12] which is an incremental extension of the popular ID3.[5]

In any case, pure incremental learning is less efficient than one step learning, i.e. when the leaning set D_L is made simultaneously available to the learning algorithm. In fact, considering one example at a time, incremental learning may forget pieces of knowledge, which have been previously learned. For this reason ID5, and several others, keep in memory a subset $D_t \subseteq D_L$ of the instances that have been previously seen, and any time they modify their knowledge base they also check that no mistakes occur on D_t.

Incremental learning algorithms are gaining popularity in data mining applications, where learning with full memory is impossible owing to the very large dimension of the learning set.

THEORY REVISION AND THEORY REFINEMENT

A generalization of supervised concept learning is supervised classification theory learning. Given a set of concepts $H=\{h_j|1\leq j\leq n\}$, defined over a set D, a classification theory T_H of H is a set of axioms, such that:

$$\frac{given: \quad d_i, \quad T_H}{follows: \quad h_j(d_i) = True}, \qquad \forall d_i \in D_j \tag{3}$$

and

$$\frac{given: \quad d_i, \quad T_H}{follows: \quad h_j(d_i) = False}, \qquad \forall d_i \in D - D_j \tag{4}$$

being D_j the subset of D for which h_j=*True*.

The task of learning a classification theory has been first approached under the name of knowledge revision in the domain of knowledge acquisition from human experts, in order to build up expert systems.[13,14] Later on the task has been approached in Machine Learning, both in propositional and in first order logic, under the name of theory refinement.

Theory revision and *theory refinement* deal with two complementary facets of the general task of learning a classification theory.[15] A theory may contain errors due to two reasons: (a) it contains contradictory axioms, so that it is inherently wrong; (b) it does not classify correctly all the items belonging to D because overly general or overly specific, or both. Removing contradiction among axioms is a task that may be accomplished without the data. This is what is done in theory revision when the knowledge elicited from a human expert is verified in order to eliminate inconsistencies.

On the contrary, in theory refinement, axioms are automatically generated in such a way that they do not contain contradiction. Nevertheless, the theory is continuously updated (generalized and/or specialized) on the basis of new data, which are not correctly classified (learning from mistake). In many cases systems designed for theory revision also include a component of theory refinement. For instance, the system SEEK2[14] has a module for eliciting knowledge from a user, which works by generating artificial cases that the user is asked to classify. On the basis of the answer from the user, a step of theory refinement is made in order to let the classification theory be able to correctly classify the new case without the help of the user. A similar method is used by the system CLINT.[16]

In the following we will review only the work done in the area of theory refinement, which is based on the learning from mistake framework. The base line for any theory refinement algorithm can be summarized as in the following. The goal is to keep a classification theory T_H consistent with all instances D_L, which are incrementally acquired. When an instance is assigned to a wrong concept, the axioms responsible for the mistake are located and, then, made more specific by adding some new condition, or some axiom is removed from T_H. On the contrary, if an instance is not assigned to any concept, some axiom is made more general by relaxing some condition or new axioms are added to T_H.

In both cases, after T_H has been modified, it is necessary to verify if the instances previously acquired are still correctly classified. If not, new refinement steps are repeated until the theory is consistent with all the instances. However, this basic reasoning scheme can be very complex and many heuristics have been presented in the literature in order to reduce the complexity.

Theory Revision in Propositional Logic

Being an easier framework, most work in this area has been done in propositional logic. One of the first systems appeared in the literature is DUCE,[17] which relies on the basic refinement scheme described above. Other systems in this line are RTLS[13] and Ductor.[18]

A more sophisticated system is EITHER,[19] followed by an interactive version, NEITHER,[20] designed for acquiring knowledge from a human expert. The novelty of this system is an extension to the basic propositional logic represented by the M-of-N construct. Very briefly, let $\Phi = \{\varphi_1, \varphi_2, \ldots \varphi_N\}$ be a set of formulas. The assertion *M-of-N*$[\Phi]$ is true if at least M of the N formulas in Φ are true. Another system, which makes use of this construct is CRLS.[21]

Another variant of the knowledge representation paradigm, that exits from classical propositional logic, consists in associating weights to the different axioms. When alternative derivation chains assign an instance to different concepts, the decision is made on the basis of the strength of the weight of the axioms involved in the competing derivations. The refinement process increases or decreases the weights of the axioms depending on the number of mistakes they generate. The first system appeared in the literature, which is based on this method, is IA-EBL.[22,23] Weights and M-of-N constructs are combined together in the system KBANN.[24]

Theory Revision in First Order Logic

Several successful attempts to develop systems for theory revision in first Order Logic have been made by several authors. However, in order to limit the complexity of the task, full first order logic has never been used but only decidable subsets of it.

One of the first systems appeared in this domain has been developed in the university of Ottawa by Matwin and Plante,[25] which combine deductive and inductive reasoning. Another family of systems of this type (KRUST, CREATOR and REFINER) has been developed at the University of Aberdeen[25] for different application fields.

In the domain of mechanical trouble-shooting it is worth mentioning the system ENIGMA,[26] which represents a real application having been used in field, both for training and for diagnosis purposes, at ENICHEM in Ravenna. An evolution of ENIGMA is represented by the system WHY,[27,28] which adopts a body of causal knowledge in order to constrain the space of the inductive search.

Finally, two very recent systems are NTR,[29] and INTHELEX.[30] NTR has been designed in order to refine numerical terms, only, in a classification theory consisting of Horn clauses. INTHLEX is an experimental system, which is able to deal with a limited form of recursion and with negation.

LEARNING CLASSIFIERS WITH NEURAL NETWORKS

As neural network it is intended a wide family of function approximators, which may be used to approximate any function, and, then, any boolean function as a special case. Neural networks are called in this way because, independently from the approach they originate from, they can be represented by some multi-layered networks in which nodes (or neurons) computing a simple function (activation function) are connected by means of weighted links. The differences among the various kinds of approximators is in the activation functions, and in the methods used to build up the network and to update the weights (learning algorithms).

The activation functions can be grouped into two major families: Open Field Functions (OFFs), and Locally Receptive Field Functions (LRFFs). The OFFs exhibit an anti-symmetry property with respect to a hyperplane splitting the function domain into two semi-spaces, whereas LRFFs have radial symmetry. A good example of the first family is the well-known sigmoid function, widely used in MLPs and in all their descendants. Examples of the second family are the multi-dimensional Gaussian functions used in Radial Basis Function Networks[31] (RBFNs), in Statistical Neural Networks (SNN),[32,33] and in Support Vector Machines.[34] Other kind of networks, instead, use activation functions representing the "potential" of a central force field, as k-NN[3] method and all its descendants. The choice of an activation function belonging to OFF or to LRFF family entails properties fundamentally different.

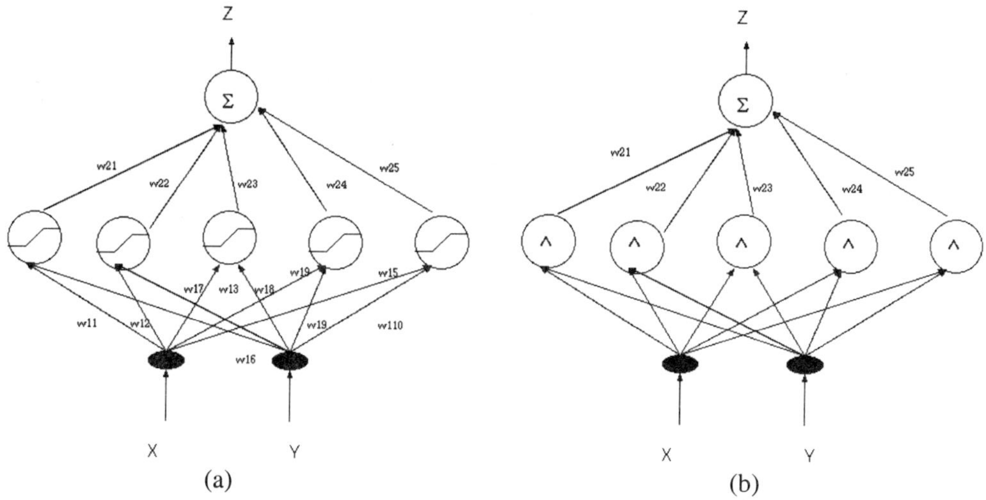

Figure 2. Multilayer perceptron (a) and Radial Basis Function Network (b) architecture. The fundamental difference is in the activation function in the hidden layer neurons. In both cases the output neuron performs the weighted sum of the output of the units in the intermediate layer, and is followed by a squashing function, such as a sigmoid.

As it has been proven by Hornik,[35] a universal function approximator can be constructed using only three layers of nodes (input, hidden and output), provided the activation function in the hidden layer be non-linear. In Figure 2 we compare the structure of two networks of this kind: one based on sigmoidal and the other on bell-shaped functions. From a topological viewpoint, they look identical. Nevertheless, the way they encode the target function is very different.

Considering the activation function of the MLPs hidden units, we see that each neuron, apart a narrow region where the sigmoid transient occurs, splits the input space into two semi-spaces where its output is significantly close to 1 or to 0. Only the semispace where the output is close to 1 contributes to the value of the target function.

Considering now the LRFN, each neuron is associated to a closed region in the input space (hyper-ellipse), where its response is significantly greater than zero, and dominates over those of every other node. The contribution to the approximation of the target function comes essentially from this region.

In order to understand the different behaviors of the two network types, suppose to modify a weight between two nodes in the MLP, as it is done by the back-propagation updating rule[2] during the training phase. The effect involves an infinite region of the input space and can affect large part of the co-domain of the target function. In practice, this network family is not suitable for incremental learning as it is required in the "learning from mistake" scenario. In fact, when the network sees too frequently instances of a given type, tends to forget the knowledge previously acquired about instances of a different type.

On the contrary, changing the amplitude or shifting the position of the activation function in a LRFN will have an effect local to the region it dominates. This locality property of the LRFN allows the network layout to be incrementally constructed (see for instance THESEO)[36] adjusting the existing neurons and/or adding new ones. As every change has a local effect, the knowledge encoded in the other parts of the network is not lost; so, it will not be necessary to go through a global revision process.

Another property related to locality is the tendency to undergeneralize exhibited by LRFNs. Depending on the way the hidden layer (the layout) is constructed, it is possible to have large regions of the input space that are not covered by any neuron. This means that,

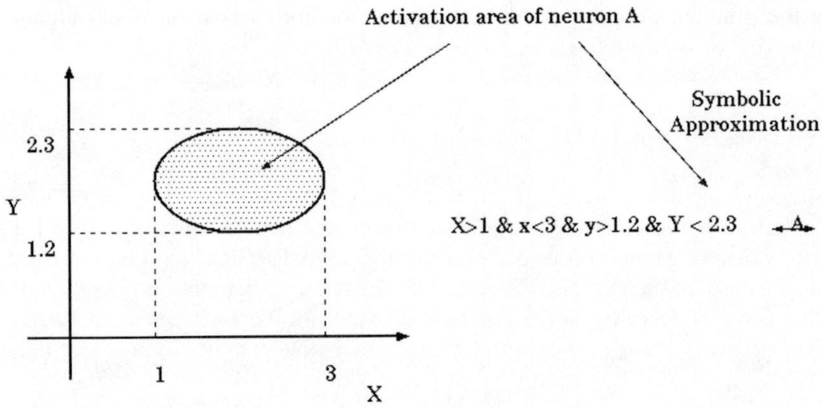

Figure 3. The closed region where LRF-Function is dominant can be roughly approximated using a conjunctive logical assertion.

when current input situation falls in such regions, the output of the network will constantly remain close to the null value (usually zero), because no neuron will fire. On the contrary, every neuron of a MLP covers an infinite region of the input space, so the network will be activated in any situation. A trained MLP has therefore the tendency to overgeneralize, whereas a LRFN always tends to undergeneralize, in the sense that its output will be null for every situation which is far from the examples processed during the training phase. As we will discuss later on, under(over)generalization can be either an advantage or a disadvantage, depending on specific cases.

Finally, an important property of the LRFNs (as opposed to MLPs, which, with few exceptions,[24] are black-boxes w.r.t. their interpretation) is the possibility of giving an immediate, symbolic interpretation of the hidden neurons. In fact, a closed region in the input space can be approximated using a hyper-rectangle which, in turn, can be described by a propositional formula, as shown in Figure 3.

As mentioned above, the basic learning rule for MLP family is the Back-Propagation rule[2] (BP), which implements a "gradient descent" on the error surface that minimizes the quadratic error of the network. When applied considering the full learning set D_L simultaneously available, BP is guaranteed to reach a local minimum of the quadratic error. However, it is also possible to use an on-line version of BP, which is quite popular in robotic applications, and considers one instance at a time. Unfortunately, in this case, the algorithm is no longer guaranteed to converge. Moreover, when trained on-line with BP, MLP tends to forget the knowledge previously learned, if it periodically does not see again all the instances in D_L. Then, in a similar way as it is done in theory revision, the learning strategy must repeat periodically a "refreshing cycle" on the full learning set, or on a selected subset of it.

Open point, with OFF networks, is the construction of the layout of the network. Up to now there is no general method for designing the network layout, which is committed to the experience of the developer of the application. On the contrary, the construction of the layout of LRFNs can be made automatically using the instances of the learning set. In the simplest formulation, every hidden unit corresponds with an instance in D_L. This is made in k-NN and in the basic version of Radial Basis Function Networks.[37] In more advanced solutions, the learning instances are first clustered and, then, the hidden units are built on the cluster centers.[38] In any case, these methods well support incremental learning algorithms: when required, a new hidden unit can be added without affecting the performance of the other parts of the network.[39] The weights on the output unit can be tuned again by performing the gradient descent,[37,38] or using other methods based on linear algebra.[31] A review of this kind of networks may be found in Baroglio et al.[40].

LEARNING FROM A REINFORCEMENT SIGNAL

In the second section, a weaker form of learning has also been considered where the feed-back from the world consists in an evaluation of the behavior of an agent in a given period of time (a chess game, in the specific example). We also discussed how it is no more possible, in this case, to clearly identify a specific mistake responsible of a bad evaluation. Actually, this form of learning is interesting because it is the most frequent scenario we encounter in our everyday life, where the evaluation of our behavior comes continuously from the world under the form of rewards (earned money, saved time, pleasure, etc.), or penalties (lost money, troubles, etc.).

This learning scenario is investigated in the machine learning approach that goes under the name of *reinforcement learning*.[41,42] The basic setting of the problem is described in Figure 4. An agent is represented as a finite state machine with N possible states and M possible actions. Every action when applied to the environment moves the agent in another state, which is detected from the agent's sensors. The agent also receives a reward (positive or negative) depending upon its behavior. The learning task consists in learning a policy Π that for each state scores the possible actions, on the basis of the chance they have of maximizing the reward cumulated over time.

As described in Figure 4, the policy can be represented as a matrix $\Pi_{N,M}$ where each element π_{ij} assigns a score for choosing action A_j in state S_i. In this scenario, the effect of an

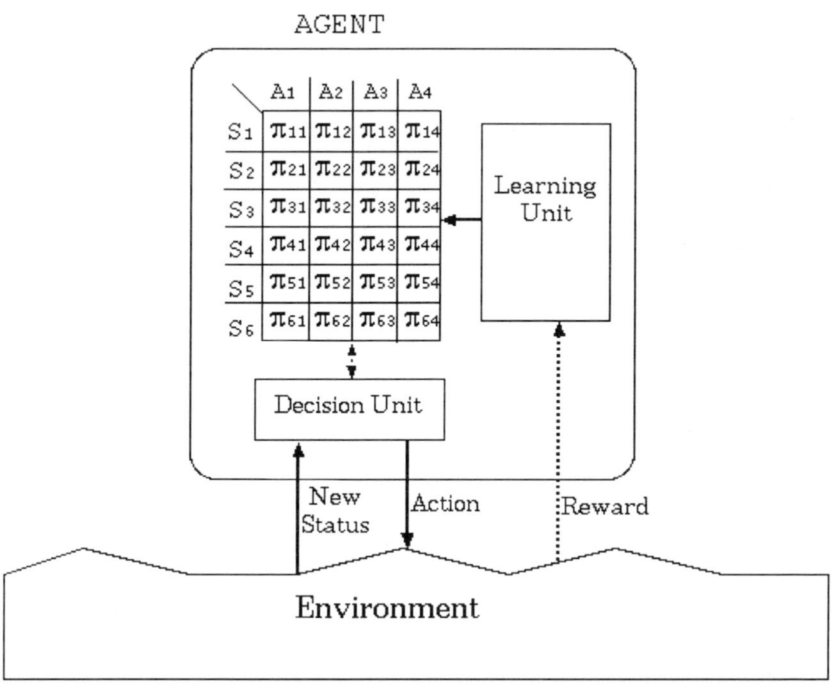

Figure 4. Basic architecture of an agent, which learns from a reinforcement signal (reward). The policy of behavior can be represented as a matrix assigning in each state a score to every action that can be selected in it. Goal is learning an optimal policy, so that, choosing the best scored action, the reward over time is maximized.

action on the cumulated reward can be discovered only by repeating the action many times and in many contexts, so that the opportunity of choosing it in a specific state can emerge. In general, reinforcement learning is set as a Markov Decision Problem in the framework of Dynamic Programming.[43] The best known algorithm for learning an optimal policy from a reinforcement signal is Q-learning.[44,45]

However, the simplified scheme of Figure 4 does not fit real scenarios for two reasons: First, the current state not always can be observed from the sensors input. This means that the agent must cope with hidden states. Second, in many cases it is required to deal with continuous states instead of discrete states. Extensions of Q-learning to continuous states have been made using neural networks for encoding the policy Π. Unfortunately, in this case the algorithm may not converge to an optimal policy. Moreover, the complexity for learning a policy in this scenario is so high that the method may be not viable in practice.

THE COMPUTATIONAL COMPLEXITY APPROACH

The task of learning a concept definition in propositional logic (or in an equivalent representation language) has also been theoretically investigated in the framework of the computational complexity approach to machine learning initiated by Valiant.[46]

The problem is set as in the following. Let h be a concept that can be correctly classified by a classifier c in a class C. Suppose moreover that the feed-back from the world is noise free, i.e. the mistake reporting is always correct. What is the maximum number of mistakes an agent will do before learning a correct definition of the concept? The answer to the question is due to Littlestone,[47] in a paper appeared in 1988, which proved that the number of mistakes for an optimal learning algorithm is bound by the relation

$$VC(C) \leq Opt(C) \leq \log_2(|C|) \tag{5}$$

In (5) the meaning of the term is the following one:

- $Opt(C)$ is the number of mistakes made for identifying the correct classifier c of concept h in the class C, using an optimal algorithm;
- $VC(C)$ is the Vapnik-Cervonenkis dimension of class C,[48] i.e., it is a parameter related to the global number of distinct concepts that can be identified by the different elements in C;
- $|C|$ is the cardinality of the class.

As an example, assuming that a concept h can be classified by a boolean expression of 10 bits, the cardinality of the class is 2^{10}=1024. Then, h can be identified doing 10 mistakes, at most.

However, the assumptions made by Littlestone are very rarely met in practice, where it is required to cope with noise and continuous attributes (then $|C| \to \infty$). Extensions of this analysis to bayesian learning can be found in Haussler et al.[49]. Other developments are found in Ben-David and Eiron.[50]

CONCLUSION

Learning from mistake is a typical form of learning, which is applied by all of us in our everyday life. In this chapter we have shown that, even if it is almost never called in this way, learning from mistake is also one of the most investigated learning scenario, in machine learning.

More specifically, we analyzed supervised concept learning, and its extension to learning classification theories, showing how both tasks are set in the typical scenario of an agent which learns from mistake, when at the same time operate on the world. This happens both in the symbolic approach and in the connectionist approach to machine learning.

We also considered an extension of the basic learning scenario, which is called reinforcement learning. Even if, in this case, the notion of mistake is no more clearly defined, the basic learning paradigm remains the same. Moreover, reinforcement learning subsumes as a special case the other two learning tasks we considered here.

The related literature in machine learning has been briefly surveyed with the purpose of offering a guide to a novice entering the field.

REFERENCES

1. R. Michalski, J. Carbonell, and T. Mitchell eds, *Machine Learning: An Artificial Intelligence Approach*, Morgan Kaufmann, Los Altos, CA, (1983).
2. D. Rumelhart, G: Hinton, and R. Williams, *Learning internal representations by error propagation*, Tech. Rep. 8506, Institute for Cognitive Science, La Jolla, University of California, San Diego, (1985).
3. P. Hart, The condensed nearest neighbor rule, *IEEE Transactions on Information Theory*, 14, pp. 515-516, (1968).
4. G. Gates, The reduced nearest neighbor rule, *IEEE Transactions on Information Theory*, 18, pp. 431-433, (1972).
5. R. Quinlan, *Induction over large data bases*, Tech. Rep. HPP-79-14, Heuristic Programming Project, Stanford University, (1979).
6. T. Mitchell, Generalization as search, *Artificial Intelligence*, 18, pp. 203-226, (1982).
7. R. Michalski, A theory and methodology of inductive learning, in *Machine Learning: An Artificial Intelligence Approach*, R. Michalski, J. Carbonell, and T. Mitchell eds., Morgan Kaufmann, Los Altos, CA, pp. 83-134, (1983).
8. Y. Kodratoff and J. Ganascia, Improving the generalization step in learning, in *Machine Learning, an Artificial Intelligence Approach*, J.G.Carbonell, R. Michalski and T. M. Mitchell eds., vol. II, Morgan Kaufmann, San Mateo, CA, pp. 215-244, (1986).
9. R. Gemello, F. Mana, and L. Saitta, Rigel: An inductive learning system, *Machine Learning*, 6, pp. 7-36, (1991).
10. T. Mitchell, *Machine Learning*, McGraw Hill, (1995).
11. J. C. Schlimmer, Incremental learning from noisy data, *Machine Learning*, 1, pp. 317-354, (1986).
12. P. Utgoff, ID5: an incremental ID3, in *Proceedings of the 5th Machine Learning Conference*, MLC-88 Ann Arbor, MI, pp. 107-120, (1988).
13. A. Ginsberg, Theory reduction, theory revision and retranslation, in *Proceedings of the 8th Conference on Artificial Intelligence*, pp. 777-782, (1990).
14. A. Ginsberg, S. Weiss, and P. Politakis, Automatic knowledge base refinement for classification systems, *Artificial Intelligence*, 35, pp. 197-226, (1988).
15. D. Sleeman, S, C., I. Ellery, S. Sharma, and P. Edwards, Machine learning, knowledge refinement and knowledge acquisition: systems which cope with incomplete and inconsistent knowledge bases, in *Proceedings of EKAW 1991*, pp. 332-342, (1991).
16. L. De Raedt and M. Bruynooghe, Interactive concept-learning and costructive induction by analogy, *Machine Learning*, 8, pp. 107-150, (1992).
17. Muggleton, S. Duce, An oracle based approach to constructive induction, in *Proceedings of the 10th International joint Conference on Artificial Intelligence*, IJCAI87, pp. 287-292, (1987).
18. T. Cain, The ductor: a theory revision system for propositional domains, in *Proceedings of the 8th International Workshop on Machine Learning*, Morgan Kauffman, San Matteo, CA, pp. 485-489, (1991).
19. D. Ourston and R. Mooney, Changing the rules: a comprehensive approach to theory refinement, in *Proceedings of the 8th National Conference on Artificial Intelligence*, Detroit, MI, pp. 851-820, (1990).
20. P. Baffes and R. Mooney, Symbolic revision of theories with m-to-n rules, in *Proceedings of the 2th International Workshop on Multistrategy Learning*, R. Michalski ed., pp. 69-75 (June 1993).
21. K. Spackman, Learning categorical decision criteria in biomedical domain, in *Proceedings of the 5th International Conference on Machine Learning*, Morgan Kauffman, Ann Arbor, MI pp. 36-46, (1988).
22. W. Choen, Learning approximate control rules of high utility, in *Proceedings of the 7th International Conference on Machine Learning*, B. W. Porter and R. J. Mooney eds., Austin, TX, pp. 268-276, (1990).
23. W. Cohen, Incremental abductive ebl. *Machine Learning*, 15, pp. 5-24, (1994).
24. G. Towell and J. Shavlik, Interpretation of artificial neural networks: Mapping knowledge-based neural networks into rules, in *Advances in Neural Information Processing*, R. Lippmann, J. Moody, and D. Touretzky eds., vol. 4. Morgan Kaufmann, (1992).
25. S. Matwin and B. Plante, A deductive-inductive method for theory revision, in *Proceedings of the 1th International Workshop on Multistrategy Learning*, MIT press, pp. 160-174 (1991).

26. A. Giordana, L. Saitta, and F. Bergadano, Enigma: a system that learns diagnostic knowledge, *IEEE Transactions on Knowledge and Data Engineering*, KDE-5, pp. 15-28 (February 1993).
27. L. Saitta, M. Botta, and F. Neri, Multistrategy learning and theory revision, *Machine Learning*, 11, pp. 153-172, (1993).
28. C. Baroglio, M. Botta, and L. Saitta, Why: A system that learns using causal models and examples, in *Machine Learning: A Multistrategy Approach*, R. Michalski and G. Tecuci eds., vol. IV. Morgan Kaufmann, San Francisco, CA, pp. 319-347, (1994).
29. M. Botta, A. Giordana, and R. Piola, Fonn: Combining first order logic with connectionist learning, in *Proceedings of 14-th International Conference on Machine learning*, ICML-97, Morgan Kauffman, Nashville, TN, pp. 157-166 (July 1997).
30. F. Esposito, G. Semeraro, N. Fanizzi, and S. Ferilli, Multistrategy theory revision: Induction and Abduction in INTHELEX, *Machine Learning*, 38, pp. 133-156, (2000).
31. T. Poggio and F. Girosi, Networks for approximation and learning, *Proceedings of the IEEE*, 78:9, pp. 1481-1497 (September 1990).
32. D. Specht, Probabilistic neural networks for classification mapping, or associative memory, in *IEEE International Conference on Neural Networks*, vol. 1, pp. 525-532, (1988).
33. D. Specht, Probabilistic neural networks, *Neural Networks*, 3, pp. 109-118, (1990).
34. C. Cortes and V. Vapnik, Support-vector networks, *Machine Learning*, 20, pp. 273-297 (1995).
35. K. Hornik, M. Stinchcombe, and H. White, Multilayer feed-forward networks are universal approximators, *Neural Networks*, 2, pp. 359-366, (1989).
36. J. Millan, Learning efficient reactive behavioral sequences from basic reflexes in a goal-directed autonomous robot, in *Proceedings of the third International Conference on Simulation of Adaptive Behavior* (1994).
37. J. Moody and C. Darken, Learning with localized receptive fields, in *Connectionist Models Summer School*, T. Sejnowski, D. Touretzky, and G. Hinton eds., Carnegie Mellon University, (1988).
38. J. Moody and C. Darken, Fast learning in networks of locally tuned units, *Neural Computations* 1, 2 pp. 281-294, (1989).
39. E. Blanzieri and P. Katenkamp, Learning radial basis function networks on-line, in *Proceedings of 13th Int. Conf. Machine Learning*, Bari, Italy, pp. 37-45 (July 1996).
40. C. Baroglio, A. Giordana, M. Kaiser, M. Nuttin, and R. Piola, Learning controllers for industrial robots, *Machine Learning,* 23, pp. 221-250 (July 1996).
41. A. Barto, R. Sutton, and C. Watkins, Learning and sequential decision making, in *Learning and Computational Neuroscience*, M. Gabriel and J. Moore eds. MIT Press, (1990).
42. A. Barto, R. Sutton, and C. Watkins, Sequential decision problems and neural networks, in *Advances in neural information processing system*, 2, Morgan Kauffman, San Mateo, Ca, (1990).
43. R. Bellman, *Dynamic Programming*, Priceton University Press, Princeton, NJ, (1957).
44. C. Watkins, *Learning from delayed rewards*, PhD thesis, University of Cambridge, U.K., (1989).
45. C. Watkins and P. Dayan, Technical note: Q-learning, *Machine Learning*, 8, pp. 279-292 (May 1992).
46. L. Valiant, A theory of the learnable, *Communications of the ACM*, 27, pp. 1134-1142, (1984).
47. N. Littlestone, Learning quickly when irrelevant attributes abound: A new linear-threshold algorithm, *Machine Learning*, 2, pp. 285-318, (1988).
48. V. N. Vapnik, *The Nature of Statistical Learning*, Springer-Verlag, (1995).
49. D. Haussler, M. Kearns, and R.E. Shapire, Bound on the sample complexity of bayesian learning using information theory and the VC dimension, *Machine Learning,* 14, pp. 83-113, (1994).
50. S. Ben-David and N. Eiron, Self-directed learning and its relation to the VC-dimension and to teacher-directed learning, *Machine Learning*, 33, pp. 87-104, (1998).

PANEL SUMMARY: THE PARALLEL BETWEEN ACTING AND THINKING: USING IT FOR BETTER PROCESS UNDERSTANDING

Bertrand Zavidovique

I.E.F., Paris 11 University
91405 ORSAY, France
e-mail: zavido@ief.u-psud.fr

The aim of the panel session was to discuss the couple ACTING/THINKING in both living and manufactured systems. Trying to avoid semantic and semiotic discussions on what thinking really means, we proposed two hypotheses to start:

- thinking deals with building and transforming representations;
- there is a fruitful parallel to draw between thinking and acting.

We based and evaluated them on progressively more abstract examples recalled here.

There were many interventions, comments and objections, during the introductory statement and in the subsequent discussion: they are reported in the last section.

MENTAL vs. PHYSICAL TRANSFORMATIONS

What is said on *acting* generally makes sense on *thinking*, and conversely. In a quasi universal way, a thought can be viewed as an abstract action through scale modeling or simulation, while acting is concrete thinking through projecting. Let us consider the example of Figure 1.

Three systems surround an object of interest. Each one has a different representation of the target that depends, among many other factors, on its own pattern (constituent matter, architecture...), on its goals, on its background, on the environment and situation including the target etc. Let us assume that elaborating the representation above and then transforming it up to a set of *actions* to fulfill the goal is what we call *thinking*.

Hence, through goals, *thinking* gets a strong influence upon *acting*.

Actions have a strong influence on *thoughts* as well.

System1, the dog, had obviously too much a basic and limited model that led to failure: it is likely to revising, may be enriching, its representation. System2, the man, did transform the initial representation to make it closer to any action (a step sometimes called the dream level), although he remains likely neither to perform any action nor to further change representation. System3, PERCEPT the robot, had a very partial view of and on the object of

Human and Machine Perception 3: Thinking, Deciding, and Acting, edited by Cantoni *et al.*
Kluwer Academic/Plenum Publishers, New York, 2001.

Figure 1a. A Thinking situation.

interest, yet a sufficient one for its goal. At least does it include dynamics. The model was obviously translated into a strategy to slow down the target and capture the interesting part, the Walkman in the present case. The successful acting process should strengthen the planning tools for being adapted in such situations.

Would that mean, in any dissertation on related topics, anything to be said on thinking (resp. acting) is likely to be true somehow for acting (resp. thinking)? In control theory and

Figure 1b. A Thinking and Acting situation.

its applications to the actual control of systems, the duality between acting and thinking was long taken advantage of. Thinking is primarily embodied as a reference to the model. A correct move is induced in minimizing some distance between the sensed situation and the model-derived one. Whonam[1] conjectured the so called "internal model principle": no system can be controlled without a model of it and of expected perturbations. Stimuli are model variables and the distance deals ultimately with the precision and robustness of the system in acting. Optimization makes it necessary then to differentiate clearly between functions, that are the thinking and acting processes, and their variables, thoughts and actions that are process results.

Thinking vs. Acting

The map of actions on to thoughts is thus *perception*, a general term to cover sensing and decision making (Figure 2). Perception can be the identity map in simple systems like automata but it becomes more and more complex a *decision* process along with system autonomy. In such theoretical frame that models *thinking* as an optimization with respect to a model, it is expected for a system to assume any unknown actor would react, first, the same way as itself or, second, according to what it knows already about its exo-system. An additional assumption remains implicit here: in case hypotheses were wrong, time and other conditions still allow for correction! The whole behavior often led man to excessive anthropocentrism. Technically, it reconciles the necessity of a model and the universal minimality principle: a body serves as a primary simulator of itself and as a natural start point for environment prediction. It is self known, close and available. Something may turn out to be more accurate and then be used, it cannot be more predictable. Note that predictable does not mean certain, since most actions remain unknown but for perception of them. Lotfy Zadeh[2] recalled that our bottom line in such discussions is system automation, whether better understanding existing systems or building new ones to last. Following his introductory talk, thinking shows eventually here a natural way to making systems precision-independent when they measure in terms of their own pattern and dimension. By the same token, (mental) representations provide avenues to exploit physical or concrete information (spatial, temporal, structural ...) for reasoning and this becomes especially true for machines too. Moreover, in supporting prediction they allow for intuition, again not a human privilege any more: in systems and computer programs, intuition is as well a start state prediction, a start state for prediction or any short circuit in the considered mental process. Hence it cannot be else than based on the above mentioned representation of the concrete world and it is an "abstract reflex" often building on analogy, may be yet another difference between men and robots at that stage. Eventually, along this "control of systems" line, we have been dealing in the subsection above with the technical duality between Physics-to-Mental or Mental-to-Mental transformations (building and manipulating a model) and Mental-to-Physics or Physics-to-Physics transformations (applying onto and manipulating objects).

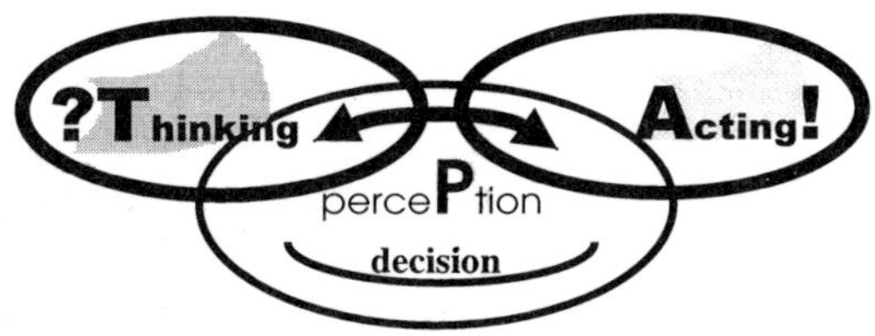

Figure 2. Set relations between Thinking and Acting with respect to Perception.

THINKING	ACTING
stimuli/deduction-abduction/conclusion	stimuli/try-move stacking/acting
answering, analyzing	responding to environment (through membrane = life)
reasoning	plan making and following
logic	action chaining
intuition	reflexes
perception for thinking vs. thinking for perception	perception for acting vs. acting for perception
mental images in retrieval, projecting	mock up and tests, simulation
naïve physics and use of structure in reasoning	experimenting, daring

Table 1 is a first attempt to summarize parallels that can be drawn between *acting* and *thinking* to be suitable for control. Most of them come from the workshop presentation (partially reported in the Preface[3]) and from the talks at this very workshop.

Actions vs. Thoughts

From the previous subsection, Perception covers both Physics-to-Mental and Mental-to-Physics transforms, that is basically building a model and applying it. While the direct way quite clearly includes *sensing*, its inverse does not split as easily from acting. Let us consider variables rather than functions and put the problem in the prospective of time, a normal step in tackling dynamic systems. It appears that a general enough map from thoughts onto actions is *learning*, of which an elementary instance is analogy-building or mimicking (i.e. when thought is neutral). Hence the sketch Figure 3.

Actions may differ from the result of thinking in so far as the system cannot physically reach the goal, and in a similar way thoughts may not be able to cover the results of acting. In both cases the system morphology or physiology had a major effect upon the process. Tom Binford[4] conjectured the equation "form = function" to simplify perception and achieve more efficient decision. Not to forget that dealing with variables and dynamics means technologically some memory to be available. Its capacity and organization, for instance the way models are stored and accessed to, cannot be irrelevant to system performances. The fundamental mechanism of optimization with respect to a model shows here more technological facets too: building it implies revising it as well. Corrections are triggered by unsatisfaction from missed actions: failure comes from the actual environment or from wrong expectations. At a higher semantic level, both natural and artificial (adaptive) systems know how to tease the environment for trial. This loop founds most processes of machine learning, that achieve another optimization at a slower pace (feature improvement by repetition). Note that efficiency often requires for more and more solutions to be stored, leading to pre-compiled thinking or acting schemes (patterned attributed graphs).

If a model is already an abstract version of the system and its universe, the latter schemes make an even more abstract version of the model. Yet technology seems a poor answer to the key question: "where, when, how does abstraction start and take place?"

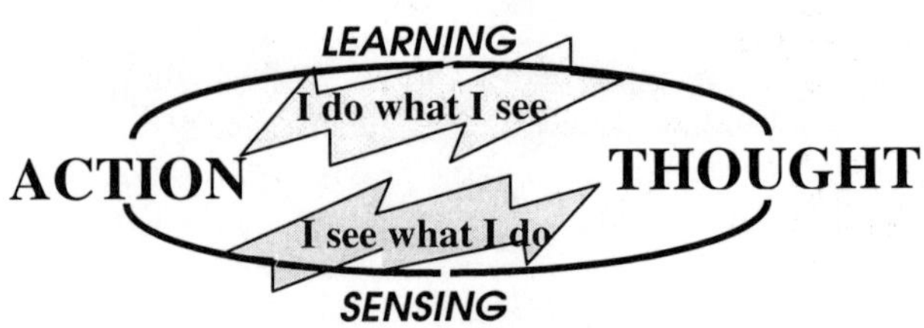

Figure 3. Precising intersections between **P**erception and **A**cting or **T**hinking.

ACTION	THOUGHT
reaction from the environment	contradiction or negation from other sources consequence evaluation
system morphology or physiology influence	memory or communication setting influence, brain organization influence
action features improved by acting, fitness in training, rehearsing	thinking ability improved by brain crunching, program cleaning by testing
graphs of actions, recipes, instructions for use	thought accumulation and mental construction mimicking, reasoning, mental schemes

Table 2 is a second attempt to summarize parallels possibly drawn between *action* and *thought* to be suitable for control. They are again mere transpositions from concrete to abstract universe of discourse.

THE ACTING-THINKING LINK FOR SYSTEM UNDERSTANDING

The previous sections suggest more than a complementarity between both concepts. Acting and Thinking may well be two facets of a same phenomenon like life itself in the necessity for adaptation (i.e. understanding and transforming). In that case they would emerge at the same time in similar conditions.

Many clues in favor of the thesis exist in both the technology and living domains. Among others: animals develop more perceptive and understanding faculties in areas where they act and relative to their environment; likewise, one reason why computers do not read well is because they do not write or draw, etc.

Putting together Figure 2 and Figure 3 leads to the complete diagram Figure 4 that describes an intelligent system provided with sensors and actuators.

Beyond the preliminary toy example in Figure 1, modeling, simulation and tests are widely used for systems when trying to prove them correct. Note that a formal proof would constitute a one-to-one map from thought (abstract system) to action (concrete system and actual making) hence an ultimate parallel. On the other hand, actions and the absence of action are widely used in animal understanding, behavior assessment, etc. under the form of reactions to stimuli.

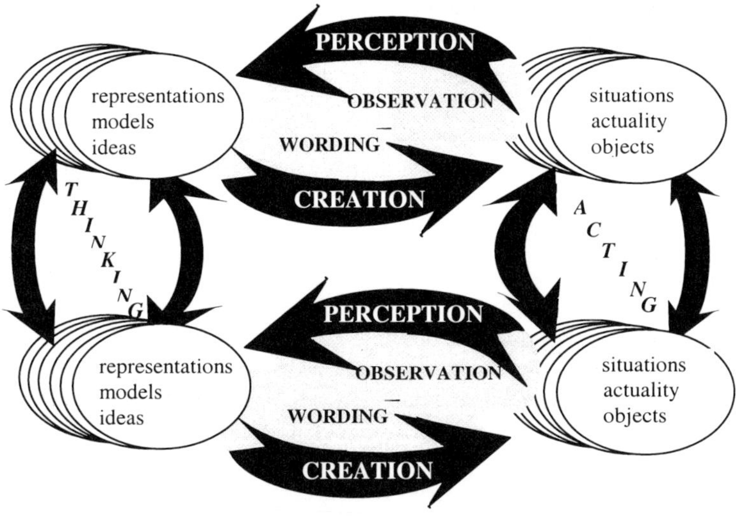

Figure 4. A potential autonomous system.

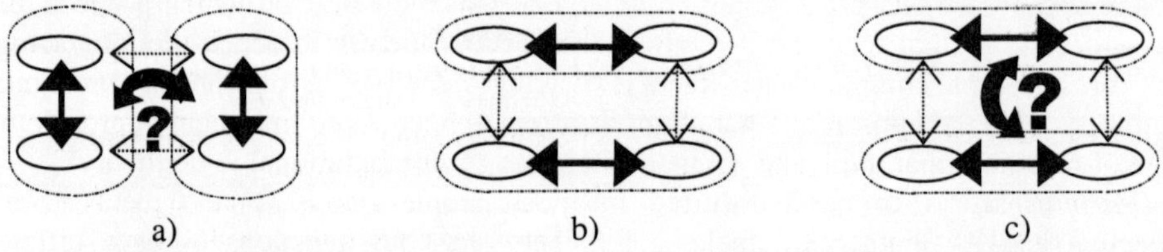

Figure 5. a) acting/thinking b) study of bio- or techno- systems c) mapping controls.

The stress so far has been put on the vertical paths, horizontal ones being considered as parameters to trigger and support respectively abstract entity handling or concrete object manipulation. It was conjectured that perception makes an efficient enough map from one to the other (see Fig. 5-a). Independently, control theory on the one hand and studies of the living on the other hand address the question of horizontal paths (Fig. 5-b). The discussion was meant to evaluate the possible maps between them (Fig. 5-c), especially when systems are of different kinds.

The audience was eventually proposed to discuss or complete the following **Table 3**, that we give a few first examples of, extracted again from the workshop presentation (partially reported in the Preface[3]) and comments above.

	ANIMAL	MACHINE
THINK	imagination, dream logic, factor analysis intuition complexity 	simulation plan making, control, graphs heuristics computability, observability
ACT	playing, projecting tactics, strategy reflexes reachability 	testing, teasing hierarchy, flowcharts, plan following automata state number, controllability

Every body on both sides could explain key words and give correspondences.

DISCUSSION

Two main points have been raised, that can be abstracted into the following questions:
- how to look at the living?
- do machines think different or should we think different?

The first one is ambiguous and was stated in various ways. For instance, is there any evidence of thinking? Edoardo Boncinelli notices: "Whatever was listed under *action* is observable and what was listed under *thinking* is not. This cannot be by chance."

Igor Aleksander goes one further: "Behavior in living organisms is generated by thought in the way in which a non-observable dynamic system generates output as a partial mapping of its internal states. This suggests that in thinking about thinking in machines, one should not have a computer in mind, but an adaptable dynamic system which learns its state structure and evolves its physical structure to represent the world on which it needs to act. This model has several features: it removes the need to attempt to couch thinking/action in terms of classical computational theory, waking adaptive, evolutionary, probabilistic dynamic systems the appropriate paradigm. It offers a much more imminent approach to the design of machines that think and act than is implied by computational logic theory!"

Learning seems to have occurred in most people minds as a typical advanced intellectual activity that would make a link between both questions or as a difference between living and manufactured systems. Sergio Cerruti further remarks: "...sometimes, it seems that learning is a one-shot process rather than an ordinate step-by-step process (at a sudden we understand a theorem, Newton developed his idea), remarkably influenced by the passed. One can speculate that underneath a deterministic chaotic model could be more realistic, but given the high *complexity* of it, it is hardly observable."

Let us underline that complexity is a crux of the matter. Most people were agreeing that dynamic system control and self state-learning is a fruitful frame. Yet artificial systems need to be implemented in the end, and for a series of known reasons ranging from technology to economics, they will be implemented on digital systems. Hence computability or recursivity theory are helpful, especially in dealing with learning. Let us underline for instance that we proved results like "there exist systems that cannot learn in one pass" thanks to formal logic (see HMP 96, pp. 155-177 (Martin)[5] and HMV 93, p. 381[6]). Here, the phenomenon of *insight* emerged along with the argument, it could be a counter example: there is no continuous solution, no computationality, no algorithm apparently involved. Same for "Gestalt" and the restructuration of percepts. If we consider insight within the more general frame of intuition and reflexes, i.e. unconventional use of a partially known model, the question becomes whether machines are capable of it and before that, whether it makes sense to them.

In wider terms, do they think different? Sergio Cerutti alleges: "Thinking is expressed in man through high-level brain functions which are not yet known from a physiological point of view. Therefore it is impossible to believe to introduce them into a machine. I am afraid that thinking, in a machine today, has to be considered a different thing than thinking in a man. Thinking is a very complex phenomenon which certainly has to do with genes and the way the subject has evolved (experience and interaction with the environment)." Although it does not help any proof either way, let us recall that machines do not act like animals either: planes vs. birds or submarines vs. fishes etc. are hackneyed examples.

For some reason, Attilio Giordana understood the proposed introductory scheme as "... a parallelism between animals able to think and reasoning , and a machine running an AI program like an intelligent agent!" He goes on "I claim that the fundamental function characterizing a thinking animal is the capability of creating on demand problem specific symbolic abstractions. This capability is actually missing in artificial agents. Even if we see abstraction mechanism "hardwired" in the programs, no one of them is able to construct an abstraction hierarchy on purpose". On top of the many objections to such statement already stressed upon in introduction and during the discussion, Igor Aleksander advocates once more for the efficiency of the dynamic-adaptable-system model: "It avoids the mistake of describing thinking (or attempting to) in algorithmic terms, and it avoids errors of saying "nobody understands thinking" in a subjective sense, by realizing that this only applies to computational logic as a way of understanding." Virginio Cantoni is more divided over this: "Some experts consider that a robot is intelligent if it is able to react in real time to external stimuli, this seems even reductive because now we have to consider industrial robots able to plan action in operative environments which are sometimes unpredictable. Moreover, in many cases the problems of auto-adaptation to the changes of the environment are pursued with limited resources (see the subsequent introduction to visual attention), in this field the performance of artificial systems are far from those of human beings. Consider also that evolution took several million years to set up the current human sensory system and dozens of thousand years to improve conscious thought and planning (see Gauthier et al.[7])".

Actually the common underlying notion in these three statements and a potential objective difference between systems is the *goal*: any system control requires its objective to be made explicit, industrial robots have a task to complete... while "high-level" when applied to brain functions often correlates wide, remote, fuzzy or even unknown objectives. The idea is confirmed by Edoardo Boncinelli: "I agree that this is a very important difference. Robots and other machines have goals to reach and purposes. Living beings are simply there. From time to time they have also some specific purposes. We tend to focus on these aspects but this is clearly a bias." Let us mention that some generic goals can be put forward in the living case like survival, species conservation. Virginio Cantoni thinks that it is even possible to build goal-oriented processes without models and conjectures on the environment or field of the solution developed. ... examples are neural network training or, even more explicitly, the Puzzle principle of Marvin Minsky:[8] "We can program a computer to solve any problem by trial and error, without knowing how to solve it in advance, provided only that we have a way to recognize when the problem is solved". It can be of interest to investigate the limits of the approach (eg. in terms of time, efficiency, training set, reinforcement and reward, costs, etc. see also Cantoni et al.[9])". And the discussion further dealt with the generality of goals: to the least, thinking allows to adapt action. A behavior in such context would result from decision cycles based upon perception, and that sent back to

- complexity: in correlating mental vs. objective parameters,
- emergence: in asking at what animal level thinking starts

Wording being definitely accepted for an action, it was underlined in contrast that pigeons are capable of abstraction, the question further becoming whether they are conscious of it (see also Virginio's comment). As consciousness is even less observable (not even its effects) than any other thinking matter already handled in that session, the audience agreed to split.

That leaves the last word to Aristoteles, who said some thing like: "I do not know if I do not know that I know". And let us conclude with two more sentences about the equal benefits from control theory, computer science and formal logic:

- systems should learn that infinity, whether H or other, is very long especially in the end;

- if perception were dependable life would be so boring that it would kill itself before existing.

ACKNOWLEDGMENTS

We are grateful to all the attendance at the panel session for their participation to the interesting discussion. Thanks to Emilien Prost for his help with the drawings.

REFERENCES

1. BA. Francis, WM. Wonham, The internal model principle of control theory, *Automatica 12*, pp. 457-465. (1976).
2. L. Zadeh, Toward a perception based theory of probabilistic reasoning, in *Human and Machine Perception 3: Thinking, Deciding, and Acting*, V. Cantoni, et al. eds., Kluwer Academic/PlenumPublishers, New York, (2001).
3. V. Cantoni, V. Di Gesù, A. Setti, and D. Tegolo, Preface, in *Human and Machine Perception 3: Thinking, Deciding, and Acting*, V. Cantoni et al. eds., Kluwer Academic/PlenumPublishers, New York, (2001).
4. Tom Binford, John McCarthy , Research in Progress in Robotics at Stanford University, *AI Magazine 2(1): Winter 1980*, 27-28 *(Research in Progress).*

5. B. Zavidovique, Perception for decision or decision for perception?, in *Human and Machine Perception: Information Fusion*, V. Cantoni et al. eds., pp. 155-177, Plenum Press, New York, (1997).
6. S. Gaglio, F. Esposito, and S. Nolfi, Panel Summary: Perceptual learning and discovering, *Human and Machine Vision: Analogies and Divergencies*, V. Cantoni ed., Plenum Press, New York, pp. 371-383, (1994).
7. G.M. Gauthier et al., Integrating reflexes and voluntary behaviours: coordination and adaptation controls in man, in *Human and Machine Perception: Information Fusion*, V. Cantoni et al. eds., pp.189-205, Plenum Press, New York, (1997).
8. Marvin Minsky, *The society of mind*, New York, p. 73, (1986).
9. V. Cantoni, V. Di Gesù, A. Setti, and D. Tegolo, *Human and Machine Perception 2: Emergence, Attention, and Creativity*, Kluwer Academic/PlenumPublishers, New York, (1999).

TIME-EFFECTIVE DETECTION OF OBJECTS OF INTEREST IN IMAGES BY MEANS OF A VISUAL ATTENTION MECHANISM

Roman M. Palenichka and Peter Zinterhof

University of Salzburg
Department of Scientific Computing
Salzburg, Austria
e-mail: palenich@coma.sbg.ac.at

INTRODUCTION

Time-effective detection and recognition of objects of interest in images is still a matter of intensive research in computer vision community because the artificial vision systems fail to outperform the detection results by a human being. The detection problem is complicated when objects of interest have low contrast and various sizes or orientations and can be located on noisy and inhomogeneous background with occlusion occurrences. In many practical applications, the real-time implementation of object detection algorithms in such conditions is a matter of great concern. The results of numerous neurological and psychophysical investigation of human visual system (HVS) indicate that the human vision can successfully cope with these complex situations because of using the visual attention mechanism associated with a model-based image analysis.[1,2] The goal of the investigation presented here was not the simulation of human visual perception but the incorporation of its advantageous features into computer vision algorithms. Besides many remarkable properties of HVS like the mentioned model-based visual attention, the HVS has also some disadvantages while detecting and identifying objects. For instance, it is prone to illusions that should be not automatically copied onto an artificial vision system.[3]

Several models of attention mechanism in HVS in the context of reliable and time-effective object detection in static scenes have been proposed in the literature. They are mostly based on the generalization of edge and line detection operators and on the utilization of a multi-resolution image analysis including the wavelet theory.[4-6] Especially good results of attention modeling have been reported by the application of symmetry operators to images.[7,8] However, a few work has been done toward designing a model-based attention mechanism based on image model of low and intermediate levels (general description of objects of interest), which could yield an optimal detection performance with respect to the underlying image model. In such a design, the invariance to size, rotation and shift in position of objects are usually required besides the feasibility of real-time implementation. The low-level image modeling is requested because the attention

Human and Machine Perception 3: Thinking, Deciding, and Acting, edited by Cantoni *et al.*
Kluwer Academic/Plenum Publishers, New York, 2001.

mechanism in its narrow sense is a *bottom-up* image analysis process based on quite general intensity properties in order to respond to various *stimuli* (objects of interest and their parts) as well as to provide a reasonable response when no object of interest is present at all. Moreover, the attentional mechanism *per se* is an instrument for formation or learning of intermediate and high level models of objects based on a representative sample of images. This is the so-called *inherited attention* that should be distinguished from the learned attention mechanism which is a feature-based attention using intermediate and high level image and scene models.

In this chapter, a new model of visual attention based on the concept of multi-scale relevance function is proposed as a mathematical representation of generally recognized results on the explanation of HVS mechanisms. The visual attention mechanism based on the relevance function provides the following advantageous features in detecting objects of interest. A model-based approach is used which exploits multi-scale morphological representation of objects and regression representation of their intensity in order to perform time-effective image analysis. The introduced concept of a multi-scale relevance function in application to object detection provides a quick location of local objects of interest with various sizes and orientations.

The chapter is organized so, that after a brief introduction, the underlying image model consisting of image intensity model and structural multi-scale model of local objects on the background is described. Next, it contains the description of the proposed model for visual attention mechanism by using the relevance function approach. At the end the experimental results of object detection using the relevance function applied to synthetic and real images from diagnostic imaging are described and concluding remarks are given.

THE UNDERLYING IMAGE MODEL

Analysis of images and detection of local objects of interest in images can be efficiently performed by using image structural properties. There are considered separately properties of object planar shape as well as intensity properties within a region of interest containing an object on the background. The available results of psychophysical investigation of HVS and other biological systems (e.g., animals) indicates also that a model-based, top-down, paradigm lays in the basis of fast and reliable perception of real world. Moreover, it confirms that the consideration of structural planar properties of objects of interest has a priority over the intensity properties. Such properties can be computed based on the so-called conformable polynomial regression (CPR) model of an image fragment under analysis. The CPR model of image fragment is an extension of a dynamic polynomial regression (DPR) model of images.[9] It is valid in every image point but may have different model parameters in different points in order to represent the non-homogeneity of image intensity. The DPR model states that the image intensity function $g(i,j)$, where (i,j) are two coordinates as the non-random explanatory variables (due to the exact discrete values of the coordinates), can be represented by a polynomial function of order q within a neighborhood region (window) $W(u,v)$ relatively to a current point (u,v) plus independent white noise term:

$$g(i,j) = \sigma \cdot n(i,j) + \sum_{r+s \le q} \theta_{r,s}(u,v) \cdot i^r j^s, \ \forall (i,j) \in W(u,v), \qquad (1)$$

where $\theta_{r,s}(u,v)$ is the (r,s) regression coefficient (parameter), $n(i,j)$ is white Gaussian noise having zero mean and unit standard deviation, σ is the scale parameter of noise residuals. The variance σ^2 is constant over the image plane whereas the coefficient vector $\theta(u,v)$ may vary from point to point. In order to ensure such a condition, noise $n(i,j)$ have to contain a relatively small fraction of outliers ε with other distribution or parameters. The outliers may arise at the object (uniform region) boundaries or on locations of insignificant small objects. The CPR model is based on the DPR model by equation (1) and considers a piece-wise constant model of an image fragment by the function $f(i,j)=F\{\theta(i,j)\}$, called the *property map*, within another window $V(u,v)$. The property map is computed as a function of the model parameter vector $\theta(i,j)=[\theta_1,...,\theta_t]$ in equation (1). It is also supposed that a zero-mean term $\lambda\cdot\nu(i,j)$ of white noise (independent and identically distribution variables) is present in the property map $f(i,j)$:

$$f(i,j)=\lambda\cdot\nu(i,j)+\sum_l \tau_l\cdot\varphi_l(i,j)\,,\ \forall(i,j)\in V(u,v)\,, \tag{2}$$

where $\{\tau_l\}$ are constant intensity values of image plane segments corresponding to objects and background, $\varphi_l(i,j)$ is the indicator function for lth segment. The function $\varphi_l(i,j)$ is equal to zero in the whole image plane except for the points belonging to lth segment. The explicit form of the function $F\{.\}$, which extracts an intensity property, depends on the problem statement in order to be computationally efficient. It is usually defined parametrically with the parameters to be learned from the image sample of objects to be detected. For example, it can be defined as a linear function of absolute component values taken from the coefficient vector $\theta(i,j)$. The simplest example is the object detection by local mean intensity where the value of $F\{.\}$ coincides with the value of the coefficient $\theta_1(i,j)=\theta_{0,0}(u,v)$ in equation (1).

However, it is not enough to have the CPR model alone as a basis for robust binary segmentation in the two-dimensional case when imaged objects have various planar shapes and sizes. The planar shape modeling is aimed at a concise representation of possible objects of interest whose intensity satisfies the CPR model. An efficient approach to describe the planar shape is the multi-scale morphological image model describing the objects of interest by using structuring elements, axes of symmetry, skeletons, as well as the contour description. One initial structuring element S_0 of minimal size as a set of points on the image grid is selected that determines the resolution of imaged objects. It possesses a symmetric disk shape in order to insure the rotation invariance as well as a reasonable approximation of different possible shapes. Two types of scales can be well distinguished: *uniform scales* and *logarithmic scales*. The structuring element at the scale k in the uniform scales is formed as a consecutive binary dilation (denoted by $\oplus$) by S_0, $S_k=S_{k-1}\oplus S_0$, $k=1,2,...,K-1$, where K is the total number of scales. The size in pixels for kth scale S_k is given by its disk radius r_k.

The generation of planar shape of a simple imaged object can be modeled in the continuous case by a growth process.[10] The formation of every isolated object starts from a

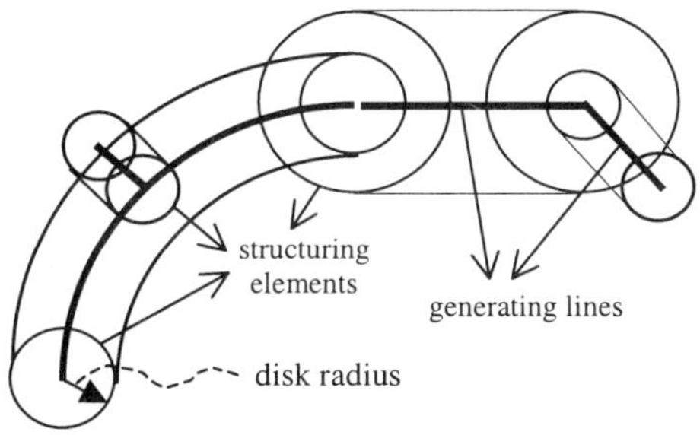

Figure 1. Multi-scale formation of a simple object of interest using concatenation of primitive objects at three different scales. The generating lines are marked by bold lines.

single point named *a seed*. The seed begins to grow along a *generating line* (straight or curve) as the growth path. In the discrete case, the growth path is represented by a generating set. The generating set is one-pixel wide connected set of points, each of them except for the end points have two neighboring points. The growth process of imaged objects starts at largest scale if used in a multi-scale representation and can be modeled by means of generation of the so-called primitive objects. A *primitive object* at scale S_k of the uniform scale system is formed using two structuring elements: object structuring element $S_a=S_k$ and background structuring element $S_b=S_{k+1}$. The domain region U_k of a primitive isolated object consists of two distinct sets of points O_k and B_k, $U_k=O_k \cup B_k$; $O_k \cap B_k=\varnothing$, which are formed by using the operation of dilation of a single generating line (set) G_k:

$$O_k = G_k \oplus S_k \text{ and } B_k = (G_k \oplus S_{k+1}) \setminus O_k \tag{3}$$

A simple model is adopted for multi-scale object formation using the primitive objects at different scales: primitive objects are just concatenated by their generating lines to form a connected domain region of a simple local object (blob) (Fig. 1). Finally, the morphological planar shape model is coupled with the CPR model of the feature intensity by equation (2) in such a way that every indicator function $\varphi_l(i,j)$ in equation (2) is represented by the morphological domain model using equation (3).

MODEL OF VISUAL ATTENTION MECHANISM USING MULTI-SCALE RELEVANCE FUNCTION

Regions of Attention and the Relevance Function

The notion of relevance function originates from the principle of edge detection by local extrema of derivative operators. More recently, the generalization of this principle has been made toward simulating visual perception in humans and animals based on the available results of physiological and neurological investigations.[3] The relevance function $R\{f(i,j)\}$ is an image local operator which takes larger values in points belonging to an object of interest rather than in the background points: $R\{g(i,j)\} \geq R\{g(m,n)\}$ for $\forall\ (i,j) \in O$ & $\forall\ (m,n) \in B$, where O and B are the object region and the background region within a window $U=O \cup B$. It is also required that the function $R\{g(i,j)\}$ takes maximal values at the supposed center location of an object of interest, i.e. ideally, the local maxima coincide with the points on object generating lines if assuming that objects are modeled by equation (3).

Here, the model-based relevance function[11] is considered as applied not to the initial image $g(i,j)$ but to the property map $f(i,j)$ modeled by equation (2). Considering a single scale S_k, let the object sub-region $O(i,j)$ be a symmetric structuring element centered at point (i,j) and the sub-region $B(i,j)$ be a ring around it generated by the background structuring element, i.e. $O=S_k$ and $B=S_{k+1} \setminus S_k$ (see Fig. 2). Assuming the morphological model of local objects coupled with the CPR model of image intensity, the relevance function $R\{f(i,j)\}$ can be defined in the case of valid model by equation (2) by the following expression:

$$R\{f(i,j)\} = \left(\frac{1}{|O|} \sum_{(m,n)\in O(i,j)} f(m,n) - \frac{1}{|B|} \sum_{(m,n)\in B(i,j)} f(m,n) \right)^2 - \alpha \cdot \left(a - \frac{1}{|O|} \sum_{(m,n)\in O(i,j)} f(m,n) \right)^2 \quad \text{for } \forall (i,j) \in A\,, \tag{4}$$

where A is the current region of interest, $|O|$ and $|B|$ denote the number of points in object and background sub-regions $B(i,j)$ and $O(i,j)$, α is a constant coefficient. The value a which is constant within A (called the object *intensity of reference*) is the mean value of intensity within an object sub-region $C(r,s)$. When analyzing in detail the same object of interest, the

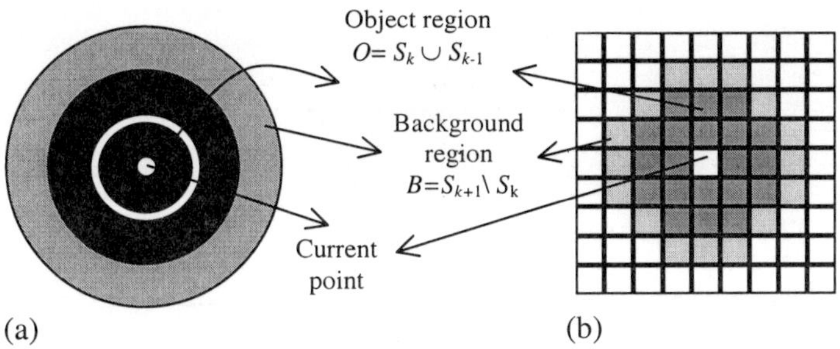

Figure 2. Sub-regions O and B in the definition of a two-scale relevance function for continuous (a) and discrete (b) cases.

sub-region $C(r,s)$ is identical to the sub-region $O(r,s)$ centered at point (r,s) which is assumed to be the object center for previous region of interest with the detected object of interest. At the beginning, when the first region of interest is considered, the region $C(r,s)$ coincides with a defined sub-region of $O(i,j)$. The difference value x in parentheses of the first term in equation (4) takes into account the region salience by object contrast, whereas the difference value y in the second term is aimed at the consideration of the object homogeneity constraint. The value of α can be determined by using the maximum likelihood (ML) approach for estimation of the most probable location. Namely, assuming Gaussian distributions $\mathbf{N}(h;\sigma_x^2)$ and $\mathbf{N}(0;\sigma_y^2)$ for object contrast x and object homogeneity y in equation (4) with different mean values and variances σ_x^2 and σ_y^2, the object location by the ML rule is reduced to the finding the maximal value of the relevance function by equation (4). At a quite reasonable assumption and after simple mathematical transformations, the coefficient $\alpha = \sigma_x^2 / \sigma_y^2$.

If the size of an object of interest to be detected is not known, a multi-scale relevance function should be applied in order to locate reliably such an object (objects) and then determine the object scale. The definition of the multi-scale relevance function takes into account all object's potential scales:

$$\frac{1}{|O|}\sum_{(m,n)\in O(i,j)} f(m,n) \leftrightarrow \frac{1}{K}\sum_{k=0}^{K-1}\left(\frac{1}{|S_k|}\sum_{(m,n)\in S_k} f(m,n)\right), \tag{5}$$

where the object mean intensity is averaged over all K mean values at K scales $\{S_k \subseteq O(i,j)\}$ for estimation of object contrast as a saliency measure for the location point (i,j), $|S_k|$ denotes the number of points in S_k. It will locate accurately different in size local objects provided a single object is present in the region of interest A or multiple objects are respectively isolated. It requires additionally an averaging with respect to K scales, but allows to avoid other steps of image analysis. The two object sub-regions S_0 and S_1 of a two-scale relevance function are illustrated in Figure 2 by dark regions.

It can be easily proved that for the case of the CPR model without present noise the relevance function takes the maximum at the seed or end point of a primitive object. Insignificant shift in the location might be introduced by the present noise depending on the noise variance. The relevance function $R\{f(i,j)\}$ have to be computed within a region of interest A and takes its maximal value in the *focus of attention* (i_f, j_f):

$$(i_f, j_f) = \arg\max_{(m,n)\in A}\{R\{f(m,n)\}\} \tag{6}$$

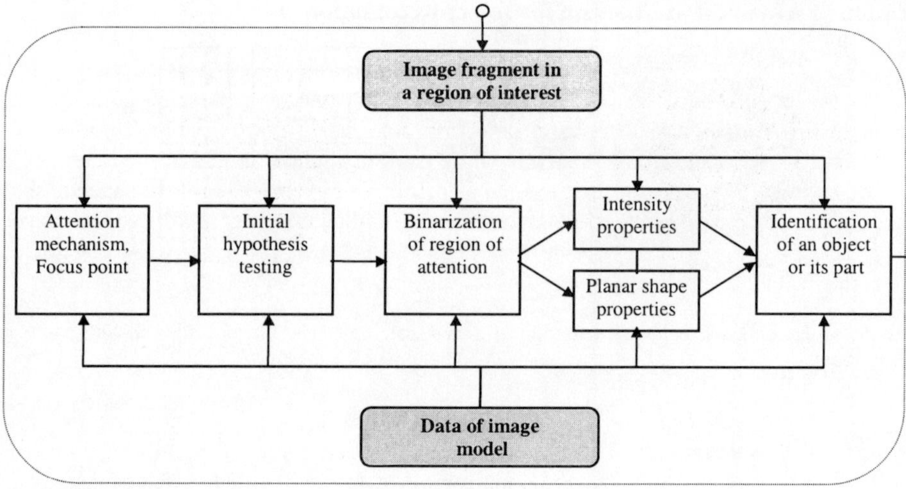

Figure 3. One stage of image binary segmentation and object identification using the relevance function approach.

The focus of attention defines a new region to be analyzed, the *region of attention*, that is centered at the focus of attention. The binary segmentation will be performed within this region if at all. If a current region of attention satisfies the condition of the CPR model, then it is segmented into two regions: object and background. Otherwise, a next region of interest will be considered in the vicinity of A.

Estimation of Object Scale and Orientation

The location of an object of interest as the focus of attention point is followed by the determination of its potential scale and potential orientation in order to ensure a size and rotation invariant recognition. On the other hand, such a preliminary estimation of scale and orientation facilitates further image analysis provided the estimation is computationally simple. This is the case of the relevance function approach because all necessary computationally complex parameters are pre-computed and stored during the determination of maximum of the multi-scale relevance function in A. The potential object scale $\rho(i_f, j_f)$ is determined as the maximal value of K single-scale relevance functions with respect to each of the K scales. When considering just one point, i.e. the focus of attention (i_f, j_f), it is reduced to the decision rule:

$$\rho(i_f, j_f) = \arg\max_k \left\{ \left| \frac{1}{|S_k|} \sum_{(m,n)\in S_k} f(m,n) - \frac{1}{|S_{k+1}|} \sum_{(m,n)\in S_{k+1}} f(m,n) \right| \right\} \tag{7}$$

The object orientation with respect to the found center can be also determined by selecting M nearest maximal values, $M \le r_k$, where r_k is the radius of the found object scale S_k of the relevance function with respect to the local maximum. In the condition of the primitive object model, M nearest maximal values will be distributed along the object generating line providing the object orientation. A preliminary testing on the distribution compactness (roundness) of M nearest maxima is made before determining the orientation.

Application of Attention Mechanism to Object Recognition

The proposed model of visual attention mechanism can be successfully applied to time-efficient detection of objects of interest and object recognition by its shape and intensity properties. In this framework, the object detection consists of several (many) consecutive stages of a multi-scale local image analysis while each of them is aimed at the detection of an object of interest or its relevant part termed as primitive pattern. The image analysis starts at larger possible scales used in the relevance function and proceeds gradually to finer scales when it is needed for an object recognition. Each of such stages (see Fig. 3) considers a region of interest determined by the results of previous stage. A statistical hypothesis, the so-called *initial hypothesis*, is first formulated and tested on object presence (absence) in the region of attention in order to know whether an object or its part is present at all. Statistically, the estimated value of actual object contrast x is tested on its significance with respect to the standard deviation σ_x in the framework of the assumed model in equation (4). Their ratio x/σ_x is a *measure of saliency* of local maximum of the relevance function in a region of interest. The explicit expression for testing estimated object contrast x is the inequality $x<\sigma_x \cdot \Delta/2$ derived from the likelihood ratio test and assuming equal prior probabilities, where Δ is a limit value for the saliency measure. In the case of a *non-salient maximum*, i.e. when the fragment is without an object of interest, this inequality is true and the current attention region is ignored as uninteresting and the next region of interest have to be considered. If an object is present, a local image binarization is made within this region. The goal of the binarization is the reduction of store space and time for recognition and analysis of primitive objects which compose the objects of interest.[11]

Once the object planar shape has been extracted from the background, a separate computation of object intensity properties and properties of planar shape is performed. In general, planar shape properties are considered more important in most applications because they insure invariance to amplification or attenuation of image intensity provided they are position, size, and rotation invariant. However, besides the shape properties, some applications require knowledge of object position, size and orientation which are considered as transformation attributes of planar shape properties. There are several different methods for planar shape description used in application to local object detection. The methods of mathematical morphology is quite a natural way to describe such objects in a concise manner. The underlying image model is based on a formation of planar object shape using the notions of structuring elements and generating sets (skeletons) and their interaction by the operation of morphological dilation.[11]

EXPERIMENTAL RESULTS

Experiments with Synthetic Images and Visual Illusions

The described approach to detection and binarization of local objects have been tested on synthetic and real images from industrial and medical diagnostic imaging. The main purpose of the testing on synthetic images was the performance evaluation during localization and binarization of low-contrast and noisy images. For example, the graph in Figure 4 shows the experimental dependence of the location bias versus noise level for a synthetic noisy image with a rectangular bar as an object of interest. For comparison, the object center has been determined in the result of a histogram-based binarization with determination of the center of mass by standard equations. The binarization threshold was determined at the middle of the two histogram modes. The method of relevance function turned out to be more tolerant to present noise, especially at higher noise variances.

Several shape and intensity illusions can be modeled (explained) by the visual attention mechanism described here. One known example of illusions concerned to planar shape is

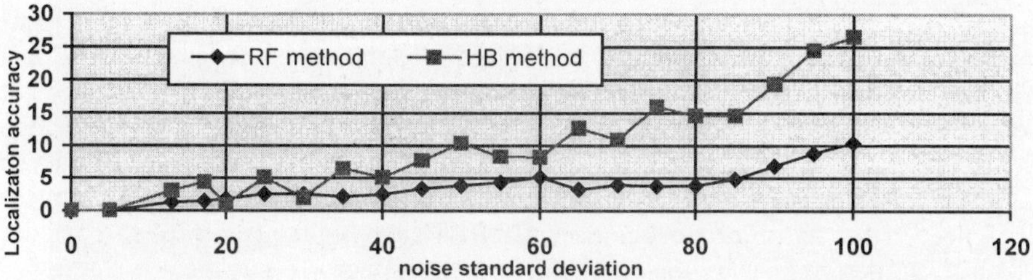

Figure 4. Localization accuracy (in pixels) vs. noise deviation for two methods: relevance function (RF) and histogram-based binarization (HB).

the Kanizsa figures[3] (see Fig. 5). The application of the relevance function and larger scales yields the focus of attention at the centers of the illusionary triangle in Fig. 5(a) at the middle of the image plane. The next stage of image analysis at finer scales reveals the first focus of attention at one of the corners of the Kanisza figures. The same image analysis started at middle scales with respect to the image size yields the first focus of attention approximately at the center of disks with cut sectors. In this case, the corner is not recognized at first because the constraint of bright object (i.e. disk with cut corner) on dark background makes the relevance function greater in the disk points.

Experiments with Real Images in Diagnostics Imaging

The proposed object detection method using relevance function has been tested on real images from diagnostic imaging where the visual attention model has its natural application area. The objects of interest are defect indications (industrial diagnostics) or abnormalities of human body (medical diagnostics) which are usually small in size, low-contrast and located on noisy and textured background. In these experiments, binarization and identification of objects of interest was made according to the flowchart in Figure 3. One sample of test images was taken from the visual inspection of photoreceptor surfaces where the sought defects in images have a round shape located on a textured background (example in Fig. 6). Another tested images are related to lesion detection from conventional radiographic and tomographic images in chest radiography. The obtained experimental

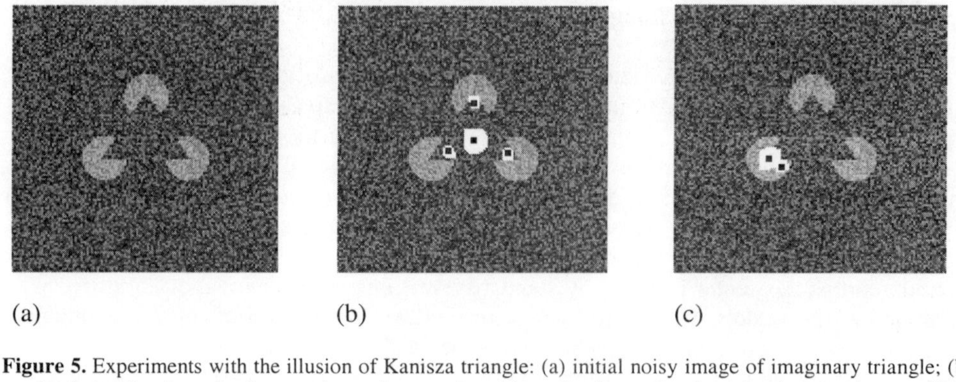

(a) (b) (c)

Figure 5. Experiments with the illusion of Kanisza triangle: (a) initial noisy image of imaginary triangle; (b) result of attentional mechanism starting at large scales; (c) result of attentional mechanism starting at middle scales. The focus of attention is marked by the black dot and the nearest maxima of the relevance function are shown white around it.

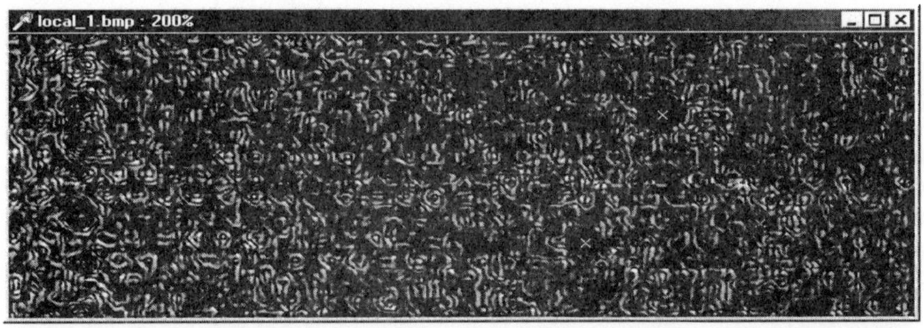

(a)

(b)

(c)

(d)

Figure 6. Results of object detection and segmentation in images with textured background by means of the multi-scale relevance function: (a) input image; (b) property map; (c) local binarization; (d) pattern recognition.

results show that the proposed model of visual attention can successfully solve the stated problems in such a wide application area as the diagnostics imaging.

CONCLUSIONS

A model for visual attention mechanism has been proposed in the context of object detection and recognition problems. A multi-scale relevance function has been introduced for time-effective and geometry-invariant determination of object position. The attention mechanism is based on the maximization of the relevance function response to image fragments (i.e. respective scene fragments) which contain objects of interest according to the assumed image model.

The theoretical investigation and preliminary testing results have shown a good matching with the results of neurophysiological and psychophysical investigation of HVS and very good performance of the relevance function approach in computer vision applications at the same time. The approach with an adequate image model of objects of interest allows to locate reliably objects of interest while eliminating visual illusions which may occur in the HVS. The experiments confirm that the visual illusions of HVS are mainly due to a prompt decision making based on an incomplete image analysis using visual attention mechanism, i.e. considering only a limited number of salient image fragments at larger scales determined by the attention mechanism.

The feasibility of real-time implementation by using fast recursive computation makes this approach attractive in application to diagnostics imaging where the defects as objects of interest are located on inhomogeneous and textured background with the occlusion possibilities.

REFERENCES

1. V Cantoni, S. Levialdi, and V. Roberto eds., *Artificial Vision: Image Description, Recognition and Communication,* Academic Press, (1997).
2. A. L. Yarbus, *Eye Movement and Vision*, Plenum Press, New York, (1967).
3. M. D. Levine, *Vision in Man and Machine*, McGraw-Hill, (1985).
4. T. Lindeberg, Detecting salient blob-like image structures and their scale with a scale-space primal sketch: a method for focus of attention, *Int. Journal of Computer Vision*, Vol. 11, pp. 283-318, (1993).
5. L. Itti, C. Koch, and E. Niebur, A model of saliency-based visual attention for rapid scene analysis, *IEEE Trans., Vol. PAMI-20,* No. 11, pp. 1254-1259, (1998).
6. J. K. Tsosos et al, Modeling visual attention via selective tuning, *Artificial Intelligence*, Vol. 78, No. 1-2, pp. 507-545, (1995).
7. D. Reisfeld et. al., Context-free attentional operators: the generalized symmetry transform, *Int. Journal of Computer Vision*, Vol. 14, pp. 119-130, (1995).
8. V. Di Gesù, C. Valenti, and L. Strinati, Local operators to detect regions of interest, *Pattern Recognition Letters*, Vol. 18, pp.1077-1081, (1997).
9. R. M. Palenichka and I. B. Ivasenko, Fast and robust parameter estimation in the polynomial regression model of images, *Proc. SPIE Nonlinear Image Processing X*, Vol. 3646, p. 28-37, (1999).
10. R. M. Palenichka and P. Zinterhof, A fast structure-adaptive evaluation of local features in images, *Pattern Recognition*, Vol. 29, No. 9, pp.1495-1505, (1996).
11. M. Palenichka and M. A. Volgin, Extraction of local structural features in images by using multi-scale relevance function, *Proc. Int. Workshop MDML'99*, LNAI 1715, Springer, pp. 87-102, (1999).
12. P. Meer et al., Robust regression methods for computer vision: a review, *Int. Journal of Computer Vision,* Vol. 6, No. 1, pp. 59-70, (1991).

EXPERIMENTS ON CONCURRENT ARTIFICIAL ENVIRONMENT

Vito Di Gesù, Giosuè Lo Bosco, and Domenico Tegolo

DMA - University of Palermo
Via Archirafi 34, 90123 Palermo, ITALY
e-mail: {digesu,lobosco,tegolo}@math.unipa.it

ABSTRACT

We show how the simulation of concurrent system is of interest for both behavioral studies and strategies of learning applied on prey-predator problems. In our case learning studies into unknown environment have been applied to mobile units by using genetic algorithms (GA). A set of trajectories, generated by GA, are able to build a description of the external scene driving a predators to a prey. Here, an example of prey–predator strategy, based on field of forces, is proposed. The evolution of the corespondent system can be formalized as an optimization problem and, for that purpose, GA can be use to give the right solution at this problem. This approach could be applied to the autonomous robot navigation in risky or inaccessible environments (monitoring of atomic power plants, exploration of sea bottom, and space missions).

INTRODUCTION

During the last years many studies, refining or extension of the Lotka Volterra model,[1,2] have been developed on the influence of predators on preys to study the instability of the natural ecosystems and to models the ecosystem around.[3]

In this chapter we will focus the attention on behaviors studies and on strategies adopted by predator to reach the prey. A simulation, based on genetic algorithms, has been taken for studying the modeling of external environment into the prey and predator context. The predator, a unit mobile robot,[4] builds an environment description both using the generated trajectories[5] and using a connectionist paradigm for the detour problem solutions.[6] Here, a prey–predator approach, implemented in the framework of GA, is considered. Because the system evolution can be formalized as an optimization problem, GAs seem useful to solve, at the best, that problem. Two basic features of predator-prey interactions have been considered: i) relationship between the prey's loss and the predator's gain; ii) in absence of prey, the predator population should die out exponentially.[7]

Moreover, predators and prey already know each other and prey has an internal model of the refuge site (they do not need learning for that). Abdul-Aziz Yakubu demonstrate the

Human and Machine Perception 3: Thinking, Deciding, and Acting, edited by Cantoni *et al.*
Kluwer Academic/Plenum Publishers, New York, 2001.

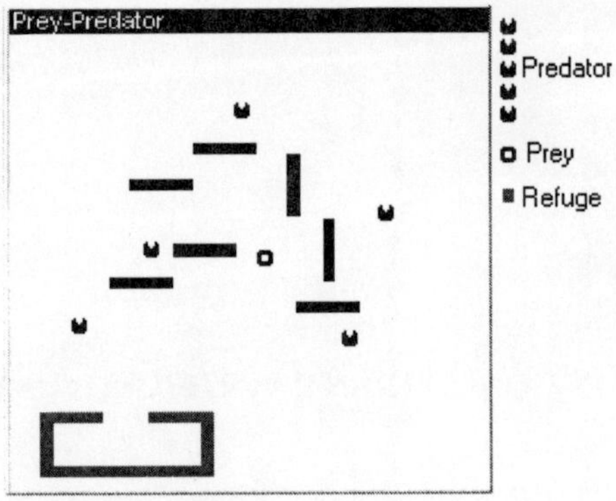

Figure 1. Prey-predators scheme.

stable coexistence of species that would otherwise exclude each other without a prey refuge.[8] We have also introduced a memory mechanism that can be useful to remember configurations, emulating mental image creation. Figure 1 shows a case of prey-predator distribution on a software simulator.

In the following will be reported some experiments regarding the relationship between the memory mechanism and the capability of the predator to find the solution for reaching the prey. We show also some complex configurations that our model is able to solve.

In the next session the concurrent environment will be presented, later the planner of the algorithm will be discussed, finally experiment and conclusion will be given.

THE CONCURRENT ENVIRONMENT

The system emulate an environment with n mice that want to reach the cheese (the goal). They are constrained in a room with obstacles and they do not know their position, the positions of the other mouse and the position of the cheese. They have only a visual system that allows them to perceive obstacles, other mice and the cheese.

Moreover they have a mental image system with memory that allows to remember for a while the directions of the cheese. Each mouse acts using three agents that are dedicated to the evaluation of the visual information (the *target recognizer*, the *mouse recognizer* and the *obstacle evaluator*) a fourth agent (*the planner*) plans the mouse movements (*action*) according to an optimization strategy[9] (see Figure 2).

The *planner/action* module can be further detailed as an evolution phase based on the *Active Information Fusion* (AIF) loop.[10] The AIF loop consists of five modules (see Figure 3): *Observe, Evaluation, Optimization, Choose-Next*, and *Action*. *Observe* is responsible for the acquisition and preprocessing of sensorial data; *Evaluation* compute, on the basis of the observed parameters and the current state of the robot the next choose; the *evaluation* uses an optimization's method (in our case a GA). *Choose-Next* selects the new movement of the robot on the basis of the evaluation module. It should be noted that all processes may run in parallel mode and they may be executed on the network of distributed processors.[11] The result of the choose module determines the *Action* (next movement in the environment). After the movement the *Observe* module drives further sensor-explorations. The *Real World* represents the environment on which the ARN operates. The information, within the system, flows in a continuous active fusion loop.

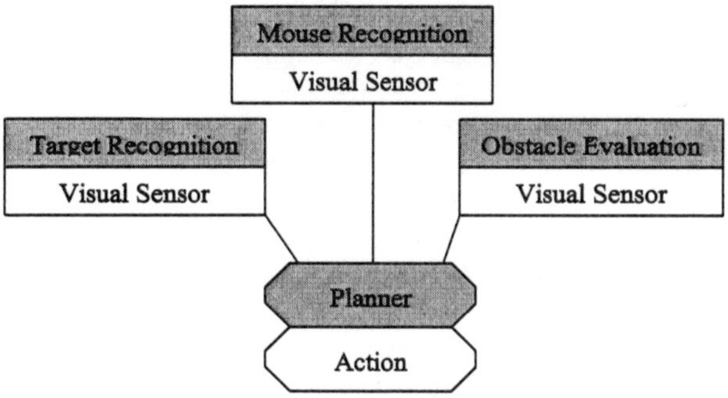

Figure 2. The agents of the ARN System.

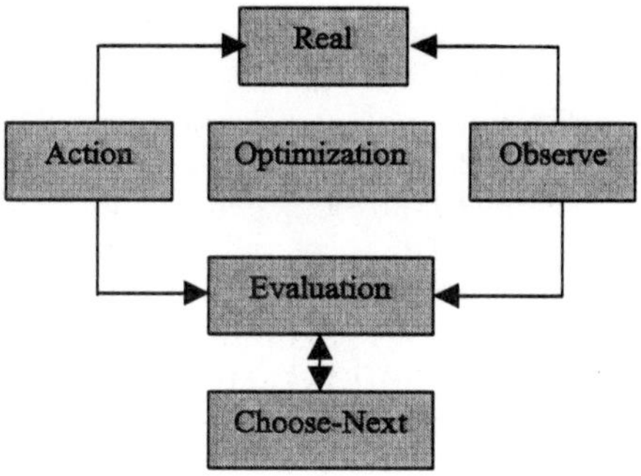

Figure 3. The Information Active Fusion Loop.

THE PLANNER

GAs have been applied to behavioral studies, simulation and optimization. They have been applied for the control of instrumentation in risky environments (atomic energy site, sea bed, space missions). GAs are useful whenever the complexity of the problem requires for approximated fast solutions.

The planning agent drives the robot toward the target.[12] The problem can be stated, informally, as follows: *find the minimum number of steps to reach the target in position* (x_T, y_T) *starting from the position* (x_0, y_0) *and avoiding obstacles and other mice in the room.*

The navigation problem, before sketched, can be formulated, in the case of one rectangular obstacle of size *a* x *b*, as a dynamic programming problem (see Fig. 4):

$$\begin{aligned} x_i &= S_x(\theta_{i-1}, x_{i-1}, y_{i-1}, x_{ob}, y_{ob}, x_F, y_T; a, b) \\ y_i &= S_y(\theta_{i-1}, x_{i-1}, y_{i-1}, x_{ob}, y_{ob}, x_F, y_T; a, b) \\ \theta_i &= S_\theta(\theta_{i-1}, x_{i-1}, y_{i-1}, x_{ob}, y_{ob}, x_F, y_T; a, b) \\ &\min(\delta(x_0, y_0, x_T, y_T)) \end{aligned} \tag{1}$$

The variables in the equations represent the coordinates of the mice (x_i, y_i), the coordinates of the top-left corner of the obstacle (x_{ob}, y_{ob}), the coordinates of the target (x_T, y_T), the angle (θ) between the direction of the robot displacement and the vector target.

Functions S_x, S_y, S_θ describe the dynamical evolution of the system (for sake of simplicity it includes the bounds conditions of the obstacle), the function δ measures the length of the path made by the robot to reach the target in step units. The problem has exponential complexity, even in the case of a single obstacle with rectangular shape.

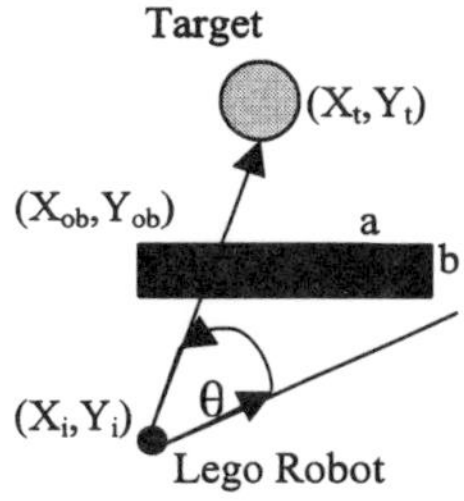

In real situation, the mouse does not know the coordinate system and only the visual sensors and the distance evaluators drive its movements. In our case it can measure the distance from the obstacles and the direction of the target with respect its displacement direction.

The GA, here described, has been designed to approximate the solution of the general optimal problem stated above.[13] The planner, on the basis of the information given by the sensors agents, provides the direction of the displacement and the speed of the robot. The GA can be sketched as follows:

1. select a starting population of genomes (random choice of 200 samples directions for example "1010101111010111");
2. perform the crossover operation on randomly chosen genomes (the crossover is performed in a single point and the probability of crossover is uniform), for example: let us be two genomes "10010110 01011101" and "11110110 01000001", the new genomes are "10010110 01000001" "11110110 0101110";
3. perform a bit-mutation with probability 0.01 to change one or more bit randomly
4. compute the fitting function, $F_{h,k}$, for each pair of chromosomes (k,h);
5. select the new population (with the tournament technique);
6. **if** the global fitting, $\overline{F}$, satisfy the convergence criteria **goto** 7) **else goto** 2);
7. end

Data Representation. One of the key points in GA is the representation of the data that constitute the *genome* on which to perform *genetic* transformations (*selection, crossover and mutation*). Our genomes are the displacement directions, α, in the interval [-π,π] (see Fig. 5). They are mapped in 16-bits word using the function: $R(\alpha) = (\alpha + \pi)(2^{15}-1)/\,2\pi$.

Fitness Function. The mathematical definition of the fitness function has been formulated by considering obstacles and target and mouse as objects that generate an electrostatic field that for the predator is attractive in the case of the target and repulsive in the case of the obstacles and mouse (see Fig. 5):

$$F_{ob}^{predator} = b_1 \times F_{prey} + b_2 \times F_{obstacle} + b_3 \times F_{clash} \tag{2}$$

with $b_1 + b_2 + b_3 = 1$.

In the case of the prey the field can be written in an analogous way:

$$F_{ob}^{prey} = a_1 \times F_{predator} + a_2 \times F_{obstacle} + a_3 \times F_{clash} \tag{3}$$

with $a_1 + a_2 + a_3 = 1$.

Where: $F_{prey} = K \times (\pi - |\alpha|)$ and $F_{predator} = -H \times (\pi - |\alpha|)$ are attractive and repulsive fields

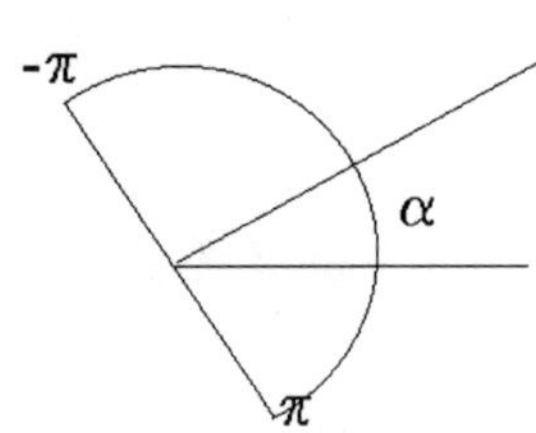

Figure 5. The direction of the displacement.

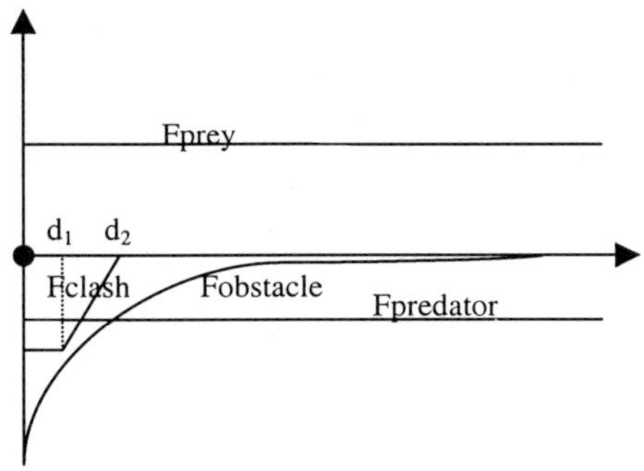

Figure 6. Components of the fitness function.

respectively; $F_{obstacles} = -\sum_{i=1}^{K} \frac{C_i}{R_i^2}$ is a Coulomb's field generated by K point sources; in our examples K=2.C_i are parameters that depend on the infrared sensors direction. They have been computed during the calibration phase, described in the next section.

The last term represents the clash between two mice or a mouse and two preys. It has been represented with a short interaction field:

$$F_{clash} = \begin{cases} -C & if\, d < d_1 \\ -C\dfrac{d-d_2}{d_1-d_2} & if\, d_1 \le d \le d_2 \\ 0 & if\, d > d_2 \end{cases} \tag{4}$$

the meaning of d_1 and d_2 are given in Figure 6.

The Memory Model. The visual memory has been implemented as a decreasing S shape expressed as follows:

$$S(p,a,b,c) = \begin{cases} 1 & p < a \\ 1-2\left(\dfrac{p-a}{c-a}\right)^2 & a \le p \le b \\ 2\left(\dfrac{p-c}{c-a}\right)^2 & b \le p \le c \\ 0 & p > c \end{cases}$$

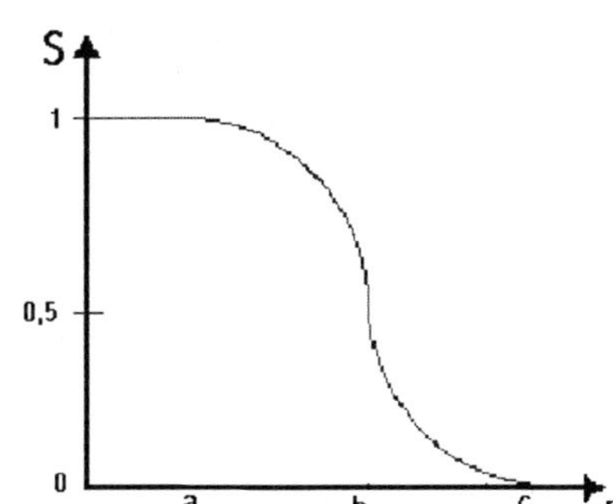

with b=(a+c)/2. This choice reflects the fact that at increasing steps (p) the mouse forgive the direction in which the cheese is if it looses the direction and do not see it. The memory is refreshed any time the mouse sees the cheese.

EXPERIMENTS

In the experiments we have used the following parameters for the predator b_1=0.05, b_2=0.45, b_3=0.50. The parameters used during the simulation of the GA are given in Table 1.

Table 1. GA Parameters Used in the Simulation

Population	200 individualsselection
Method	binary tournament
Crossover	single point
$Prob_{crossover}$	0.8 (uniform distribution)
Mutation	bit-mutation
$Prob_{mutation}$	0.01 (uniform distribution)

The simulator, see Figure 7, has been implemented in C++ and to perform experiments a visual interface has been designed.

Figure 7b shows the trajectory in the case of memory step c=30 (the cheese is reached after 29 iterations) and in the case of memory step c=10 (the cheese is reached after 96 iterations). This result shows the effect of the memory in reaching the prey.

Figure 8 shows the number of steps to reach the prey as a function of the directional memory length. The curve has been obtained by performing the experiments 100 times with the same initial conditions (see Fig. 7a,b), but different randomization.

CONCLUDING REMARKS

In the previous paragraph we analyzed, by using a simulator, the strategy of a set of predators to reach the prey. Theoretical and empirical studies have examined how patterns of prey might reflect a balance between the conflicting demands of feeding and avoiding predators, see also the dissertation in Yakubu.[8] A refuge has been included into the

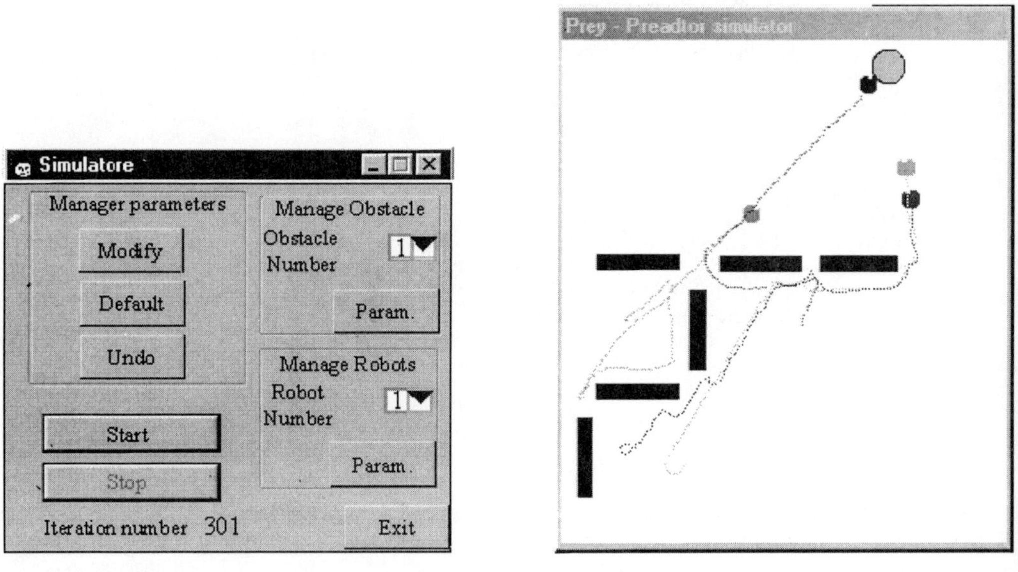

(a) (b)

Figure 7. (a) User interface of the simulator; (b) visualization of the predator trajectory.

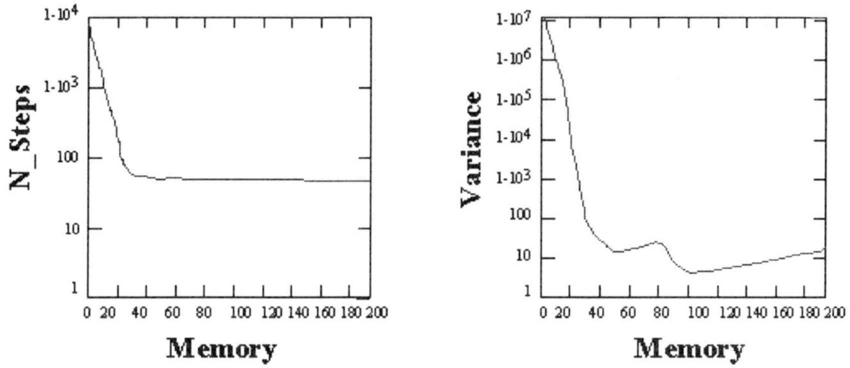

Figure 8. Steps to reach the prey versus the memory length.

simulator to allow a way of flight to the prey, otherwise the predator find the prey without exchange of information among them.

The n predators and the prey have been distributed on the environment in a random way, on the contrary obstacles and refuge have been arranged by the user.

Stability of trajectories can be seen as an optimization planning problem,[14] on this item there are a large number of different approaches, that can be classified in the following categories: classical methods, based on calculated derivatives, and evolutionary methods, based on initial values and its estimation.[9] In our approach we have used GA as evolutionary algorithms.[7] Crossover, mutation and translation allow the variation on the starting trajectories and on the next generation, moreover its mutation is guided by a fitness function based on the theory of attraction and repulsion of particles, for a strong improvement of the trajectory a memory has been introduced on the predators unit.

Further work will be addressed to the analysis of co-operating systems (for example, neural networks and genetic algorithms combination) and the comparison of different learning and problem solving strategies.

REFERENCES

1. V. Volterra, Sulle variazioni e fluttuazioni di specie animali conviventi, *Memorie della R. Accademia dei Lincei S.*, 6, 2, pp. 31-113, (1926).
2. A.J. Lotka, *Proc. Nat. Accad. Sci.*, USA 6, 410, (1920).
3. R. Levins, *Some demographic and genetic consequences of environmental heterogeneity for biological control*, Bull Entomol. Soc. Am 15, pp. 237-240, (1969).
4. I. Harvey and P. Husbands, Evolutionary robotics, in *Proc. Of IEE Colloquium on Genetic Algorithms for Control System Engineering*, London, 8 May, 1992.
5. S. Yamada, Learning Behaviors for Environmental Modeling by Genetic Algorithm, in *Evolutionary robotics*, Husbands, Meyer eds., Lecture Notes in Computer Science, pp.179-191, Springer Verlag, (1998).
6. H. Hautop Lund, D. Parisi, Preadaptation in Populations of Neural Networks Evolving in a Changing Environment, in *Artificial Life*, Vol.2, N. 2, pp.179-197, (1995).
7. V. Rai, V. Kumar, L.K. Pande, A New Prey-Predator Model, *IEEE Transaction on Systems, Man, and Cybernetics*, Vol.21, no.1, (1991).
8. A. Yakubu, Pery Dominance in Discrete Predator-Prey Systems with a Prey Refuge, *Mathematical Biosciences*, vol. 144, pp.155-178, (1997).
9. S. Burrerat and J.E. Cooper, Evolutionary Methods for the Optimization of engineering Systems, *Optimisation in Control: Methods and Applications* (Ref. No. 1998/521), 1/1–110 (1998).

10. V. Di Gesù, Multi-agent System and Artificial Vision, in *Computer and Devices for Communication*, pp. 393-402, (1998).
11. B. Lenzitti, G. Lo Bosco, and D. Tegolo, An integrated Environment for Dynamic Processes in Distributed Image Analysis System, in P. Biljanovic, K.Skala, S.Ribaric, and L. Budin eds., *Proc. MIPRO'98*, IEEE Computer Soc., Opatija (Croatia), pp.14-21, May 1998.
12. P.E. Agre and D.Chapman, What Are Plans for?, *Robotics and Autonom. Systems*, Vol.6, pp.17-34, (1990).
13. H. Muhlenbein and D. Schilierkamp-Voosen, Predictive Models for the Breeder Genetic Algorithm – Continuous Parameter Optimization, *Evolutionary Computing*, v1 n1, pp. 25-49, (1993).
14. S.R. Sutton, Planning by Incremental Dynamic Programming, in *Proc. of the Eighth Int. Workshop on Machine Learning*, Morgan Kaufmann, pp. 353-357, (1991).

RATIONAL DECISION: A MATTER OF COMMON SENSE

Mariagrazia Semenza

Computer Vision Laboratory, DIS
Pavia University
Via Ferrata 1, 27100 Pavia - Italy
e-mail: lella@vision.unipv.it

All of us have to make decisions every day of our lives. Many empirical studies have been made which required extensive surveys and considerable laboratory work, including investigations on simple decision-making processes. They mostly appear to show that human beings are incoherent.

Every now and then, however, we have to make decisions which we recognize as important and to which we devote a great deal of thought.

In particular, in the field of the risk analysis of dangerous industrial plants, wrong decisions can have heavy consequences. Risk analysts have a precise responsibility towards companies from the economical point of view and moreover they have a heavy responsibility towards colleagues and people in general as regards safety, health and environmental pollution. This kind of decisions compels the risk analyst to use a scientific approach, which reduces the possibility of mistakes as much as possible.

But what is meant for being wrong about the prevision of events which are uncertain and, luckily, in most cases, rare? According to the subjective probability theory, being wrong does not mean not guessing the "true" probability; being wrong means being "not rational", that is incoherent in one's previsions.

Then, in these cases, we need coherent, that is rational, decision-making. We refer to coherence as a necessary property in dealing with uncertainties, and thanks to the fact that uncertainty can be described numerically in terms of probability, and probability obeys laws, these laws are the essential devices that secure coherence.

They are:

$$0 \leq P(E) \leq 1 \qquad \text{the convexity law} \tag{1}$$

$$P(E_1 \cup E_2) = P(E_1) + P(E_2) \qquad \text{the addition law for exclusive events} \tag{2}$$

$$P(EH) = P(E/H) \cdot P(H) \qquad \text{the multiplication law} \tag{3}$$

Human and Machine Perception 3: Thinking, Deciding, and Acting, edited by Cantoni *et al.*
Kluwer Academic/Plenum Publishers, New York, 2001.

About the above mentioned laws Dennis Lindley[1] wrote:

"The convexity, addition and multiplication laws are the three basic laws of probability. No one of them can be deduced from the other two, yet every other result in probability can be deduced from them. Probability is an enormous tripod built on these three legs. The key point here is that the ways in which statements of uncertainty can be combined are by no means arbitrary. One cannot sit down and think up apparently reasonable rules. The only ones are the three we have discussed: no more, no less.

The laws ensure that several statements of uncertainty cohere.

The laws of probability constrain one's statements of uncertainty yet still leave a lot of freedom for personal views.

The constraints typically provide the most useful way of assessing probabilities."

Before showing how the probability theory can be applied to a risk assessment, we must stress the importance of another concept by quoting Dennis Lindley[1] about the nature of probability:

"The probability is subjective: it depends on the subject making the judgements.

Probability expresses a relationship between a person and the world he is contemplating. Some have regarded this as a disadvantage. In fact, the subjective nature of probability is its great strength, for it describes a real situation of a subject, or observer, contemplating a world, rather than talking about a world divorced from the observers living in it.

Recognizing that two observers may have different probabilities from the same event, the suggestion has been made that the difference arises not because the people are different but because they have different amounts of information.

For a given subject contemplating an event the probability will depend on the information he has.

Indeed one of the most important tasks is for us to see how the probability changes with information"

That means that, since observers with a different amount of information make different previsions, a fundamental requirement of an analysis methodology is the fact that information and uncertainties connected with it are to be ordered and elaborated in a transparent way. This methodology in facts makes it possible to compare the states of information of two different subjects with the utmost precision.

According to this approach a tool to aid a decision maker does not try to replace him; it provides some guide lines which enable a complicate decision process to be broken down into smaller, and then simpler, parts the separate analyses of which can be combined to provide a solution to the whole.

BUILDING A PROBABILITY MODEL

Let us see how we can collect and rationally organize information to solve our complex problem: the aim is to build a logical model, for example, of the behavior of an industrial plant under all the relevant accidental conditions.

First of all we have to give a representation of the basic element of the model: the event. We can describe an event[2] either as a question requiring a single answer: "true" or "false" (we can call this particular event "fact") or, more generally, as a question the answer to which is uncertain: to every possible answer we have to relate a probability which represents the weight of our expectation.

Notice that the answers to every event must be an exaustive set of exclusive possibilities, so that the related probabilities, summed up, give as a result 1, that is certainty.

The questions, therefore, may be of this kind:

- Is the phenomenon X present or absent?

- Will the parameter Y reach the value A or B or C or D at time T?
- Will the equipment Z act effectively on demand?

Each answer defines a situation, a change of state or the adequacy of the intervention of an equipment of the plant that we consider relevant to understand the problem.

The learning process, in fact, begins by drawing up a list of the events defined as we said above. We call them "universe of events" and their order is never arbitrary. In the case of analysis of physical or chemical processes, the most intuitive order is generally the temporal one because it makes it possible to describe the evolution of the phenomena during the time. In other cases, instead, in which we must model a situation independent of time, the order of events is determined by a typical mechanism cause-consequences.

The choice of the starting point is however critical: we often realize we have to study the problem under a different point of view, and we must build up the structure of the model again; we all know the useless difficulties deriving from having faced a problem the wrong way round.

At this point we must link the events with one another to build the structure of the logical model which, unlike the well-known event tree, is very articulated and flexible to fit the process of improvement of knowledge in the best way.

As a matter of fact, as a consequence of the choice of a specific answer, we must define:

- the direct links between events: we declare the event which follows;
- the variation of probabilities: we can change the probabilities associated with one or more events which follow; we can also force an event to become certain (P=1) or impossible (P=0)
- the variation of path: we can change the links between following events; that is we can change the order and the number of the events that will be analyzed afterwards.

This way we obtain a dynamic representation of our process of collecting and organizing knowledge.

IMPLEMENTING A SOFTWARE TOOL

All this information, expressed in an elementary and precise syntax, can now be processed by a software,[3] which follows all the possible paths, choosing each time, for each event, one of the possible answers, taking into account, at the same time, the constraints that each of the answers chosen imposes on the following questions.

In detail a syntactic language that tries to simulate the human thought in the construction of logical paths, defined in accordance with the laws of logic, has to adopt the dynamic conditional capabilities we listed above. And these are the possibilities that the syntactic language of this methodology provides to the user. Surely the human thought has a flexibility, an imagination freedom that cannot in any way be simulated. But the limitation of human memory and the freedom in regarding all at once semantic as well as syntactic aspects can lead to logical errors.

A purely syntactic elaboration made by a tool with a large permanent memory capability gives some more guarantee against such logical errors, even if in a forcefully limited universe of events.

The complexity due to the branching of logical paths is pruned in our mind when we judge some of them insignificant, that is when we perceive either probability or relevance as below a subjective threshold; in this case the whole path is dismissed and erased from the memory. A similar pruning criterion can be adopted by our methodology that can dismiss the constituents the probability of which is below an established threshold. A definite advantage is that the software takes under control the total probability lost because of this dismission and gives us notice of its value. We can therefore decide on the adequacy of the threshold and, if the case, we can change it.

Since the software does a purely syntactic elaboration, we have to act as interface between semantic and syntactic both at input and at output stages. We have to define the problem to be solved as we said above and to translate it into the software own language: we have to operate the semantic-syntactic translation. The software does the syntactic elaboration necessary to individuate, to build and to store into memory one by one all the constituents of the partition, that is the exaustive set of the exclusive sequences, which can be built on the basis of the universe of the events and of their logical interrelationships. Each one of these constituents represents just one of the logical paths, that is a possible alternative sequence of events coherent with the input constraints.

Notice that since the rules imposed to describe the problem are purely syntactic, the software we use can be applied to a wider range of situations besides the risk analysis field.

MODEL COHERENCE AND CORRECTNESS VERIFICATION

Now we have the task of inspecting the constituents that have been generated in order to verify:

- the coherence of the logical sequence of events. Any incoherence that can be found is due to lacks or mistakes in the initial translation: lacks must be filled, mistakes must be corrected;
- the completeness of the universe of events that have been taken into consideration, as far as the problem to be solved is concerned:
    * any event that conditions the events of the universe either from a logical or a probabilistic point of view must have its very place in the sequence of events, in any case before the events it conditions;
    * any event necessary to define significant physical behaviors of the system along each logical path has to be explicitly present in it;
- the consistency with our experience both of the described courses of events and of the probability evaluations.

If these conditions are satisfied, our understanding of the problem can be regarded as sufficient and the model is correct; if not, we have to think again about our problem and to make the necessary modifications. This trial and error procedure repeated a few times brings to a full understanding of the problem to be solved, without shades or concealed doubts, and therefore it brings to a satisfactory and settled model.

When at last we have a satisfactory model available, our task is limited to the use of the tools supplied by the methodology to get any answer we need about our problem: the problem can be completely solved without any further effort. Any top event (a dangerous and generally complex event, according to the definition used in fault tree methodology) which is relevant to the problem, can be completely defined by simply selecting, among the whole of the stored sequences, all the ones that the examined top event implies. They represent in fact all the alternative ways in which the top event can happen and so the sum of their probabilities gives the probability of the top event itself directly and correctly. Their selection and examination directly supply the combinations of occurrences which are critical from a probabilistic or from a risk point of view and which we try, therefore, to avoid, if necessary.

Lastly, experience shows that an alternative logical path, with its sequence of chained occurrences, very similar to a movie picture, is very easily understood by experts of the behavior of the system, mainly if it is integrated with the physical representation: so experts can be easily and usefully involved in the model verification, even if they are not familiar with these methods of analysis. In this way, if we offer exclusive sets of dynamic consequent occurrences, we have a satisfactory solution of the problem of technological transfer more easily than adopting the well known usual methodologies, which are static and deductive.

Integration of logical and physical models compels us to take into consideration system behaviors and success criteria when we assign probabilities and we compare our event definitions with actual physical occurrences: this often helps avoid that the analysis may become a sort of parlor game rather than a true breakthrough in the problem of understanding.

Bruno De Finetti[4] stated: "Probability rules are nothing artificial or odd or sublime, but they are conditions of obvious common sense". He was certainly right. However, my problem is that it is difficult to use common sense, particularly in complex decisions. It is fundamental to have an adequate aid in reasoning. If I use the methodology described above I do not obtain guarantees of perfection, but I obtain guarantees of coherence and freedom of representation, unlike that does not happen if I use traditional methodologies.

Every sequence elaborated is the precise and explicit (that is unequivocal) description of my complete amount of knowledge about the problem, and even more of my degree of ignorance. Personally, the fact that my arguments are made so explicit, so transparent, gives me the uncomfortable feeling that my professional qualities can be checked every time, for every work, which is necessary.

Every new problem is a challenge for my ability in understanding reality. I cannot hide my doubts behind the formal precision of elegant but rigid structures, such as the one of fault trees, for example, which only represent the logical conclusions I have reached, but do not help me improve and verify my knowledge. On the contrary, in practice they prevent me from critically analyzing the results. Every time I finish a work, I am sure that some expert will be likely to find it not perfect; but I also know that it is certainly the best I could do. In fact, the methodology has helped me in the process of identification and understanding of the most obscure points, and the conclusions I have reached will surely be consistent with what I have understood.

REFERENCES

1. D. V. Lindley, *Making Decisions*, John Wiley & Sons, London (1985).
2. R. Galvagni, *Problemi Decisionali e Logica dell'Incerto*, Scuola di Specializzazione in Sicurezza e Protezione Industriale, Facolttà di Ingegneria, Università degli Studi "La Sapienza", Roma (1997).
3. L. Galvagni, *Integrated Dynamic Decision Analysis: User's Guide*, S.O.S.E., Roma (1996).
4. B. De Finetti, *Teoria delle Probabilità*, G. Einaudi ed., Torino (1970).

A HIERARCHICAL ENTROPY BASED REPRESENTATION FOR MEDICAL SIGNALS

Cecilia Di Ruberto,[1] Riccardo Distasi,[2] and Sergio Vitulano[3]

[1]Dipartimento di Matematica,
Facoltà di Scienze,
Via Ospedale, 58 - 09124 Cagliari, Italy.
e-mail: dirubert@vaxca1.unica.it
[2]Dipartimento di Matematica e Informatica,
Università di Salerno,
Baronissi (SA), Italy
e-mail: ricdis@unisa.it
[3]Dipartimento di Scienze Mediche,
Facoltà di Medicina,
Via S. Giorgio 12 - 09124 Cagliari, Italy.
e-mail: vitulano@vaxca1.unica.it

ABSTRACT

In this chapter, we present a Hierarchical Entropy-based Representation (HER) for 1-dimensional signals. In fact, many 2-dimensional signals can be effectively encoded in a 1-D form that will allow the use of HER. This method represents the signal by a vector containing the energy values related to its local maxima and their locations along the time axis. Such a representation has been utilized as the feature extraction engine in an image retrieval system for medical image databases, with a K-d-tree based spatial access method. The data used for the experiments were contours and textures from various medical sources. The experiments show that HER performs very well in this respect.

INTRODUCTION

The development of new imaging methods for medical diagnosis has significantly widened the scope of the images available to physicians. The traditional X-ray photograph is now only one of the possibilities: other options include Computer Tomography (CT), Positron Emission Tomography (PET), Nuclear Magnetic Resonance (NMR), Thermography and more. The volume of data generated by an average sized medical imaging facility is such that both retrieval and classification of the images become essential issues.

Human and Machine Perception 3: Thinking, Deciding, and Acting, edited by Cantoni *et al.*
Kluwer Academic/Plenum Publishers, New York, 2001.

The problem of image classification and retrieval by content, that is, based only on the actual content of the pictorial scene, is a hard one. As it turns out, human beings are extremely good at recognizing shapes and textures independently from their position and orientation, much less at programming a machine to do so. Therefore, finding an automated technique to solve the pattern recognition problem by computer is a daunting task and no general solution is yet available. There are, however, several results that solve, at least partially, particular problems in specific areas.

The techniques proposed in the scientific literature fall almost invariably in the category of feature extraction methods. The key idea is that of analyzing the pictorial scene in order to obtain *n* numerical features. By doing so, an image is mapped from image (or pixel) space into a single point in *n*-dimensional feature space, where traditional - and exact - spatial access methods may be used to retrieve points (i.e., images) that are close to the query image. This type of user interaction paradigm is called 'query by example,' because the user supplies an example (the query image) and the system looks for images that are 'near' in some sense.

Another possible paradigm for user interaction is called 'query by sketch.' In this case, the user draws a sketch of some object which should appear in the retrieved images. Depending on the nature of the feature extraction technique, one or both paradigms might be applicable.

The underlying assumption with all feature extraction based methods is that proximity in feature space implies proximity in image space. Under certain hypotheses about the feature extraction process and the definition of distance in feature space, this is indeed the case.

Image invariants, that is, features that do not change under certain families of image transformations such as rotations and skewing, are useful in many cases. The method of Fourier descriptors, for instance, is known to work reasonably well for the recognition of object contours independent of position, orientation and size.[1] There are works that show the results of the Fourier approximation of polygons for different numbers of Fourier coefficients.[2] As it turns out, it is possible to achieve a good approximation of a polygon by using 15-30 coefficients. Even with few coefficients, the Fourier series obtains an acceptable approximation to the original curve because the low frequencies contain the most significant information about the object. However, for highly irregular or jagged contours, even 30 coefficients are not enough to characterize the shape adequately.

In order to place the present work into some perspective, we now briefly review some of the techniques already published in the literature that bear some resemblance, one way or another, to the technique we are going to describe.

A query-by-sketch-only approach to the problem of image retrieval is described in Del Bimbo and Pala.[3] The user draws a sketch of one object that should be featured in the pictorial scene and the system tries to 'fit' this sketch with the objects in its database image. The feature-space distance between the user sketch and a candidate image depends on the elastic deformation energy necessary for the fitting. This method, however, is computationally much heavier than the other ones we are illustrating and therefore might be argued to fall into a different 'horsepower category.'

Still query-by-sketch only, the system described in Del Bimbo et al.[4] is computationally lighter and has the interesting property of integrating query by sketch with a progressive query approach: the user is presented with results and can then refine his query in a friendly way. The feature extraction engine relies on a fixed sampling of the closed contours under exam (360 samples, one per degree). This does yield an index of fixed, manageable size, but leads to difficulties with certain types of contours; concave shapes are a specific problem.

Lastly, a method that, similarly to our technique, was originally designed for 1-D time series (e.g., stock price data). It is described in Agrawal et al.[5]. The discrete Fourier transform is used to extract features, and only the first few coefficients are retained. This is appropriate if one wishes to detect long-period and medium-period trends (low and mid-low frequencies).

The technique presented in this chapter is called HER (Hierarchical Entropy-based Representation). It can help both the expert (i.e., the physician) and the non-expert user in retrieving or classifying images in a database. As will be shown, the technique is general and flexible enough to be useful in a wide range of applications. The experiments have been performed on a variety of databases; however, medical image databases are the majority.

HER is based on a subset of the 1-D signal samples (namely, the signal's local maxima) and on their associated energy. The interesting peculiarity of this representation is that it allows us to reconstruct the signal's energy distribution along the time axis by using only a few coefficients.

Morphology and structure play a key role in image-based medical diagnosis, whether it is performed by a human physician or by a computer-controlled automated system. Morphological features, such as size, presence or absence of lumps, contour jaggedness or smoothness and so on, can hint at pathologies or other clinical peculiarities.

The use of HER for medical image contour analysis can provide information of several types:

- linear variations in an organ's macroscopical size or in the cells' microscopic size is detected by HER as an enlargement or contraction in the signal domain, but the number and position of maxima does not change;
- 'lumps' or other variations in the shape under exam are detected both for an enlargement of the signal domain and for the presence of a larger number maxima along with an increase of the associated value and entropy;
- the jaggedness of an organ's contour (e.g., cirrhosis) is detected for the presence of several additional maxima, as if there was some added noise. This could not be possibly detected by a Fourier analysis unless we are willing to consider a large number of harmonics in the signal (about 50). Moment-based methods or simple Euclidean distance are not able to detect this kind of anomalies at all.

When a physician performs the structural analysis of a specimen (organ, cell or whatever), he/she looks for macroscopical 'order' in the distribution of grey levels in the specimen. The presence of any disorder that cannot be justified (blood vessels, morphological anomalies induced by surgery etc.) implies the presence of some kind of pathology. As an example, the variation in jaggedness or in the distribution of cell agglomerates in a liver CT scan can point to the existence of cirrhosis in the patient.

The structure of the chapter is as follows. In the next section, we present the theoretical foundation underlying HER and describe the correspondence between the input signal and the representation vector. After that we show the results of our experiments in terms of well-known objective measures. Finally, a few concluding remarks are expressed.

HOW HER WORKS

HER for Contours

Let $x(n)$ be a monodimensional signal, time-discrete and finite in the time domain, i.e., $x(n) \neq 0$ for $n \in [0,N-1]$. Let us define the *energy* of the i-th sample as $E(i)=|x(i)|^2$. The total energy of $x(n)$ can be then defined as the sample by sample sum of the individual energies:

$$E = \sum_{i=0}^{N-1} E(i) = \sum_{i=0}^{N-1} |x(i)|^2 \tag{1}$$

Now, consider the difference operator $\Delta_i=x(i)-x(i-1)$. A local signal maximum occurs whenever Δ becomes negative:

$$x(k) \text{ local max} \Leftrightarrow \Delta_k \geq 0 \text{ and } \Delta_{k+1} < 0 \tag{2}$$

If we have several adjacent samples $x(k),\ldots,x(k+l)$ such that $\Delta_k = \ldots = \Delta_{k+l} = 0$, then we have a signal plateau. In this case, if $\Delta_{k+l+1}<0$, we pick the plateau midpoint $x(i+l/2)$ as the local maximum.

Suppose $x(i)$ is a local maximum for the signal under study; we compute its relative energy, weighted by total signal energy, as

$$E_r(i) = \frac{E(i)}{E - E(i)} E(i) \tag{3}$$

We now consider $x(i)$ as the midpoint of a Gaussian distribution. The choice of associating a Gaussian function with each maximum has been made after considering the following facts:

- given a set P of points, for large $|P|$, if the number of local maxima m is such that $m \ll |P|$, then any distribution tends to a Gaussian;
- if the contour is corrupted by 0-mean Gaussian noise, then a Gaussian distribution is obviously the best fit for the resulting distribution;
- the experiments have shown that choosing a distribution other than a Gaussian has little effect on both the energy associated to each maximum and the final normalized recall achieved;
- the symmetry of Gaussian distributions allows us to associate with each maximum the energy contained in an interval that has the maximum as its midpoint in a natural way.

The standard deviation of the Gaussian associated with maximum in $x(i)$ is then

$$\sigma(i) = \frac{1}{\sqrt{2\pi (E_r(i))^2}} \tag{4}$$

We then calculate the entropy relative to the maximum $x(i)$ as the quantity

$$S(i) = \frac{1}{x(i)} \sum_{k=-\sigma(i)}^{\sigma(i)} |x(i+k)| \tag{5}$$

$S(i)$ can be considered as a relative measure of the signal energy in the range $[i-\sigma(i), i+\sigma(i)]$ with respect to the energy in $x(i)$. If a signal has m maxima, its total entropy is therefore

$$S = \sum_{i=1}^{m} S(i) \tag{6}$$

We now consider a vector $\hat{x}$ containing the m maxima of $x(n)$ in decreasing order:

$$\hat{x} \equiv (x(i_1), \ldots, x(i_m)), \text{ with } x(i_1) \geq x(i_2) \geq \ldots \geq x(i_m) \tag{7}$$

Then, the representation of the signal x is ultimately obtained by the value, position and energy of all the relevant maxima in $\hat{x}$.

$$\hat{y} = (x(i_1), S(i_1), i_1, \ldots, x(i_k), S(i_k), i_k), \tag{8}$$

where $k \leq m$ depends on the value assigned to a threshold parameter. A summary of the whole process is presented in Figure 1, which shows a high level description of the algorithm to obtain the HER form of a given signal, while Figure 2 shows a specific example of how the HER representation obtained starting with a real-life input signal.

The HER Algorithm.

Given an input signal $x(i)$, $i \in [0,N\text{-}1]$:

1. Compute the total energy E of $x(\bullet)$ using Eq. (1);
2. Find the signal maxima and put them in a queue Q in decreasing magnitude order, along with their x-axis positions. If the number of signal maxima is m, the queue Q now contains $((i_1, x(i_1)), \ldots, (i_m, x(i_m)))$;
3. Compute the relative energy $E_r(i_k)$ of the first maximum, say $x(i_k)$, in Q using Eq. (3). Remove $x(i_k)$ from Q.
4. Compute the standard deviation $\sigma(i_k)$ relative to the current maximum $x(i_k)$ using Eq. (4);
5. The energy in the neighborhood of $x(i_k)$ has been accounted for; therefore, set the input signal $x(i)$ to zero in the range $i \in [i\text{-}\sigma(i_k), i+\sigma(i_k)]$ and update the total energy E accordingly;
6. Go back to Step 3 until the fraction of the total energy remaining in the signal $x(\bullet)$ falls below a given threshold.

Figure 1. A high level sketch of the HER algorithm.

In order to perform an approximate comparison between two given signals $x_1(.)$ and $x_2(.)$, we can compute the distance between their HER representations $\hat{y}_1$ and $\hat{y}_2$, where the distance is defined as follows:

$$D(\hat{y}_1, \hat{y}_2) = \sum_{i=0}^{\infty} \left| \hat{y}_{1i} - \hat{y}_{2i} \right| \tag{9}$$

The distance between any signal and itself is obviously equal to zero.

It should be stressed that HER is not meant to exactly reproduce the input signal, as would be appropriate for strict signal compression applications. Rather, HER extracts a

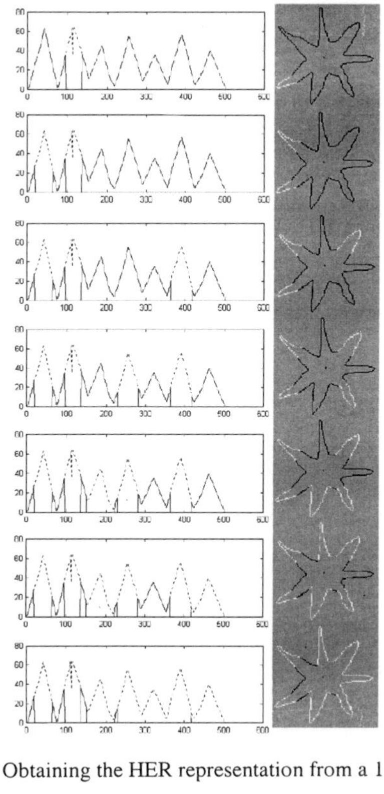

Figure 2. Obtaining the HER representation from a 1-D input signal.

number of features that can be used to retrieve signals similar to the one under consideration. Even the case $\sigma = 0$ does not imply that every single signal point will be sampled by HER: the algorithm always limits itself to the signal maxima. If, on the other hand, we select all the points (not only the maxima) and we define the instantaneous energy density as

$$\rho_E(n,l) = \frac{1}{2l}\sum_{i=-l}^{l} x(n+i) \tag{10}$$

then ρ_E does converge to the original signal:

$$\lim_{l \to 1} \rho_E(n,l) = x(n) \quad (\text{in } l_2) \tag{11}$$

Finally, in order to attain a more efficient search, we utilized a spatial access structure. Spatial access methods are techniques that allow for faster access to spatially organized data.[6-11] HER utilizes K-d-trees[12] as its spatial access method.

HER for shape contours has several nice invariance properties. First, HER is obviously invariant to any translation of the object in the image space, since the reference system gets translated together with the center of the object.

Then, HER is invariant to object rotation for any integer multiple of $\pi/2$. Indeed, in the case of continuos signals, it is invariant for any angle: a rotation of the object only causes a change of phase on $\gamma(s)$ and therefore on the HER representation.

Additionally, since mirror reflection is equivalent to a change in the direction of the contour scan used to build the 1-D signal, there's no change in the vector of ordered maxima. Therefore, HER is reflection-invariant too, because the maxima are ordered according to their magnitude, not their order of occurrence in the scan.

As for scaling invariance, suppose that an object A has two adjacent maxima m_1 and m_2 in the curvilinear coordinate representation of its contour C. Let d be the distance between m_1 and m_2. Now perform a scaling transformation on A obtaining A' (and m_1', m_2', C' and d'). Then, there is the following relation between the length of the contours $|C|$ and $|C'|$ in the original object and its scaled version:

$$\frac{|C|}{|C'|} = \frac{d}{d'} \tag{12}$$

If we represent C and C' in curvilinear coordinates, then the following is also true:

$$\frac{|\gamma|}{|\gamma'|} = \frac{d}{d'} \tag{13}$$

where d and d' represent the distance between any two maxima resp. in A and A'.

Summing up, the system can be made invariant to scaling, by replacing the test for approximate equality between $\hat{y}_1$ and $\hat{y}_2$ with a test for approximate linear dependence.

A few words about the 'approximate equality' test are in order. Techniques for approximate matching utilizing feature extraction usually adopt the distance in feature space as the main indicator of similarity.

Let T be a mapping from image space I to feature space F. T is said to be a *complete* mapping iff any element $X_I \in I$ has exactly one image $X_F = T(X_I) \in F$. Non-complete mappings may still be usable for pattern recognition tasks if they meet the weaker (and more fuzzily defined) condition of separating clusters in feature space. Let X_I and Y_I be two objects in I

and let $X_F=T(X_I)$ and $Y_F=T(Y_I)$ be their images in feature space; let $D(.,.)$ and $G(.,.)$ be distance metrics resp. in I and F.

When the transformation T represents, for instance, a Fourier transform with truncation of all but the first α harmonics, Parseval's theorem tells us that

$$G(X_F, Y_F) \leq D(X_I, Y_I) \tag{14}$$

$$\lim_{\alpha \to N-1} G_\alpha(X_F, Y_F) = D(X_I, Y_I) \tag{15}$$

where N is the number of points in the input signals - in this context, supposing we are dealing with contour signals, the number of pixels in the contours of X_I and Y_I.

When T denotes the HER representation, we have a similar result:

$$G(X_F, Y_F) \leq D(X_I, Y_I) \tag{16}$$

$$\lim_{\alpha \to N} G_\alpha(X_F, Y_F) = D(X_I, Y_I) \tag{17}$$

only that in this case α represents the number of maxima utilized in the HER representation rather than the number of Fourier harmonics.

The value that provides the best tradeoff between ease of computation and effective results is dependent on the specific shapes of the objects we are dealing with. This is true for both HER and Fourier transform based methods.

HER for Textures

In this chapter, the method we propose for texture structural analysis is that of the spiral. In a previous work,[13] we utilized a low level segmentation technique (merge) that for each region in the input image provided the user with the size of the partition element and its associated entropy. The evaluation function was the entropy associated to the image histogram. The main steps of that technique can be briefly summarized as follows:

1. associate the whole image domain to the root of a quad-tree (or other k-ary complete tree);
2. generate the histogram for the domain under exam and represent it through HER (the histogram is a 1-D signal);
3. partition the domain into its 4 children and proceed to Step 2;
4. compare the children's histograms with the parent's and consider the one that is closest (according to HER);
5. go back to Step 3, iterating down to the leaves;
6. select the minimum in the signal given by parent entropy and child entropy. Select the level where the minimum occurs.

In this way, we are able to associate to each region the partition element size and its entropy.

In the split phase, we obtain a 1-D representation of a 2-D texture. The 1-D representation is obtained by drawing a spiral-shaped path on the partition element and considering the pixels in the order they are touched. The spiral starts at the upper left corner of the tile and goes inward clockwise. Figure 3 shows an example of this procedure.

When applied to textures, HER enables us to detect even minimal structured variations among the tiles, thus realizing a split operation on the region under exam. Furthermore, the system provides topological information about the distribution of pixel values along a well-defined path.

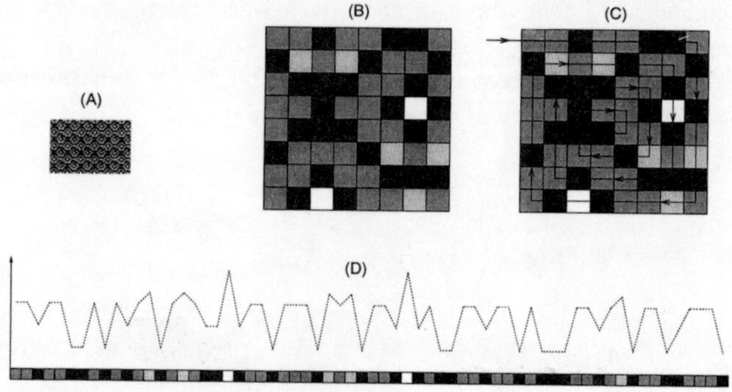

Figure 3. How HER obtains a 1-D signal from a 2-D texture. (A): the texture; (B): the partition element (texture tile); (C): the spiral; (D): the resulting signal.

The proposed model has been applied to real-life signals in order to analyze and classify the textures present in the different regions of several images found in medical databases.

EXPERIMENTAL RESULTS

In order to test the performance of HER, we implemented it on a 600 MHz Pentium-III system using Matlab under Windows 2000. Our experiments have been performed on a number of databases of different size. As a general result, it appears that five maxima are generally enough to capture a significant amount of signal energy for the signal sizes we dealt with. However, it must be said that the percentage of energy needed to describe the contours with adequate precision is strongly dependent on the type of signal examined and the type of application.

Before showing the detailed results, we briefly review the evaluation criteria usually adopted for the testing of retrieval systems. The *recall* measures the system's ability to retrieve all relevant objects, while the *precision* measures the system ability in retrieving only relevant objects. Another indicator that is often used is the *normalized recall* (NR),[14] which is defined as follows.

Suppose we have a database D of $|D|$ objects where the number of objects relevant to our query is $N<|D|$. Besides, suppose that the relevant objects are sorted *a priori* so that the most relevant object is X_1, down to the less relevant object X_N. We perform a query that returns an ordered answer set A. Let r_i be the rank of i-th most relevant object X_i in the answer set A. The *ideal rank* (IR) is then defined as

$$\mathrm{IR} = \frac{1}{N}\sum_{i=1}^{N} i = \frac{N+1}{2} \tag{18}$$

(note that it does not depend on A); the *average rank* (AR) of A is then

$$\mathrm{AR} = \frac{1}{N}\sum_{i=1}^{N} r_i \tag{19}$$

The difference AR-IR gives a measure of the effectiveness of the system. It is usually normalized in order to obtain a value between 0 and 1 known as *normalized recall* (NR):

$$\mathrm{NR} = \frac{\mathrm{AR}-\mathrm{IR}}{|D|-N} \tag{20}$$

Figure 4: An example of contour-based retrieval using HER.

The definition of Normalized Recall requires a human operator to manually rank the images for relevance. In all cases where such ranking required high-level competence, we have recurred to the assistence of experts in the field. What this practically means is that for the CT scans and microscopic images, it has been a specialist physician to determine which images were showing signs of a pathology and which were not.

Results of HER for Contours

We have evaluated HER's effectiveness at retrieving images based on contour data with the use of 10 heterogeneous query time series selected from a database of 1500. For each of the 10 query time series, we manually ranked the 20 most similar objects in the database in order to compute the NR. The number of maxima utilized for our tests is 10. We also performed the same queries both using Euclidean distance (ED) as a similarity measure. Additionally, we also performed comparative tests against the moments-based technique[15,16] considering the first two moments and their ratio. As shown in Table 1, this is the worst performing technique, having many false dismissals.

Table 1. **Contour-Based Retrieval: Comparison between HER, Euclidean Distance (ED) and a Moment-Based Technique (MBT) in Terms of Normalized Recall**

Measure	Value	HER	ED	MBT
Size of DB	1500			
Number of queries	10			
Normalized recall		0.984	0.971	0.960

A visual example of HER in action (see Figure 4) shows the effectiveness of HER, which is able to capture the 'thin-shaped fish' quality in the answer set. Although not illustrated here, HER's ability in retrieving rotated versions of the query should also be noted. The rotated replicas of the query are the first items to appear in the answer set, since they are at distance 0 from the query - this is a consequence of HER's invariance to integer rotation discussed earlier. Details and experimental results about invariance to rotation can be found in Distasi et al.[17]

Table 2 shows the results of a different set of queries over a slightly larger database (1600 objects). For this experiment, answer set size was 70 throughout. Here is a discussion of the data in the table: '% Enrg.' is the fraction of the total signal energy utilized for constructing the index. The value 10 for HER means that the threshold in Figure 1, Step 6 has been set to 90%. The 'Avg. dist.' column contains the average distance between all database objects in feature space. In image space, the average distance is 5399.5, as can be deduced from the 'ED'row (ED is calculated on the unmodified contours). The 'Avg. A.S. dist.' column shows the average inter-object distance (in feature space) inside the answer set only. This quantity can give some insight on the average cluster size. Finally, the 'NR' column shows the normalized recall obtained.

Table 2. Contour-Based Retrieval: Detailed Comparative Results for Euclidean Distance (ED), HER, and Moment-Based Technique (MBT)

Technique	% Enrg.	Avg. dist.	Avg. A.S. dist.	NR
ED	100	5399.50	15.86	0.74
HER	10	611.07	7.73	0.84
MBT	-	5960.43	16.02	0.62

As can be seen, HER achieves good results while using only a little fraction of the total energy present in the signal. The energy utilized has a raw relation to the number of actual signal samples used in the index. What this ultimately means is that the whole index is much lighter, since it contains less information. We have found that in many cases, 50% (or even 10%!) of the signal energy is enough to achieve good retrieval results. Euclidean distance, on the other hand, is calculated by taking into account all of the signal's samples. However, the optimal threshold to be used for the representation, and therefore the actual amount of signal energy used, is strongly dependent on the data contained in the database. Another interesting fact emerging from the table is that the moment-based mapping of image space into feature space is really an expansion rather than a contraction: the average distance in feature space is more than in image space.

Results of HER for Textures

HER's ability to classify, retrieve and retrieve images based on structural texture features has been evaluated on three different datasets. The first dataset is the classical reference of Brodatz textures. The other two datasets came from real-life medical applications: a set of liver CT scans and a set of microscopies of thyroid specimens. In the latter two cases, we had some specialist physicians sort out the images that implied a

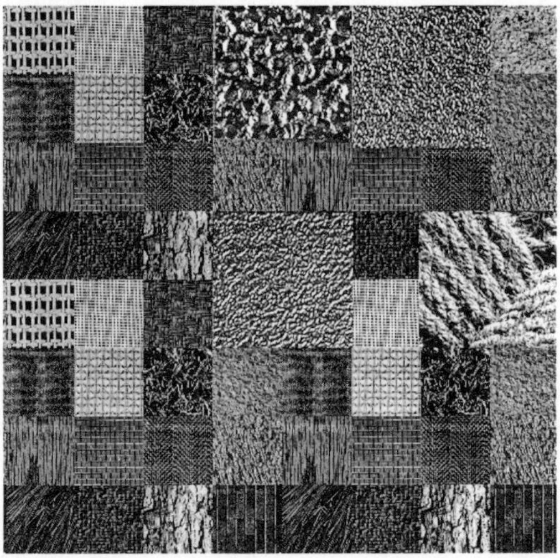

Figure 5. The 256 Brodatz textures utilized for our texture-based retrieval experiments.

pathology from the images that showed no sign of pathologic situations. The results are shown in tabular form in Table 3, while the datasets are shown in Figures 5 and 6. The medical dataset pictures only show the textures extracted, not the whole original image.

Table 3. Texture-Based Retrieval: Normalized Recall Comparative Results for HER and Euclidean Distance (ED) for the Retrieval of Images from 3 Databases

dataset	size	# queries	HER	ED
Brodatz	256	10	0.97	0.75
Liver CT	625	15	0.95	0.88
Thyroid	625	15	0.91	0.86

The normalized recall results shown in Table 3 have been obtained by averaging the results over a number of queries. A physician has been the a priori arbiter for evaluating the accuracy and discriminating power of the retrievals performed by sorting the textures that displayed signs of a pathological state from the ones that didn't.

It can be seen that HER is very effective for texture-based retrieval. The Euclidean distance based method's performance is much more modest. Furthermore, the index in the Euclidean distance case is made up by the whole signal, while HER's index is significantly smaller. As stated earlier in the section about contours, often as little as 10% of the total signal energy is enough to achieve a highly discriminant index.

CONCLUSION

This chapter presents HER, a novel technique for content based retrieval of time series. HER is based on a Hierarchical Entropy based Representation for 1-dimensional signals that utilizes the local signal maxima as well as the entropy associated to each of them. The experiments show that this representation is both effective and very general. This model can effectively be utilized whenever it is possible to obtain a 1-D representation of the patterns to be retrieved by content.

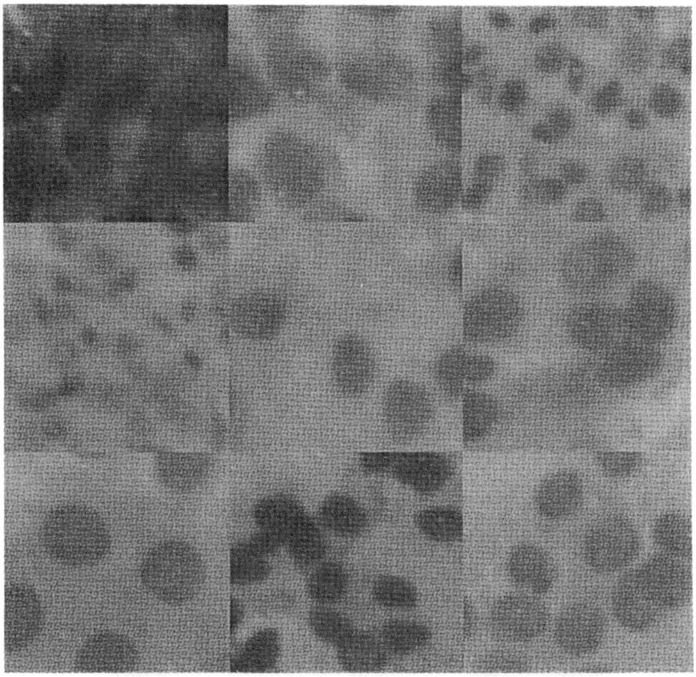

Figure 6. The 625 thyroid microscopy textures utilized for our texture-based retrieval experiments.

REFERENCES

1. G. H. Grunlund, Fourier preprocessing for hand print character recognition, *IEEE Trans. Computers*, Vol. C21, (1972).
2. C. T. Zahn and R. Z. Roskies, Fourier descriptors for plane closed curves, *IEEE Trans. Computers*, Vol. C21, (1972).
3. A. Del Bimbo and P. Pala, Visual image retrieval by elastic matching of user sketches, *IEEE Trans. Patt. Anal. Mach. Intell.*, Vol. 19, No. 2, pp. 121-132, Feb. 1997.
4. A. Del Bimbo, M. De Marsico, S. Levialdi, and G. Peritore, Query by dialog: and interactive approach to pictorial querying, *Image and Vision Computing 16*, pp. 557-569, Elsevier, (1998).
5. R. Agrawal, C. Faloutsos, and A. Swami, Efficient similarity search in sequence databases, *Proc. FODO: 4th Int'l Conf. on Foundations of Data Organization and Algorithms*, pp. 69-84, Evanston, IL, Oct. 1993.
6. N. Beckmann, H. P. Kriegel, R. Schneider, and B. Seeger, The R*-tree: an efficient and robust access method for points and rectangles, *Proc. ACM SIGMOD*, pp. 322-331, May 1990.
7. A. Guttman, R-Tree: A dynamic index structure for spatial searching, *Proc. ACM SIGMOD*, Boston, pp. 45-47, June 1984.
8. H. V. Jagadish, Linear clustering of objects with multiple attributes, *Proc. ACM SIGMOD*, pp. 332-342, Atlantic City, May 1990.
9. I. Kamel and C. Faloutsos, On packing R-trees, *Proc. CIKM*, 2nd International Conf. on Information Knowledge Management, Nov. 1993.
10. H. Samet, *The Design and Analysis of Spatial Data Structures*, Addison Wesley, (1989).
11. J. D. Ullman, *Principles of Database and Knowledge-based Systems*, Computer Science Press, Rockville, (1989).
12. J. L. Bentley, Multidimensional binary search trees used for associative searching, *Comm. ACM*, Vol. 18, No. 9, pp. 509-517, Sept. 1975.
13. S. Vitulano, C. Di Ruberto and M. Nappi, Different methods to segment biomedical images, *Pattern Recognition Letters 18*, pp. 1125-1131, (1997).
14. G. Salton and M. J. McGill, *Introduction to Modern Information Retrieval*, McGraw-Hill, (1983).
15. A. K. Jain, *Fundamentals of Digital Image Processing*, Computer Science Press, Rockville, (1989).
16. E. G. M. Petrakis and C. Faloutsos, Similarity searching in medical databases, *IEEE Trans. on Knowledge and Data Engineering*, Vol. 9, No. 3, May-June 1997.
17. R. Distasi, D. Vitulano and S. Vitulano, A hierarchical representation for content based image retrieval, *Journal of Visual Languages and Computing*, to appear.

ADAPTATION AND LEARNING FOR PATTERN RECOGNITION: A COMPARISON BETWEEN NEURAL AND EVOLUTIONARY COMPUTATION

Claudio De Stefano[1], Antonio Della Cioppa[2], and Angelo Marcelli[2]

[1]School of Engineering, Università del Sannio
Piazza Roma, Palazzo Bosco Lucarelli, 82100 Benevento - ITALY
[2]DIIIE, Università di Salerno
Via Ponte Don Melillo, 84084 Fisciano (SA) – ITALY
e-mail: marcelli@unisa.it

ABSTRACT

The work presented in this chapter is aimed at developing self-governing artificial systems that are able to operate in complex, uncertain and dynamic application domains, by mimicking the learning and adaptation capabilities exhibited by biological systems. For this purpose, we are exploring the possibility offered by the artificial neural networks and evolutionary computation paradigms for automatically extracting the set of prototypes describing the variability present in a data set. In particular, this chapter reports the results of an experiment designed for comparing the performance exhibited by a Learning Vector Quantization network and an Evolutionary Learning System using a Breeder Genetic Algorithm. For the sake of generality, the comparison has been performed on a complex classification problem obtained by generating a synthetic data set according to the distribution of distributions model.

INTRODUCTION

The ability to learn is the most important property of the living systems and constitutes the fundamental process through which the living species deal with changes in the environment. This *learning* process is strictly related to that of *adaptation*, which plays a fundamental role for the survival or the extinction of species. Therefore, studying the learning and adaptation mechanisms of biological species can suggest effective ways of building self-governing artificial systems that, by exploiting both the information embedded in a suitable set of training patterns and that provided in input during their normal use, are able to operate in complex, uncertain and dynamic application domains.

In this framework, two main categories of approaches have been followed in the last decades: the ones based on the simulation of biological neural networks (Artificial Neural

Human and Machine Perception 3: Thinking, Deciding, and Acting, edited by Cantoni *et al.*
Kluwer Academic/Plenum Publishers, New York, 2001.

Networks) and those based on the simulation of the natural evolution of biological species (Evolutionary Algorithms).

In the first studies on artificial neural networks, the goal was to demonstrate the possibility of solving problems without explicit programming and by using very simple processing elements. The network was supposed to learn from examples and store this knowledge in a distributed way by assigning weights to the connections among the processing elements. Thus, the original methodology was exactly opposite to the goal-driven or top-down design of statistical classifiers, which make use of explicit error functions. As far as the research activity in this field has proposed more and more sophisticated network architectures (by adding non linear fire functions, embedding more layers of neurons working in parallel and allowing different ways of interactions among neurons and layers), it has been possible to perform complex tasks such as, for instance, speech synthesis or pattern recognition.[1-3]

In the current neural network research, the original views are becoming vague, as some of the most fundamental neural networks, such as the Multi-Layer Perceptron or the Learning Vector Quantization, have shown to have properties very similar to those of statistical techniques. However, the main goal of artificial neural network research, i.e. that of building effective systems in complex application domains, in which it is difficult or eventually impossible to exploit any kind of *a priori* knowledge, still remains unchanged.

The research activity in the field of Evolutionary Computation started about thirty years ago. In that period, a number of researchers came up with the idea of mimicking mechanisms of biological evolution in order to develop powerful algorithms for problems of adaptation and optimization. Since many optimal natural structures have emerged through biological evolution, the idea of using the underlying mechanisms for the solution of optimization problems has motivated a considerable amount of research. The result of this activity is the definition of several approaches that have proved their effectiveness and robustness in a variety of applications such as machine learning, pattern recognition, immune systems, function optimization, and that have been used in different domains, such as ecology, engineering, economics, management, etc.[4-6].

Evolutionary algorithms have been developed to deal with complex applications where analytical information is not available. They are based on the concept of a population of individuals which, by applying probabilistic operators such as mutation, selection and recombination, evolve towards new populations of individuals having better fitness values. The individuals represent potential solutions and the fitness measures their effectiveness in solving the problem at hand.

The work reported in this chapter is related to a larger project aimed at developing a novel pattern recognition system that exploits both the learning and the adaptation processes. The basic assumptions of the project are the following: i) the output of the system is a set of prototypes; ii) there is no *a priori* knowledge about either the location of the prototypes in the feature space or their number. Basically, the system should exhibit both the ability of modeling the variability present in the training set by means of a suitable set of prototypes, and the capability of modifying such a set if, during normal operation, the performance drops too much.

In this framework, the aim of the chapter is to compare artificial neural networks and evolutionary algorithms in extracting the set of prototypes from a training set. In particular, following the basic assumption of our project, we have considered a Learning Vector Quantization (LVQ) network and an Evolutionary Learning System (ELS) using a Breeder Genetic Algorithm. For the experimental work, we have used synthetic data generated according to a distribution of distributions model, because it allows controlling the overlap among classes and the amount of noise introduced. Thus, it is possible to compare the LVQ network and the ELS performance as the complexity of the task increases.

The chapter is organized as follows: the second section illustrates the LVQ network, while the third section illustrates the ELS. Subsequently the experimental results obtained for both algorithms are presented and discussed, while some conclusions are eventually left to the final section.

LEARNING VECTOR QUANTIZATION CLASSIFIER

The LVQ network belongs to the so-called prototype based category. The set of weights associated with the connections between the neurons of the input layer and each of the output neurons represents a prototype. During the training phase the network generates the prototypes of the classes by performing a clustering of the samples included in the training set. In the operative phase, an input sample is assigned to the class corresponding to the nearest prototype. The Euclidean distance is commonly used.

The learning algorithms can be either supervised or unsupervised and belong to the competitive learning category,[7] i.e. they have the property that a competition among the neurons takes place before each learning step. At each step, the neuron winning the competition is allowed to modify its weights in a different way from that of the non-winning units. In the supervised case each output neuron is associated with one class before the training starts. In the unsupervised case, the output neurons must be labeled after training in order to allow the classification: this can be done only after the whole training set has been examined, by associating a neuron to the class for which it obtained the highest winning frequency. In both cases identifying the prototypes gives rise to a Voronoi tessellation of the feature space. Each region of this partition is associated with a prototype, and the samples located in a region are attributed to the class of the associated prototype.

One of the problems typical of this architecture is neuron *under-utilization*: for particular configurations of the points representing the samples in the feature space, some neurons cannot modify their weights during the training and remain unused. Algorithms substantially based on a modified distance calculation, which takes into account the number of times each neuron wins the competition, make it possible to overcome this problem.[8-10] In order to guarantee the convergence of the training algorithm,[11] the learning rate has to be a decreasing function of the number of learning cycles (for instance, by varying the learning rate on the basis of the Robbins-Monro stochastic approximation[12]). This makes it possible to avoid the problem of choosing when the training has to be stopped. However, the results obtainable during the training are significantly dependent on the initial value of the learning rate.[13]

EVOLUTIONARY LEARNING SYSTEMS

An ELS learns by examples, by inducing descriptions for the concepts to be learned from a set of samples of such a concept (positive samples) and samples of other concepts (negative samples). The samples are represented as points in the feature space, while concepts are represented as subsets of points (regions) in the same space. The learning is achieved by means of interactions between a population of individuals representing the behavioral rules of the system, and an environment, made of the positive and negative samples.

The internal structure of an ELS is based on a collection of individuals (prototypes) similar to the population of a simple evolutionary algorithm. Each individual is assigned a positive real-valued score, called *strength*. During the evolution, the individuals are subject to be matched with the environment and their strength is computed by means of a fitness function that takes into account the number of positive and negative samples matched in the environment.

An ELS works by iterating a cycle made up of an evaluation phase and a discovery phase. The evaluation phase consists of a number of interactions between the system and the environment and the discovery phase is performed by an evolutionary algorithm which modifies the population trying to discover new strong individuals, by selection, recombination and mutation of the strongest among the extant ones. As a whole, at the end of the learning, the population provided by the system should represent the behavioral patterns, which optimize the system response to the environmental stimuli for accomplishing a given task.

Evolutionary algorithms, however, are not able to deal with cases when a class is described by a set of different prototypes rather than by a single one. In fact, they do not allow the evolution of such a set of prototypes, because the fitness of each prototype is evaluated independently on the fitness of the others, and the task of the evolutionary process is that of optimizing the average value of the fitness in the population. Consequently, the optimal solution provided by such a process consists entirely of genetic variations of the best prototype. To overcome this intrinsic limitation, and to allow the formation and the maintenance of sets of prototypes (*niches*), each modeling a subset of the class to be learned, niching methods have been proposed.[6,14-16] By coupling niching methods and evolutionary algorithms is therefore possible to implement a parallel and adaptive search for prototypes in the feature space.

EXPERIMENTAL RESULTS

For our experimental work, we have considered a complex classification problem in which the data are generated by a statistical process based on the distribution of distributions model,[17] using Gaussian and uniform distribution functions with different statistical moment values, so as to model the complexity of a large number of real applications. Since in many real applications more than one prototype may be required to adequately model the variability exhibited by the sample belonging to a given class, we have introduced a set of clusters for each class. The samples belonging to each cluster have been generated according to a Gaussian distribution, the mean and the variance of each cluster have been randomly generated by means of uniform distribution functions. The random generation of the mean representing the centroids of the clusters simulates the independence of the prototypes and causes each cluster to be randomly located in the feature space. Moreover, the random generation of the variance effectively models the variability introduced by the description algorithms, generating clusters that are more or less spread over the feature space and more or less overlapped. The results reported in this chapter refer to a two-dimensional classification problem with three classes, each one made of three clusters.

As neural networks, we have considered a LVQ network using a supervised Frequency Sensitive Competitive Learning Algorithm.[9] According to this learning strategy, the problem of neuron under-utilization is faced by taking into account the winning frequencies of the competing neurons. The learning parameters of the net have been dynamically changed, during the training phase, according to the Robbins-Monro stochastic approximation procedure[12]. The actual values of the parameters have been set as follows: $\alpha = \beta = \gamma = 0.05$ and 300 iterations were performed in each trial.

As regards the implementation of the ELS, we have used a Breeder Genetic Algorithm[18] (BGA) using stochastic universal sampling, extended intermediate recombination with $p_{eir} = 1$, and BGA mutation with $p_m = 0.5$. As for the niching mechanism we have adopted a Resource Sharing[6] with σ = 60% of the population size. A number of generations equal to $\lceil 10^5/\text{population size} \rceil$ have been performed for each run. We have chosen a BGA in that it deals with real-coded genes. This property is very useful for the considered application because the samples to classify are represented by means of pairs of real numbers.

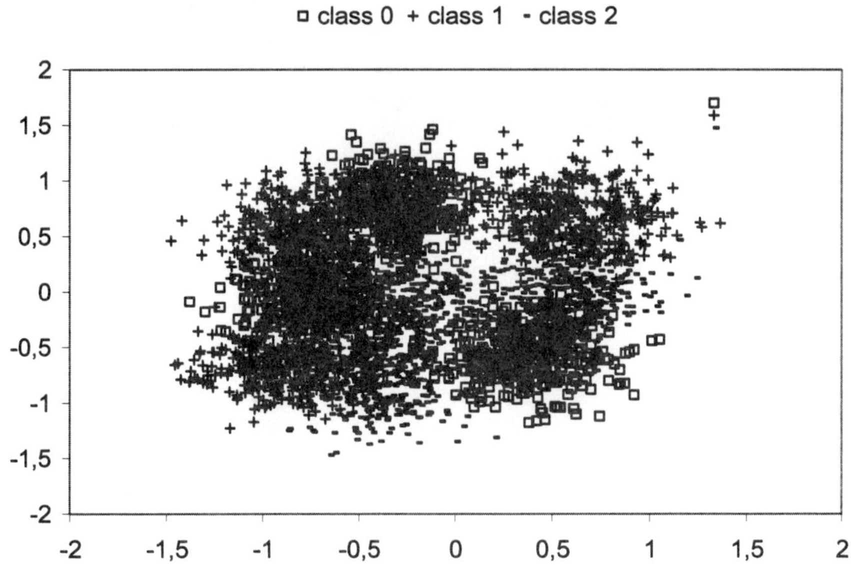

Figure 1. The training set TR3 used in the first experiment.

A first experiment aimed at evaluating the ability of the learning systems to deal with problems of increasing complexity has been performed by using three data set (TR1, TR2 and TR3), and a test set (TS), each one of 3,600 samples. The mean and the variance of the clusters have been generated according to a uniform distribution in the range [-1, 1] for the mean, [0.08, 0.13], [0.13, 0.23] and [0.23, 0.33] for the variance of the three training sets, respectively. In this way, the overlap among the clusters increases as far as the variance increases. Figure 1 shows the distribution in the feature space of the samples of the training set TR3. The clusters of the test set have the same centroids as the training sets (otherwise they would represent a different classification problem), but the variance is 20% greater than that used to generate TR3, so as to obtain larger overlaps among clusters.

Figure 2 reports the recognition rates obtained by a nearest neighbor classifier using the prototypes sets provided by the two learning systems as a function of the number of neurons (LVQ) and of the population size (ELS). The data show that both the systems were able to achieve 100% recognition rate when the clusters do not overlap, while the ELS performs better than the LVQ when the complexity of the task, as measured by the overlap among the clusters, increases. Such a behavior can be explained by noticing that, in case of the LVQ, at the beginning of the learning each neuron receives a label representing the class it is meant to represent and that such a label cannot be changed during the whole training phase. As a detailed analysis of the experimental results has shown, there are several prototypes that cover very few samples (none in the worst case) belonging to the class represented by those prototypes and many samples belonging to other classes, thus contributing to increase the error rate and, as a consequence, to reduce the recognition rate. Better performance, therefore, could be obtained by eliminating these prototypes from the prototypes set by means of an ad hoc procedure. The ELS, on the contrary, assigns the label to each prototype dynamically, during the learning. This labeling is such as each prototype receives the most frequent label among those present in the subset of the training samples that are closer to that prototype than to any other one.

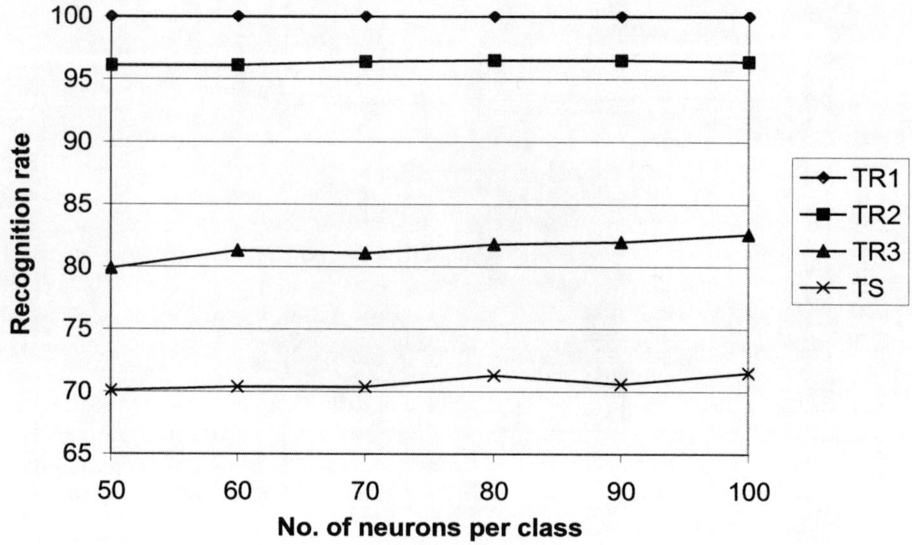

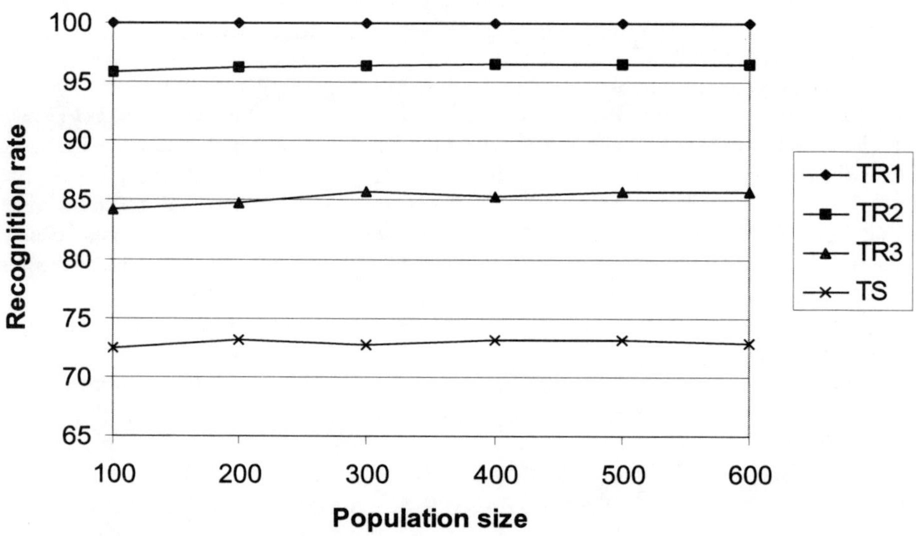

Figure 2. Recognition rates obtained by a NN classifier using the prototypes sets provided by the LVQ (top) and the ELS (bottom) as function of the size of the training set.

A second experiment has been designed to evaluate the power of generalization of the two learning systems, i.e. their ability to extract the relevant information from the training set. For this experiment, four different training sets TR900, TR1800, TR2700 and TR3600 were generated, made of 900, 1,800, 2,700 and 3,600 samples, respectively. The mean and the variance of the clusters are the same as in the set TR3 of the first experiment. The same test set of the previous experiment was also used.

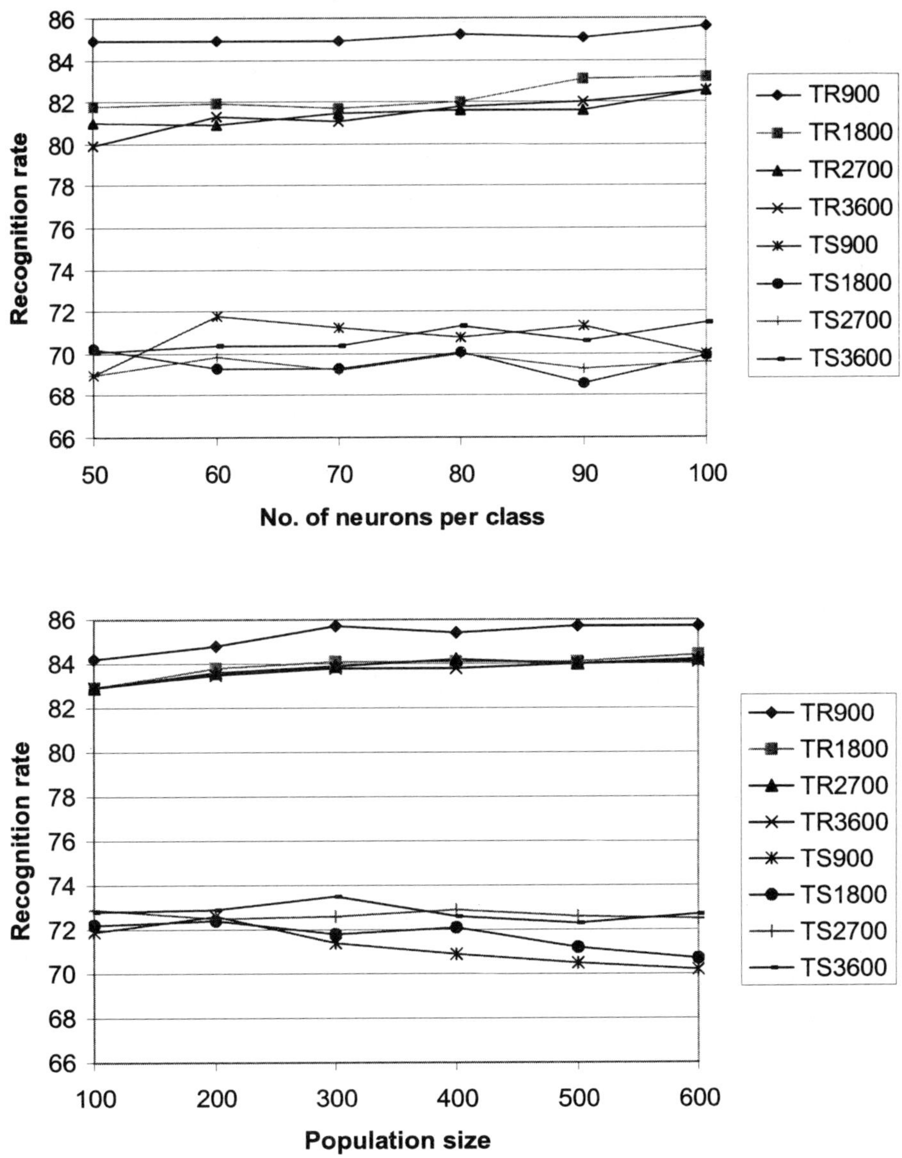

Figure 3. The recognition rates as a function of the number of neurons per class (LVQ, top) and the population size (ELS, bottom), obtained on the four training sets and on the test set while using the set of prototypes obtained by using as training sets TR900, TR1800, TR2700 and TR3600.

Figure 3 reports the recognition rates on both the training sets and the test set as a function of the number of neurons per class (LVQ) and the population size (ELS). The plots denoted by TS900, TS1800, TS2700 and TS3600 reports the results on the test set obtained by using the set of prototypes provided extracted from the training sets TR900, TR1800, TR2700 and TR3600, respectively. They show that the ELS performed consistently better than the LVQ on both the training and the test sets.

The plots reported in Figure 3 show that both the systems exhibit the same behavior on the training set TR1, with the ELS performing slightly better than LVQ. They also show that the performance on the training sets of the ELS does not vary significantly with either the training set size or the population size. This is not true for the LVQ, because its performance is significantly better on TR1800 than on TR2700 and TR3600, and it improves with the number of neurons per class. Looking at the performance on the test sets, the plots show that the ELS performance depends on the population size in such a way that there exists a value for the population size corresponding to the best tradeoff between completeness and consistency of the prototypes. In the current experiment, this value can be estimated at about 300 individuals, in that larger populations do not provide prototype sets that perform significantly better, but rather they can exhibit worst performance. This suggests that the population in excess has been used to produce prototypes covering outlying samples of the training set, thus not contributing significantly to improve the performance, or even being responsible for errors when they overlap more general prototypes. Moreover, the plots show that, as expected, the best performance is obtained by using the set of prototypes extracted from the largest training set. On the contrary, the LVQ exhibits its best performance while using the set of prototypes extracted from the smallest training set.

As regards the number of prototypes, the LVQ net requires that such a number must be determined *a priori*, while the ELS automatically provides the number of prototypes needed to maximize its performance. In particular, at the end of the learning phase, the ELS provided the following number of prototypes: 106 for the Class 0, 93 for the Class 1 and 101 for the Class 2. On the contrary, in case of the LVQ a large set of experiment is needed to decide the number of prototypes to be used for each class, because, as it may be derived from the plots of figure 3, it is difficult to foresee the performance of the LVQ as a function of the number of neurons per class.

CONCLUSIONS

In this chapter we have presented a comparison between a Learning Vector Quantization network and an Evolutionary Learning System aimed at evaluating their capability to automatically extract the set of prototypes describing the variability present in a given data set.

The experiments have been performed with reference to a complex classification problem, obtained by generating a synthetic data set according to the distribution of distributions model, and characterizing the performance of the two systems by means of their recognition rate, as a function of the complexity of both the task to accomplish and the system internal structure.

The results have shown that both the approaches achieve satisfactory performance, but that the ELS exhibits better performance on both the training sets and the test set, and a more consistent behavior with respect to the complexity of the task than that exhibited by the LVQ. Moreover, the LVQ net requires that the number of prototypes needed to describe each class must be fixed *a priori*, and such a number must be equal for all the classes. On the contrary, the ELS automatically provides, for each class, the number of prototypes needed to maximize its performance.

Finally, all the plus exhibited by the ELS are counterbalanced by a computational cost for both the learning and the classification phase higher than that of the LVQ net, because of its more complex learning mechanism and the higher number of prototypes per class.

REFERENCES

1. S. Haykin, *Neural Networks: A Comprehensive Foundation*, Macmillan College Publishing Co., New York: NY, (1994).
2. J.A. Anderson and E. Rosenfeld eds., *Neurocomputing: Foundations of Research*, MIT Press, Cambridge: MA, (1988).
3. I. Sethi and A. K. Jain eds., *Neural Networks and Statistical Pattern Recognition*, North Holland, Amsterdam, (1991).
4. J. Holland, *Adaptation in Natural and Artificial Systems*, University of Michigan Press, Ann Arbor: MI, (1975).
5. D. E. Goldberg, *Genetic Algorithms in Search, Optimization, and Machine Learning*, Addison-Wesley, Reading: MA, (1989).
6. S. Forrest, B. Javornik, R. E. Smith, A. S. Perelson, Using genetic algorithms to explore pattern recognition in the immune system, *Evolutionary Computation*, 1(3): 191 (1993).
7. R. Hecht-Nielsen, *Neurocomputing*, Addison-Wesley, Reading: MA, (1990).
8. D. DeSieno, Adding a conscience to competitive learning, in: *Proc. 2nd Annual IEEE Intern. Conference on Neural Networks*, IEEE Computer Society Press, Los Angeles: CA, (1988).
9. S. C. Ahalt, A. K. Krishnamurthy, P. Chen, and D. E. Melton, Competitive learning algorithms for vector quantization, *Neural Networks*, 3:277, (1990).
10. L. Xu, A. Krzyzak, and E. Oja, Rival penalized competitive learning for cluster analysis, RBF net and curve detection, *IEEE Transactions on Neural Networks*, 4:636, (1993).
11. T. Kohonen, The self-organizing map, *Proceedings of the IEEE*, 78(9):1464, (1990).
12. H. Robbins and S. Monro, A stochastic approximation method, *Annals of Mathematical Statistics*, 22:400, (1951).
13. C. De Stefano, C. Sansone, and M. Vento, Evaluating Competitive Learning Strategies for Handwritten Character Recognition, *Proc. IEEE Intern. Conference on Systems, Man and Cybernetics*, IEEE Computer Society Press, Los Angeles: CA(1994).
14. L. B. Booker, D. E. Goldberg, and J. H. Holland, Classifier systems and genetic algorithms, *Artificial Intelligence*, 40:235, (1989).
15. K. Deb and D. E. Goldberg, An investigation of niche and species formation in genetic function optimization, in: *Proc. Third International Conference on Genetic Algorithms*, J. D. Schaffer ed., Morgan Kaufman, San Mateo: CA, (1989).
16. J. Horn, D. E. Goldberg, and K. Deb, Implicit niching in a learning classifier system: nature's way, *Evolutionary Computation,* 2 (1):37, (1994).
17. J. de Villiers and E. Barnard, Back-Propagation neural networks with one and two hidden layers, *IEEE Trans. on Neural Networks*, 4(1):136, (1993).
18. H. Muhlenbein and D. Schlierkamp-Voosen, The science of breeding and its application to the breeder genetic algorithm (BGA), *Evolutionary Computation*, 1(4):335, (1993).

DECISION MAKING IN EVOLVING ARTIFICIAL SYSTEMS

Richard Walker,[1] Maurizio Cardaci[2]

[1]II University of Naples, Italy
e-mail: richard.walker@unina2.it
[2]University of Palermo, Italy
e-mail: Cardaci@unipa.it

INTRODUCTION

The theme of this workshop is artificial perception. In this chapter we will argue that the *ecological* function of perception is to serve decision-making. If this is so the mechanisms chosen to implement perception, in natural or artificial systems, will be constrained by the requirements of decision-making and theories of decision-making will inevitably influence theories of perception. In what follows we will look at decision-making from what we hope is a new perspective, applying concepts and techniques developed by what we will call "new artificial intelligence". We will begin, in the second part of the chapter, with a review of traditional, "normative" theories of decision-making and of the mounting body of experimental evidence, showing the inadequacy of these theories. In the third part we will outline an alternative ecological/biological view of decision-making, inspired by computer models from the fields of "Artificial Neural Networks", "Artificial Evolution" and "Evolutionary Robotics. Finally, we will briefly examine how far this new view may be considered to be biologically realistic and examine some of the possible consequences for theories of perception and of individual perceptive systems.

TRADITIONAL THEORIES OF DECISION-MAKING

Traditional theories of decision-making have their roots in classical and neo-classical economics and are essentially *normative* in nature.[1] The goal of these theories is to describe how rational decision-makers should take their decisions and to prescribe criteria for what should, and should not be considered, as *rational* behavior.

The theory of decision-making has been traditionally formulated as a "theory of rational choice". In this view a rational decision-maker "scores" the expected consequences of a set of alternative policies. Policies leading to favourable outcomes receive a high score and policies leading to unfavourable outcomes a low one. When all possible policies have been examined the decision-maker makes a *rational choice,* choosing the policy with the highest score.

Human and Machine Perception 3: Thinking, Deciding, and Acting, edited by Cantoni *et al.*
Kluwer Academic/Plenum Publishers, New York, 2001.

This view of decision-making has a number of implications. In particular: rational choices are based, by definition, on objective facts; they should therefore be *description invariant*. Given a set of possible policies and their known consequences two rational decision-makers, with the same preferences, should calculate the same scores and arrive at the same decision, regardless of the way in which the problem is formulated.

Preferences should exhibit *transitivity:* if a decision-maker prefers policy A to policy B and policy B to policy C she should prefer policy A to policy C.

Classical experiments conducted during the 1970s and 1980s, have shown that these predictions are false.

In one experiment[2] two groups of volunteers were asked to consider alternative plans to fight a viral epidemic expected to cause 600 cases of a potentially fatal disease. The first group was told that Policy A would save 200 lives whereas Policy B would either completely fail, leading to 600 deaths, or prove 100% effective, saving all 600 patients. The majority of volunteers in this group chose Policy A.

The second group of volunteers was told that with Policy A there would be 400 deaths; policy B was described in the same way as for the first group. This second group chose Policy B.

In reality the two policies are identical. With 600 cases two hundred successes imply four hundred deaths and vice-versa. The difference in the reaction of the two groups seems to depend on the different ways in which the problem is formulated. This is a clear *violation of description invariance*.

A second classic experiment tested the prediction of transitivity.[3] Here volunteers were asked to simulate the selection of candidates for admission to university. The selection was to be based on candidates' scores for intelligence, mental stability and social skills. The experiment was designed so that candidate B received a higher score for intelligence than candidate A, candidate C had a higher score than candidate B and so on, whereas the scores for stability and social skills showed the opposite trend with the lowest scores going to E and the highest to A.

During the experiment candidates were presented two at a time and volunteers were asked to indicate which of the two candidates they would select. In general volunteers showed a preference for high stability and good social skills: A was preferred to B, B to C, C to D and D to E. When however the volunteers were asked to choose between A and E a large majority chose E. Here again the experiment showed that real human decision-makers fail to satisfy the predictions of the theory of rational choice, in this case the prediction of transitivity.

Psychologists have presented a number of theories to explain the conflict between the predictions of normative theory and observed behavior. So-called "Prospect Theory", for instance,[4] argues that decision-makers tend to ignore small differences (i.e. the relatively small differences in the intelligence scores of A and B, B and C etc.) making their choices on the basis of major differences (such as the large difference in the intelligence scores of A and E). A number of alternative explanations have also been offered. These explanations are theories of *imperfect rationality*. Human beings, we are asked to believe, are not clever enough or do not have enough information to achieve the higher rationality of the economist.

A BIOLOGICAL/ECOLOGICAL VIEW OF DECISION-MAKING: INSIGHT FROM "NEW ARTIFICIAL INTELLIGENCE"

From a biological or an ecological viewpoint a "*rational*" *decision-making procedure* is a procedure, which, when applied repeatedly, benefits the reproductive fitness of the decision-maker. In what follows we use theoretical and experimental results from "New Artificial Intelligence" to suggest that there exist organisms whose decision-making is

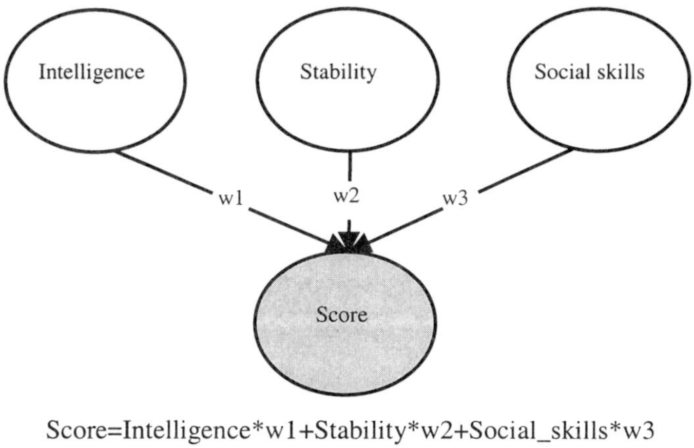

Figure 1. Design for a neural network displaying the predicted behavior on the "student selection" task.

perfectly rational in this biological sense yet whose behavior fails to satisfy the predictions of the theory of rational choice. This in turn leads us to reconsider our definition of rational behavior.

As a thought experiment, imagine the design of two Artificial Neural Networks. The first of these networks is designed to reproduce the *predicted* behavior of volunteers on the "student selection" task described earlier; the second network is designed to reproduce their actual, observed behavior.

The task of reproducing the predicted behavior is a simple one. Any algorithm which computes a single valued score for each student as a function of her intelligence, stability and social skills will automatically demonstrate transitivity. If the results of the tests are coded as integers with maximum value k even the most complex scoring function can be represented as a look-up table with k^3 entries. Applying a number of simplifying assumptions we can model our network as the "single layer perceptron" depicted in Figure 1.

Modeling the actual behavior is rather more complex. Here the selector has to compare couples (or other groupings) of students. The network will therefore require (at least) twice as much input (intelligence, stability, social skills for student1, intelligence, stability, social skills for student2). For an arbitrary scoring system this requires a look-up table of (at least) k^6 entries. If we try and train a network to reproduce the behavior observed in the laboratory we discover that the problem is not "linearly separable" and cannot be resolved by a single layer perceptron. The simulation of non-transitive choices requires the kind of multi-layer network shown in Figure 2. Unexpectedly we discover that "imperfect rationality" is harder to model than the "perfect rationality" predicted by the theory of rational choice. Why, one asks, should nature choose a complex solution with poor performance when there exists a simpler solution which would work better? Other work in "New Artificial Intelligence" may help to provide an answer.

The conceptual roots of Artificial Neural Networks are grounded in an "*information processing*" view of cognition. Neural networks take an input and process this input to produce an output.[5] More recent work in the "New Artificial Intelligence" has tended, on the other hand, to take an *ecological stance,* considering not just the neural network but the way in which organisms interact with the environment.[6,7] In this interaction the only data which the organism can use is the data provided by its sensors. Sensor input depends on the position of the organism and thus on its own actions. Organisms take an *ego-centric view* of their environment; rational behavior consists in making the best possible use of this view.

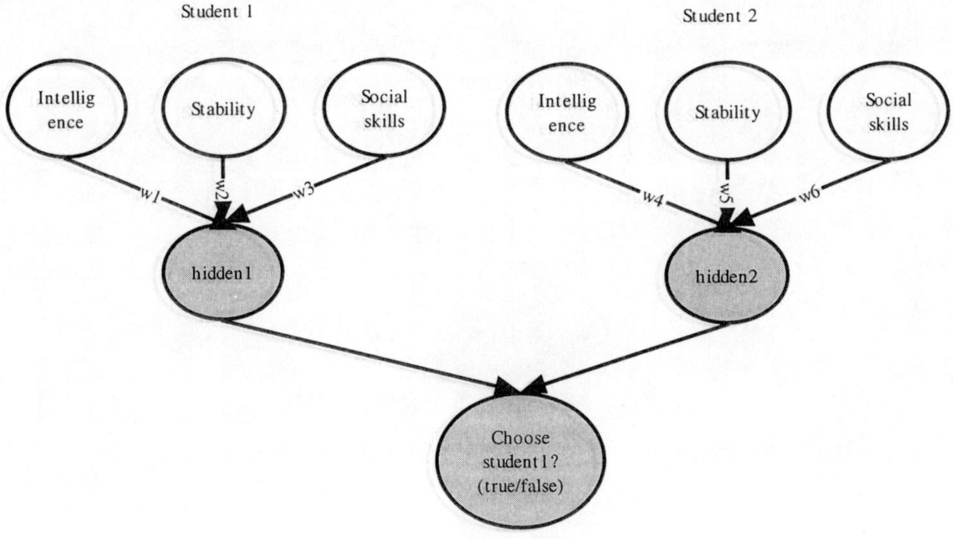

Figure 2. Design for a neural network reproducing the observed behavior on the "student selection" task.

Ecological models raise a series of new issues for models of cognition and decision-making. One of these is the issue of so-called "Perceptual Aliasing".[8] In an *ecologically grounded* model different environmental situations, requiring different decisions by the organism, may, at a particular moment, generate identical input to the organism's sensors. A large object, viewed from a distance, may, for example, produce the same image on the retina as a smaller object viewed from closer to. This has implications for our theories of decision-making. In particular it means that *for a physical organism, inhabiting a physical environment there exists no simple mapping from input to output capable of producing biologically optimal behavior. A biologically optimal decision-making mechanism has to be context-sensitive, to react in different ways at different times.* This leads to new questions. How can we (or nature) design an adaptive decision-making mechanism? What are likely to be the characteristics of such a system? Here again "New Artificial Intelligence" can provide us with useful insight.

The decision-making capabilities of natural organisms are the fruit of an evolutionary process. "New Artificial Intelligence" has produced models which simulate this process to evolve "artificial organisms" with specific cognitive abilities.[9,10] This work has made it possible to understand some of the constraints which apply to evolutionary processes. One of these constraints is that evolution is necessarily proceeds "step by step". In our experiments we have discovered that it is impossible to evolve a complex behavior in a single step. The most effective strategy is to evolve relatively small modules, with relatively simple behaviors, and then to evolve *connections,* between these modules so as to produce more complex, *emergent* behaviors. A second constraint is the need for redundancy. In work currently in progress we show that populations of artificial organisms evolve most effectively when individual organisms in the population are "redundant", that is when they include many more components than are strictly needed to do the job. Last but not least experimenters have demonstrated that evolution works most effectively when it is combined with *learning*. D. Floreano of EFL Lausanne,[11] for example, has produced models which use evolution to design the *learning rules* for individual synapses allowing the weights of the synapses to change continuously during interactions between the organism and the environment.

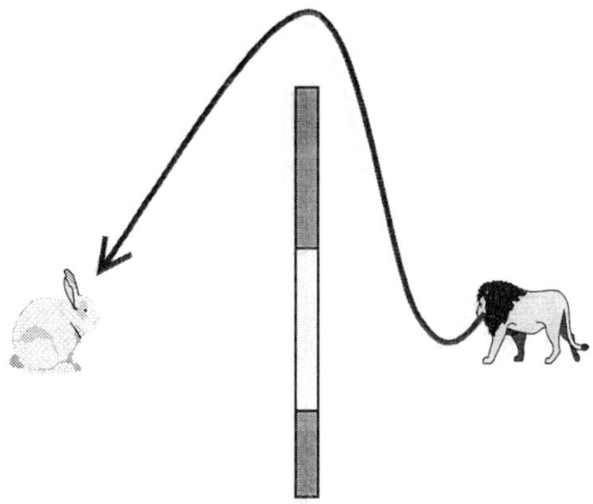

Figure 3. Detour behavior.

Putting these constraints together we can derive a number of consequences for a theory of decision-making. The requirement that evolution should proceed "step by step" implies that it is very hard to evolve big, complex rules in a single step and that it is easier to evolve hierarchies of simple rules, perhaps with complex emergent properties. The requirement for redundancy means that a system governed by a large, functionally inefficient, set of rules, will have a greater ability to adapt to a changing environment than a smaller, more efficient system. Most important of all the requirement for learning signifies that effective decision-making procedures have to be capable of changing over time – perhaps on a very rapid time scale. In short a "biologically rational" decision-making system should make its decisions using large numbers of simple rules based on "local input"; it should, what is more have the ability to change its rules rapidly in response to changes in context. Systems which meet these criteria are likely to benefit from an important side-effect. It is precisely this kind of system which is likely to prove most robust when faced with environmental change.

Two examples can help to illustrate the kind of robust decision-making system which evolution can produce. The first of these systems, developed by one of the authors, in collaboration with O. Miglino,[12] is a system designed to simulate so-called detour behavior when an animal (see Fig. 3) the animal "detours" round an obstacle in order to reach a target. In the experiment the authors "evolved" a population of artificial organisms with the ability to perform the task. The behavior of a typical organism is shown in Figure 4. This behavior can be summarized in three rules:

•Main: Move forward turning first clockwise (slowly) and then anti-clockwise (more rapidly).

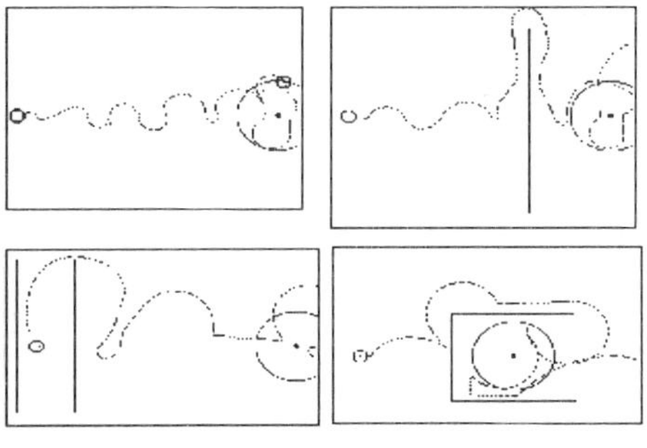

Figure 4. Solutions to the detour problem by an artificially evolved organism.

•Taxis: On visual contact with target turn sharply towards the target. Return to Main

•Wall following: If left proximity sensors active turn right. If right sensors active turn left. Move forwards until obstacle out of view. Return to Main

At no time in this experiment does the "animal" have a complete view of its environment; at no time therefore is it in a position to evaluate the consequences of its future actions. The information used by the organism comes completely from its sensors and depends not just on the objective characteristics of environment but also on the animal's position within the environment. This in turn depends on the animal's previous actions. The "rationality" of the animals behavior derives, in the last analysis, from complex interactions between extremely simple rules, which guarantee that the detour will function robustly in a broad range of different environments.

The second example of evolved rationality is based on the work by D. Floreano, referred to earlier in this chapter. In this experiment[11] Floreano evolved an "artificial mouse" with the ability to solve a simple, context-sensitive decision problem. The "mouse" inhabited the environment shown in Figure 5. In this environment the dark grey area was lit by an electric light controlled by a switch (the small rectangle at the bottom of the diagram). The mouse was rewarded for the amount of time it spent in the illuminated area *when the light was on.* When the light was *off* the optimal behavior for the mouse was to go to the switch and switch it on; when it was on the optimal behavior was to go to the illuminated area and stay there. Floreano successfully evolved mice with the ability to rapidly change their decision rules as the context rewarded different behaviors. The result is the behavior illustrated in the diagram. For these mice the mapping between input and output was a dynamic one. Decision-making was *adaptive.*

Bearing in mind the need to use simple, multiple, highly adaptive rules of decision-making we can now try and model an ideal student selector. Assuming that the selector's goal is to produce a functional, productive academic community it is easy to see that the application of a static set of decision rules is likely to prove counter-productive. If all students are chosen using the same criteria the result is likely to be a "mono-culture" of very similar students all of whom will be very intelligent, or very stable or very sociable, according to the actual decision rules chosen. A rational decision-making procedure should, on the other hand, display adaptivity. If, for example, the current rules are producing an over-concentration of very intelligent, unstable, unsociable students the selection criteria should shift so as to increase the input of stable individuals with better social skills. The skilled decision-maker will, in practice, apply many different rules adapted to particular

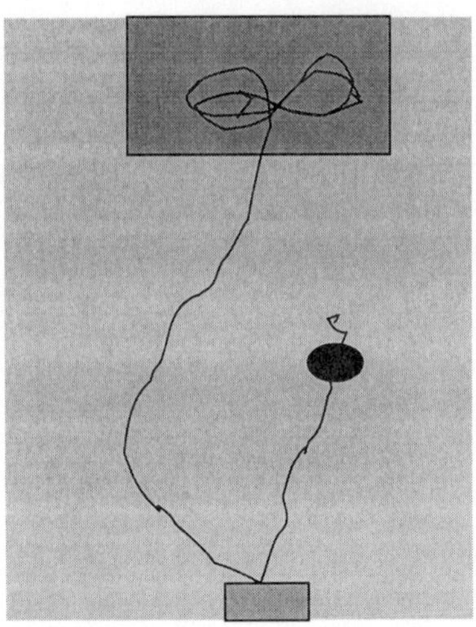

Figure 5. Context-sensitive decision-making with an artificial mouse.

circumstances. In a purely static laboratory setting the application of such rules may not meet abstract criteria for perfect rationality. A decision-maker, taking her decisions in this way, is likely however to be far more adaptable and to produce far better long-term results than a colleague, applying a single mechanical scoring system.

CONCLUSIONS: THEORIES OF DECISION-MAKING AND PERCEPTION

On the basis of what has been said so far it can be argued that the application of traditional criteria for rationality is likely to produce ecologically sub-optimal decision-making. If this is so adaptivity is a fundamental aspect of "rational" decision-making and the "imperfect rationality" observed in the laboratory is a symptom not of *insufficient intelligence* or *imperfect information* but of an artificially static test environment where adaptivity is of unusually low value.

A dynamic theory of rationality and of decision-making has important implications for theories of perception. The goal of "traditional theories", such as those proposed by Marr,[13] is to build systems with the capability to construct a complete, objective representation of external reality. In this framework perception is seen as a computation generating a high level "symbolic" representation of reality from low level sensorial input.

In an ecologically grounded theory of perception on the other hand the goal of perception is to support biologically rational behavior and the generation of rational behavior is seen as an on-going process involving continual adaptation to context via the use of multiple, redundant, very simple computational rules. The kind of theories produced by researchers such as Marr, or traditional theorists in the field of Artificial Neural Networks assume that perception is essentially a static computation. Translated into neuro-biological terms this means that information is transmitted from neuron to neuron by neuro-transmitter release across synapses. Once a network had been "trained" synapses were taken to be static. Perceptive sub-systems of the brain act as a computers calculating output from the motor system as a function of sensorial input. More recent biological research shows however that this view is a serious distortion of reality. Even extremely primitive biological animals display a surprising degree of synaptic plasticity. During the life of the organism the strengths of synapses change continuously on a number of different time scales; the pattern of transmission of information between neurons is subject to continual modification via the action of a large family of neuro-modulators which inhibit or excite the action of particular neural circuits.[14] The biological evidence supports the concept of perception as a dynamic system serving ecological goals. The challenge today is to translate this point of view into concrete theories and models which can provide inspiration both for the biologist and the engineer.

REFERENCES

1. E. Shamir and A. Tversky, Decision Making, in E. Smith and D.N. Osherson, *An Invitation to Cognitive Science: Thinking*, Volume 3, MIT Press, 1995.
2. A. Tversky and D. Kahneman, The framing of decisions and the psychology of choice, *Science,* 211, 453-458, 1981.
3. A. Tversky, Intransitivity of Preferences, *Psychological Review,* 76: 31-48, 1969.
4. D. Kahneman and A: Tversky, Prospect Theory: an analysis of decision under risk, *Econometrica*, 47: 263-291, 1979.
5. D. Rumelhart, J.L. McLelland et al., Parallel Distributed Processing: explorations in the microstructure of cognition, MIT Press, 1987.
6. A. Clark, *Being there. putting brain, body and the world Together Again*, MIT Press, 1997.
7. D. Parisi, *Mente: i nuovi modelli della vita artificiale*, Il Mulino, 1999.
8. S. Nolfi, Adaptation as a more powerful tool than decomposition and integration in T. Fogarty and G. Venturini eds., *Proceedings of the Workshop on Evolutionary Machinery and Machine Learning, 13th*

International Conference on Machine Learning, University of Bari, Italy, (available at http://kant.irmkant.rm.cnr.it/nolfipub.htm) 1996.

9. J.H. Holland, Adaptation in Natural and Artificial Systems, MIT Press, 1975.
10. M. Mitchell, An Introduction to Genetic Algorithms, MIT Press, 1999.
11. D. Floreano and J. Urzelai, Evolution of Neural Controllers with Adaptive Synapeses and Compact Genetic Encoding in Lecture Notes in Artificial Intelligence 1674, *Advances in Artificial Life,* Springer Verlag, 1999.
12. R. Walker and O. Miglino, Replicating Experiments in "Detour Behavior" with Artificially Evolved Robots: an A-Life Approach to Comparative Psychology, 1999.
13. D. Marr, *Vision*, W.H. Freeman and Company, 1982.
14. P.S. Katz, *Beyond neurotransmission*, Oxford University Press, 1999.

CASE STUDIES IN COGNITIVE ROBOTICS

Luigia Carlucci Aiello, Daniele Nardi, and Fiora Pirri

Dip. di Informatica e Sistemistica
Università di Roma "La Sapienza", Italy
e-mail: {aiello|nardi|pirri}@dis.uniroma1.it

INTRODUCTION

Cognitive Robotics is concerned with the theory and implementation of embodied intelligent agents. Cognitive Robotics poses a unique challenge to researchers: it deals in fact with manufacts that are inherently autonomous, situated in a not structured environment, and whose behavior is characterized by a cycle "Perception-Reasoning-Action" or "Sense-Plan-Act" (sometimes called SPA architectures). Of course this cycle has to be performed reliably, and in real time. Sensing, planning and acting are difficult problems to be solved, per se. Their integration within a single system poses additional challenges.

Sometimes researchers call systems endowed with the above capabilities "mobile robots" or "mobots" (see for instance Kortenkamp[1]), or even "autonomous robots", thus emphasizing that, in order for a robot to be able to move in the real world, it is essential that it is capable of performing the SPA cycle in an autonomous way, i.e. without "being told" what to do by an external (human) operator.

We do not stress mobility here. We stress instead that, in order for a robotic system to make autonomous decisions about what to do in a real world environment, it must be endowed with cognition. Cognition about the world and the general rules that regulate it. Cognition about its own position and its abilities to perform both reasoning and acting. In fact, the robot has to draw conclusions, often instantaneous ones, about how to act to achieve goals set by an external agent (e.g. a human operator) or by itself (for instance, as an intermediate step in achieving another goal).

Note that the environment in which the robotic system is supposed to act is not a structured artificial setting, but the real world. Hence, the environment is dynamic (if for no other reason, because of the effects of the robot's actions). Unfortunately, the evolution of the world is in general not predictable, both because the effects of robot's actions are (sometimes) not deterministic, and because there are events that happen independently of the robot's control (i.e., exogenous actions). In addition, the robot's sensors can, in general, perceive only part of the world. For these reasons we cannot make any kind of close world assumption in the description that the robot possesses of the world.

Human and Machine Perception 3: Thinking, Deciding, and Acting, edited by Cantoni *et al.*
Kluwer Academic/Plenum Publishers, New York, 2001.

A cognitive robot is called to interact with its environments intelligently: this means that it must be capable of autonomous reasoning, i.e. autonomously synthesizing a plan to achieve its goal. A plan in this context is a course of actions that brings a robot from the situation where it currently is to a goal situation. Of course, achieving the goal is not in general sufficient, we also want to have precise specifications of the robot's behavior and some sort of "certification" with respect to some desired properties. For example, the robot does not destroy other fellow robots on its way, it does not fool around, etc.

The SPA paradigm is not at all new. It is the paradigm of artificial intelligence. At the same time, it is the paradigm of robotics. In fact, each robotic system operates under the control of some software that loops around a sense-plan-act cycle.

The First Cognitive Robot

The first cognitive robot is SHAKEY,[2] developed at SRI International in the sixties. SHAKEY was in fact endowed with autonomous planning capabilities with the STRIPS system (a LISP based formalism to represent actions, their preconditions and effects). Plans were executed by SHAKEY, and it had also some capabilities to recover from failure, and re-plan.

Even though the work done around SHAKEY has been seminal and very influential on the following developments of both artificial intelligence and robotics (it is sufficient to mention the A* algorithm),[3] in some way it has also had some detrimental consequences. The goals of its designers were in fact very ambitious, while both the theory and the technologies of the sixties were inadequate. The success of SHAKEY was largely tied to the fact that it was confined to act in a small and structured domain, with no possibility of scaling up to real size applications.

As a consequence, after SHAKEY, artificial intelligence and robotics split. Still based on the same perception-reasoning-action paradigm, they put emphasis on different aspects of the cycle. Artificial intelligence concentrated on knowledge representation and reasoning; robotics concentrated on perception and action. To bring the split to an extreme, we may say that artificial intelligence has forgotten embodiment of the intelligent artifact, while robotics has forgotten the representation of the world.

The split has been made sharper after several (negative) complexity results of the planning problem. Brooks brings this position to an extreme in Brooks[4] and later in two very influential papers.[5,6] Their title is significant of the disillusion: "Intelligence without Representation" and "Intelligence without Reason". Brooks claims that intelligent behaviors can be achieved by reactive architectures based on a sense-act cycle, patterned after animal behaviors.

The Cognitive Robotics Manifesto

Brooks's proposal created much discomfort among the researchers in the knowledge representation and reasoning community, who had been motivating their research with the construction of the "ultimate" intelligent robot. Their claim that there is no intelligence without cognition, together with the poor performance of purely reactive architectures at the robotic competitions organized by the AAAI (American Association for Artificial Intelligence) since the early nineties, brought Cognitive Robotics into being.

The name "Cognitive Robotics" was introduced in the Fall of 1998, at a AAAI Symposium, together with the "Cognitive Robotics Manifesto".

The manifesto can be summarized as:

- interesting agent behaviors require reasoning;
- reasoning with and about explicitly represented knowledge leads to decisions about how to act.

The second item is also known as the Knowledge Representation Hypothesis: it brings the attention back to the empowerment that an explicit representation of knowledge gives a robot.

It is evident that the manifesto is a way to draw a sharp boundary between the Cognitive Robotics research and the "Reasoning without Representations" à la Brooks, but - very important - it awaked the "Pure Knowledge Representation and Reasoning" community reminding that reasoning is only an aspect of the SPA cycle, and after all, not even the most critical one. The manifesto has reminded that robotics is "Theory and *implementation* of *embodied* intelligent agents", where the words "implementation" and "embodied" play as crucial a role as "theory" does.

Many research groups started cognitive robotics projects, most notably the Toronto Group,[7-15] Lakemeyer,[16,17] Sandewall,[18] and Shanahan.[19] The languages used for representing actions and change go from the Situation Calculus to the Event Calculus, to Feature and Fluents, and Dynamic Logic. These formalisms are strongly connected. Research also started at our Department (Dipartimento di Informatica e Sistemistica, Università di Roma "La Sapienza"), building on a strong tradition of research in knowledge representation and reasoning.[20-24] Two of the projects will be reported in the sequel of this paper, based on Situation Calculus and Dynamic Logic, respectively.

The basic idea underlying all the cognitive robotics efforts is to have a formal language to represent the knowledge about the world and a high level description of actions and how they affect the world. Then, more fine-grain defined actions are associated with complex low level behaviors that affect the robot actuators. Sensing is dealt with as affecting the representation of the world with the appropriate updates.

The various systems differ in the knowledge representation language, in the action theory and in the way of synthesizing plans, but all of them comply with the above paradigm.

Research Challenges

The research challenges that cognitive robotic poses are many.

First, a robust theory of action has to be developed, capable to deal with changes, in particular with continuous events and concurrent processes. It has to be able to update its internal representation of the current state of affairs also as an effect of sensing actions. It has to be able to revise its beliefs in the presence of conflicting observations.

Second, the robot has to be endowed with the ability to perform actions, to reason about what to do and when to act. In addition, the robot behavior is resource bounded.

Third, the world is uncertain. The robot has to cope with nondeterministic and probabilistic events and action effects, which means that it has to be endowed with execution monitoring capabilities and ability to recover from failure.

Fourth, the reasoning component has to be embedded in actual mobile robots that can suitably combine numeric and symbolic representations, as well as reactivity and deliberation.

The robot's reasoning capabilities should include the possibility to reason about goals, preconditions and effects of actions, when to perceive, what to look for. In addition, it has to reason about changes in the environment and the cognitive state of other (human and robotic) agents in the scenario.

Logic plays a major role, in particular the logics for commonsense reasoning and for reasoning about actions and changes that have been developed in the last couple of decades. Nevertheless, the solution to these challenging problems can only come from the integration of logics with vision and control theory, together with concurrency theory, probability theory, utility theory, among others.

Building cognitive robots is an effort being carried on in several institutions, now, and for a variety of applications that go from space exploration to underwater exploration and repairs, to museum guides, to soccer players.

We do not go here into many details about the state of the art in all the aspects of the construction of cognitive robots. We only illustrate what a planning problem is and why logic plays a major role in solving it. Then, among the various researches being presently carried on at our Department, we illustrate two. They can be considered two case studies in cognitive robotics, or in "logic at work".

LOGIC AND PLANNING

The field of planning is a research area in artificial intelligence aiming at the construction of algorithms that enable an agent (a robot or a "softboat") to synthesize a course of actions that will achieve its goals. We cannot illustrate the vast literature in the field; the interested reader may see the recent survey by Weld[25] for a comprehensive presentation of the state of the art. We limit ourselves to present the definition of a planning problem, to comment on the main issues arising with its solution, and to give enough background to understand the challenges it poses in the realization of cognitive robots.

A planning problem is a representation, in some formal language, of the following three aspects:

1. a description of the initial state of the world;
2. a description of the goal state the agent has to reach;
3. a description of the possible actions that can be performed by the agent. This is often called a *domain theory*.

The solution of the problem (if one exists) is a *plan*, i.e. a sequence of actions that, when executed in any world satisfying the initial state description, will achieve the goal.

In addition to the description of a planning problem, a planner has to be provided with a *background theory*, i.e. a description of generally known properties about the world.

In classical planning problems a number of simplifying assumptions are made, some of them are absolutely not trivial to be removed: atomic time (or, equivalently, instantaneous duration of action execution), no exogenous events, deterministic action effects, omniscience on the side of the agent. As stated above, all of them have to be removed in order to come to cognitive robots capable to act in real world situations.

A fundamental problem in planning is finding the appropriate and adequate representation language: in fact we look for a concise representation of dynamic situations that lends itself to an easy search for plans.

Specialized planning formalisms for representing and solving planning problems have been proposed. Logic has been proposed as one of such formalisms; though, classical first order logic is not a good candidate for such a language, as it is not expressive enough to easily represent change.

MacCarthy and Hayes[26] in 1969 proposed the *Situation Calculus* as a formalism for planning, and Cordell Green[27,28] has first proposed the deductive approach to automatically synthesize plans. The Situation Calculus, as recently revisited by Reiter,[7,8,15] will be illustrated here to clarify the issues raising in stating and solving planning problems.

The Situation Calculus

The Situation Calculus uses a sorted first order language, where predicates (functions) may be decorated with an extra parameter s, of sort *Situation*, to indicate that the truth value of the predicate (the value of the function) may change over situations. Predicates with a situation argument - usually the last one - are called *fluents*, (or *relational fluents*, to distinguish them from *functional fluents*). For example, $Has(money, s_1)$ and $\neg Has(money, s_2)$ indicate that the agent possesses money in situation s_1 and does not in situation s_2. As an example of functional fluent, in a domain where objects are colored and may be painted, consider *color*(x, s_1), which denotes the color of x in situation s_1, and may be different from *color*(x, s_2).

We distinguish the initial situation by the constant symbol S_0. Of course, the following holds for the initial situation:

$$\forall s \neg (s < S_0) \qquad [AxS_0] \tag{1}$$

The description of the initial situation is a set of sentences stating what holds in S_0. For example:

$$At(home, S_0) \wedge \neg Has(milk, S_0) \wedge Has(money, S_0) \tag{2}$$

could be the description of the initial situation of a planning problem for a robot that's supposed to go out buy milk.

Actions can be performed in situations and their result is a new situation, i.e.:

$$s' = do(a,s) \tag{3}$$

where *do* is a binary function symbol which, given an action a and a situation s, evaluates to its successor situation s'. Actions are denoted by function symbols, for example, if $put(x,y)$ is the action of putting object x on top of object y, $do(put(A,B), S_0)$ denotes the situation resulting from putting A onto B in the initial situation S_0.

In general, a situation is a first order term of the following form:

$$do(a_n, do(a_{n-1},\ldots, do(a_1, S_0)\ldots)) \tag{4}$$

Hence, situations can be seen as *histories*, or *traces*, of action executions.

Reiter, in addition to $[AxS_0]$, introduces the following axioms for situations, called *foundational axioms*.

$$\begin{array}{ll} do(a_1,s_1)=do(a_2,s_2) \equiv a_1=a_2 \wedge s_1=s_2 & [Dis] \\ s < do(a,s_1) \equiv s \leq s_1 & [Lin] \\ P(S_0) \wedge (\forall\, as)\,(P(s) \rightarrow P(do(a,s))) \rightarrow (\forall s)P(s) & [Ind] \end{array} \tag{5}$$

[*Dis*] states that distinct traces correspond to distinct situations; [*Lin*] is a linearity axiom for histories; [*Ind*] is an induction axiom (that makes the calculus more expressive than first order predicate logic).

Actions modify the current status of the world. This is described by stating the "causal laws", i.e. how actions affect the values of the fluents, in the form of the so called *effect axioms*. For example, the effect on a relational fluent *Broken* of a robot r dropping a fragile object x in a situation s is described by the causal laws:

$$Fragile(x,s) \rightarrow Broken(x,do(drop(r,x),s)) \tag{6}$$

the effect of a robot repairing an object is described as:

$$\neg\, Broken(x,do(repair(r,x),s)) \tag{7}$$

or, the effect of a robot painting an object with color c is described as:

$$color(x,do(paint(r,x,c),s)=c \tag{8}$$

Actions may have *preconditions*, i.e. requirements to be satisfied in order for the action to be executable. The predicate symbol *Poss* is used to indicate that all the preconditions are satisfied, hence an action is possible. *Poss(a,s)* reads as: action *a* is possible in situation *s*. For example:

$$Poss(repair(x),s) \equiv HasGlue(s) \wedge Broken(x,s) \tag{9}$$

meaning that the (only) preconditions for a repair action, by a robot on an object *x* in a situation *s*, are that the robot has glue and *x* is broken.

In the Situation Calculus, an axiom - called *action precondition axiom* - is associated with each action $A(\vec{x})$. It has the form:

$$Poss(A(\vec{x}),s) \equiv \Pi_A(\vec{x},s) \tag{10}$$

where $\Pi_A(\vec{x},s)$ is a first order formula with free variables $\vec{x}$ and *s*, and no occurrence of the function symbol *do*. Note here that we make a sort of "closure" assumption, i.e., all the necessary and sufficient conditions for the execution of action $A(\vec{x})$ are listed in $\Pi_A(\vec{x},s)$.*

In addition to the specification of the action precondition axioms, we have to state the "laws of inertia" for our domain, i.e., actions change the value of some fluents, all the others are unaffected, so their truth value persists through the action execution.** As an example, dropping an object onto the floor does not change its color, the position of other objects on a table, their weight, color, etc.

Reiter's proposal merges causal laws and laws of inertia into the *successor state axioms* for both relational and functional fluents. The intuition is that a successor state axiom states all the necessary and sufficient conditions for the value of a fluent to change as the result of performing an action.

A successor state axiom for a relational fluent *F* has the form:

$$F(\vec{x},do(a,s)) \equiv \Psi_F(\vec{x},a,s) \tag{11}$$

where $\Psi_F(\vec{x},a,s)$ is a first order formula with free variables $\vec{x}$, *a*, *s*, and Ψ_F specifies all the conditions for *F* to be true or false in *do(a,s)*. As an example, consider the following successor state axiom for the fluent *Holding*:

$$\begin{aligned} Holding(r,x,do(a,s)) \equiv\ & a = pickup(x,r) \vee \\ & (Holding(r,x,s) \wedge \neg (a = drop(r,x))) \end{aligned} \tag{12}$$

Successor state axioms for functional fluents can be introduced in an analogous way; we do not illustrate them here, the interested reader may see Reiter.[15]

In summary, a domain theory in the Situation Calculus consists of a sorted first order language with a unique name assumption:***

- a set of actions;
- a precondition axiom for each action;
- a successor state axiom for each fluent.

With this domain description, it is clear that if a property holds in a generic situation *s* then, by looking at the trace of actions that leads from S_0 to *s*, either at some intermediate

---

* In other words, we ignore all the other qualifications for the execution of $A(\vec{x})$, by considering them irrelevant. In the artificial intelligence literature, this is known as the *Qualification Problem*.

** This is known in the artificial intelligence literature as the *Frame Problem*.

*** The unique name assumption says that distinct symbols denote distinct objects.

Figure 1. The robot ArmHand0

situation an action has made the property true and no successive action has made it false, or it was already true in S_0 and no action in the trace ever made it false.

Given a situation

$$s_{fin} = do(a_n, do(a_{n-1}, \ldots, do(a_1, S_0)\ldots)) \quad (13)$$

we call the sequence of actions $a_1, \ldots, a_n$ the *plan* to reach s_{fin}.

In the Situation Calculus, given the foundational axioms, a domain theory mentioning also a description of the initial situation (let us call the set of all these formulas *D*), solving a planning problem to reach a situation that satisfies a goal statement *Goal*, means proving that:

$$D \vdash (\exists\, s)Goal(s) \quad (14)$$

Any binding for *s* obtained as a side effect of a proof of the above is a plan guaranteed to lead to a world situation satisfying the goal *G*.

ARMHAND0

In this section we present the manipulator ArmHand0. The robot ArmHand0 is composed by:

1. a 4 d.o.f. (degree of freedom) mobile manipulator;
2. an end-effector with two fingers;
3. a control environment where it executes its tasks.

ArmHand0 is a toy - see below for a description of ArmHand0's anatomy - nevertheless it is an autonomous agent. Our main concern in the design of ArmHand0 has been the analysis and synthesis of an efficient planning system for a robotic manipulator with a complex geometry.

We have defined a cognitive framework (see Finzi et al.,[22] Pirri et al.,[24] Pirri et al.,[23] Piazzolla et al.[29]) for the robotic manipulator ArmHand0 ensuring an autonomous and efficient behavior for the execution of complex tasks in a micro world, the well known blocks world. The blocks world has been spoiled by an overuse in the artificial intelligence community, still it exemplifies many of the real problems a manipulator has to solve when executing one or more tasks in order to achieve a goal.

The main contributions of ArmHand0 are the following:

- a basic theory of actions and perception, in the Situation Calculus, in which the representation of the domain and the specification of the different kind of actions that ArmHand0 can perform are defined;
- a task planner in GOLOG (alGOl in LOGic)[11] that can manage complex actions;
- a global planner for calculating safe paths around the towers of blocks that it builds;
- a control system that can reason about failures of actions.

The above framework, which has been initiated with the simulation of the 7 d.o.f. SPIDER arm, in the domain JERICO (see Finzi et al.[30]), consists of a planning module with several levels of abstraction. The real scientific advancement that ArmHand0 brought into the architecture relies in the implementation of an embedded system.

We distinguish among different levels of control. The high-level planner (Task Planner) is based on a symbolic model of the domain, defined within the Situation Calculus. The low-level planner(Local Planner) makes use of a geometric model of the world based on a metric description of the domain. The geometric model is connected to the symbolic model via a one-to-one correspondence between the qualitative representation, in which objects have properties and denotations, and the metrical representation of shapes and positions (Global Planner). An advantage of this three-layered architecture (first envisaged in Cameron[31]) is that goal-planning is delegated only to the upper level that can manage, in this way, a relatively simple representation of the world.

An Architecture for Planning and Control

The Task Layer. The role of the Task layer is to endow the robotic agent with the ability to reason about its domain. This ability is used to coordinate a sequence of actions, emulating an intelligent behavior that allows the robot to achieve the required goals on the basis of its beliefs and knowledge of the world. The Task layer is the highest level of abstraction and is used to compute, through the task planner, a course of tasks. At the task level, the dynamic domain is described purely in qualitative terms by means of a symbolic description of the domain in the Situation Calculus, disregarding any metric description concerning the shape or the position of the real world's objects. The task level is concerned only with the causal laws of the domain and with the amount of reasoning needed to fulfill them.

From a formal point of view, we assume that when a plan is computed by the task planner, all the control action preconditions compiled at the high level are necessary and sufficient to the solution of the corresponding physical problems, such as the evaluation of the forces that act on the end-effector or the existence of a free path. This assumption is possible because of the structure of actions, which at the task level are complex actions that will be expanded at the lower levels. In fact, while the task is searching for a complex action to execute next - in order to achieve the goal -- the other planning levels control whether the sequence of actions obtained by expanding it is executable or not. A situation is executable when each action mentioned in the sequence is possible, that is, for a given situation $s=do(a_n, \ldots, a_0, S_0)$ it is verified:

$$Poss(a_0, S_0) \wedge \ldots \wedge Poss(a_n, do(a_{n-1}, \ldots, a_0, S_0)) \qquad (15)$$

The Task Planner has been implemented as a concurrent process. It interacts with the lower module (Global Planner) in order to evaluate all the operative conditions needed for the physical execution of the actions.

The Global Layer. The role of the Global layer is to reason about distances, volumes of objects and free paths. The global planner, in fact, computes a free manipulation path for the end-effector, that has to satisfy all the post-conditions associated with a generic sequence of Task actions.

In our hierarchy, the Global level operates concurrently with the processes activated by the Task level. This is achieved simply by a spatial transposition of the abstract concepts defined by the latter. At this level of abstraction, information is managed globally, and the end-effector is supposed to be like a point free to fluctuate in the 3D-space. Movement actions, obtained by a suitable expansion of the task actions, are sequences of manipulation actions needed to reach all the key positions without worrying about the evolution of each robot joint.

The manipulation sequence is formed by a sequence of Global actions such as

$$getFreePos(),\ getPos(cubeID),\ getPosTarget(cubeID),\ moveToPos(posXYZ) \qquad (16)$$

The above first three kinds of actions have exclusively the job of finding spatial information. The two latter ones have the job of producing all the necessary data for the physical realization of the movement of the robotic arm (which will take place at the Local level). The Global planner operates in two distinguished phases. The first, as stated above, operates in parallel with the Task Planner to produce a plan in which high-level actions are translated into the corresponding Global actions, as soon as they are generated. The second one has the main task of processing all the geometric information, defined at the global level, which is related to Motion Planning.

In this second phase, it is necessary to compute an admissible path consenting to position the end-effector at the required points of the workspace, avoiding collision with the objects of the world. To attain such goals, we have realized two specific algorithms: a heuristic search (based on graphs) and a collision detection/avoidance. At the end of such processes, we get a sequence of actions containing spatial information. This sequence, after a "smoothing" phase of the representing trajectory, is then ready to be routed to the Local Planner which will translate it into basic commands acting directly on the robot joints.

Local Planning Layer. The Local Planner is the lowest level of the hierarchical planning architecture and is devoted to the resolution of the "physical aspects" related to Motion Planning. Its main task consists of finding a path that allows the manipulator to operate safely during the execution of the high level tasks. The intermediate sequences of actions computed by the Global Planner are now expanded into basic commands, such as

$$moveJoint(jointID,\ amount),\ openHand(),\ closeHand() \qquad (17)$$

operating directly on the motors controlling the movement of each joint. We have implemented a particular kind of Local Planner, called "One-Shot", to give an analytical solution to both the direct and inverse kinematics. It can compute a sub-goal configuration starting from simple information about the current situation and the sub-goal. For the description of the kinematics of the manipulator see Piazzolla et al.[29]

Anatomy of the Robot ArmHand0. The construction of the arm has been possible thanks to the technology offered by the LEGO MindStorms Robotic Invention System (RIS). The core of the MindStorms system is the RCX (Robotic Command eXplorer), a programmable brick equipped with an LCD, and based on a micro controller that can manage three little motors, three sensors and a communication serial interface. The basic system is the RCX itself, an infrared transmitter, and a computer. The control system of the robotic manipulator ArmHand0 consists of several software modules, realized with the Microsoft Visual C++ 6.0, with an embedded part of source code written with Eclipse Prolog and GOLOG. To build the robot ArmHand0 we used two packages of the Robotic Invention System. The reason is that each RCX module can control only three motors while to realize the manipulator we needed five of them (for the moving base, the shoulder, the arm, the wrist and the end-effector).

Recalling that ArmHand0 is a toy, it is impossible to exactly control the forces and the torques applied to the joints, because the system is not provided with high precision components such as step-by-step motors or sophisticated analogical sensors. Therefore, we cannot appreciate small rotations or the pressure made by the arm during the taking process; this has made it impossible to implement a module to correct those errors that inevitably occur and propagate along the kinematic chain. In this situation, the amount of failures during the execution of an entire plan is proportional to the length of the plan itself. This can be a limiting factor when evaluating the correctness of the simulation because all these problems can confer an uncertain behavior to the system.

It has also to be added that the small number of d.o.f. of the anthropomorphic structure does not allow an effective emulation of a human arm (at least 6 d.o.f. are needed). The operative capability of the robotic arm resides in the mobility of the joints of its subsystems: base, shoulder, arm, and wrist. Basing on the angular values that the joints can reach and the length of the links of the robot ArmHand0, it is possible to obtain, geometrically, the associated workspace, defined as the set of points of the three-dimensional space that the end-effector can reach while the joints change their configuration.

Taking Care of Failures. The execution of a task in the blocks world requires a certain accuracy which cannot be always achieved by ArmHand0 because, as we argued above, failures are possible. In this sense, although all actions are intrinsically deterministic, we have developed a new architecture which allows the robot to reason about failures.

To reason about failures we formalize the possibility that each action can either fail or succeed. A probability of success is given for each action. Therefore, a stochastic process is associated with each task. The stochastic process generates a measure of belief for each property of the domain according to the probability of success of each action. Successor state axioms for the measure of belief are provided for each fluent, likewise conditional probabilities manage the action preconditions. A belief state, with respect to a stochastic situation, is thus defined. To confirm or reject a certain belief state we make use of perception. Perception (see Pirri and Finzi[23]) is achieved via complex sensing actions like *ObserveScene*. The complex action *ObserveScene* recovers the values of all perceptibles (a perceptible is a term representing a property of the scene that has been sensed by the agent); these values are then used to form a hypothesis on the real configuration reached at the end of each task. The hypothesis is then compared with the current belief state of the agent. To deal with errors we have redesigned the Task part by decomposing the high level control into several tasks that can non-deterministically be chosen until the goal configuration is reached.

COGNITIVE SOCCER PLAYERS

In this section we present the approach to cognitive robotics first developed on the mobile robot "Tino",[20] and then applied in the design of robotic soccer players for RoboCup.[32] We start by providing some information on RoboCup, then sketch the player's architecture and, in particular, the representation of actions adopted in our cognitive mobile robots, and finally present the Italian national team of robotic soccer players, Azzurra Robot Team, whose players can successfully coordinate their behavior, based on a fully distributed approach.

RoboCup

RoboCup[33] has been launched a few years ago[34] as a framework for testing research ideas in the design of cooperative multi-agent systems.

Since then, RoboCup has been very successful in organizing 3 World Championships,[35-38] and is nowadays regarded as a *landmark* project[39] which identifies "playing soccer" as a standard problem to be faced by research in the field of intelligent robotics and has the rather challenging long-term goal of building robots that play soccer as humans. RoboCup thus provides an interesting scientific framework for addressing several problems in artificial intelligence and robotics.

RoboCup organizes several kinds of competitions: computer simulated soccer and soccer played by mobile robots, in which there are further differentiations based on the size and on the equipment of the robots.

The *Middle-Size* league is played within a 5x9 meters field and the body of the robot must be within a cylinder of 50 cm diameter and 80 cm height. The teams are composed by 1 goal keeper and 3 middlefield players and are distinguished by cyan and magenta markers. The ball is a size 4 orange (size 4 is the standard for indoor soccer). All sensing devices must be onboard the robots, in particular global vision as well as other external sensing devices are not available.

The distinguishing feature of the Middle-size league is the autonomy of the players, which is imposed by the constraint that all the information on the playing environment must be acquired through sensors that are positioned on board of the robot. Therefore, a Middle-size team can be regarded as a multiagent robotic system, each of whose robots can process the data coming from sensors and select the most appropriate playing actions, coordinating them with those performed by the teammates.

Architecture and Reasoning Component

The development of cognitive mobile robots requires not only a suitable representation of the knowledge of the robot, but also reactive capabilities. The ability to combine both the above features is addressed through the design of software architectures.

In our robots we adopt a heterogeneous layered hybrid architecture,[40] with two levels: the *deliberative* (cognitive) level, in which a high-level state of the agent is maintained and decisions on which actions are to be performed are taken, and the *operative* (reactive) level, in which low-level conditions on the world are verified and actions are actually executed. The central issue in devising hybrid architectures is to provide an effective integration of reasoning and reactivity.

The deliberative level is constituted by a knowledge base containing a description of the dynamic system, together with information about the state of both the robot and the environment. The operative level is based on a geometrical representation of the environment and a set of values characterizing the internal data of the robot, and is represented as a data base of objects' properties.

The main features of the architecture we adopt for the implementation of reasoning and reactivity are *heterogeneity* and *asynchronism*. Heterogeneity refers to the use of many different techniques, for example a fuzzy reactive controller, together with a knowledge representation system and low-level image processing routines. Asynchronism is essential for an effective integration of reasoning and reactivity. Indeed, asynchronous architectures, allow time-critical modules to have processing time available when needed. In our case this has been obtained by forcing the system to give control to low-level modules every 100 ms.

The different levels of the representation are characterized by the degree of abstraction of data representation. Thus, Knowledge Acquisition and Task Execution are performed over the two levels. Two kinds of data run across these levels: the first one, representing *perceptual data*, goes from the sensing devices to the deliberative level through a high-level interpretation process, and the second one, representing *action data*, goes from the deliberative level to the actuators. In addition, conclusions drawn at the deliberative level may require the update of the world model of the operative level.

Actions at the operative level are expressed as control programs in the Colbert language, that can be directly executed by the Saphira system,[41] which controls the mobile base. Such programs are generated by the planner (possibly off-line) and turned by the Colbert translator into sequences of actions, or, more generally, into execution structures (including behaviors and low level control actions), that are handled by the Colbert interpreter.

The basic feature of this architecture is thus the presence of different representations (at different degree of abstraction) of the same information on the environments. We consider the relationships between the two world models to be a critical element for the development of cognitive agents, since updating a logic-based knowledge base can require high computational time and affect reactive abilities.

In addition, we developed a formal framework, both for representing the actions of a robot and for reasoning, to derive an executable plan to achieve a given goal, which has been used in the cognitive level of our mobile robots. Our approach [20] originates from Propositional Dynamic Logics and exploits their formal correspondence with Description Logics. Its main features are the ability to represent the knowledge of the robot and, consequently, allowing for the generation of executable plans. In Iocchi et al.[21] an extension of such a framework is presented, including both concurrency on primitive actions and autoepistemic operators for explicitly representing the robot's epistemic state. The resulting formal setting allows for the representation of actions with context-dependent effects, sensing actions, and concurrent actions, and properly addresses both the presence of exogenous events and the characterization of the notion of executable plan in such a complex setting.

The proposed framework has been implemented in a system which is capable of generating plans that are actually executed on mobile robots operating in office-like environments. The system has also been used to describe the knowledge of robotic soccer players. In the implementation, the output of the planner is used to generate Colbert activities (control programs), that can be directly executed on the mobile base.

Specifically, we have formalized at the logical level several situations arising in the RoboCup scenario and generated, through the planner, a significant portion of the control programs that were executed on our soccer players.

The introduction of a system that generates plans with concurrent actions requires the robotic architecture to be able to schedule and manage concurrent behaviors and to provide synchronization among such behaviors. Concurrent plans are actually executed on a single player by making use of the Saphira built-in mechanisms for activating concurrent behaviors and for monitoring their completion before starting the next action in the plan. The execution of global concurrent plans (that concern more than one robot) is instead realized by means of explicit communication among players. Observe that in this case all the players share the same plan, and each player is able to identify the actions that must be executed. For example, the execution of the action $A_1||B_2$ is obtained by performing A on $P1$ and B on $P2$ and by a broadcast notification when actions terminate. In this way, all the robots involved in the global plan can detect when it is possible to start the next action in the plan.

ART: Azzurra Robot Team

ART is the national Italian team for RoboCup, the Middle-size league, that participated in the RoboCup championships held in July 1998[42] and August 1999.[43] ART qualified for the quarter finals in RoboCup 1998 and obtained the second place in Stockholm (1999). Figure 2 shows the ART team in action.

Figure 2. ART playing in RoboCup 99

ART is formed by players that have been developed by 6 research groups, operating in various Italian universities.[42,43] The basic player of the ART team is built on the mobile base the Pioneer 1, where we added a conventional PC, running LINUX, for onboard computing. We have reached a compromise between weight and power consumption, where the player has enough autonomy to play games. We also have a wireless high bandwidth connection that is used during development to obtain accurate information about the situation onboard, and supports the exchange of information among the players during the game, but it is not used to transfer raw data among the players. Each robot has a vision system with color cameras in different settings. Finally, the players are equipped with a kicking device driven by air pressure, with two actuators differently arranged on the players, that enable different types of kicks (left, right or both). The software has been developed autonomously by each member of the team, using different environments for the development.

ART thus forms a multi-agent, robotic system, which is heterogeneous both from the hardware and from the software viewpoint, and in which the robots coordinate their behavior through a distributed coordination protocol.

The hypotheses underlying the coordination problem we have faced are:

1. *communication-based coordination*: exploit the use of communication among the players to improve team performance;

2. *autonomy in coordination*: the players are capable to perform their task, possibly in a degraded way, even in case of lack of communication;

3. *heterogeneity in the multi-agent system*: the players are heterogeneous both from hardware and software viewpoints.

Besides the above constraints, coordination in ART has been designed to deal both with roles (defender, attacker etc.) and with strategy (defensive, offensive). While the strategic level is currently demanded to an external selection (the human coach).

Consequently, the requirements a robot of the ART team must fulfill to be able to coordinate itself with the teammates are to have the same communication apparatus (net card) and to correctly handle a set of protocols, namely: UDP-IP protocol, broadcast communication protocol, message exchange protocol, and coordination protocol.

The coordination protocol includes two steps: formation selection and role assignment. Formation selection can be either explicitly accomplished or activated in special situations (for example, if the goal keeper is not responding, then a defensive formation is chosen). For the selected formation, roles are dynamically assigned (see Stone and Veloso[44]) to the various team elements during the game, depending on the configuration present on the field. The assignment of the roles is based on the evaluation of a utility function, which provides an indication of the possibilities of the player of successfully accomplishing the task associated with the given role. A critical aspects in dealing with numerical functions for expressing the intentions of a robot is the experimental work needed to find suitable values.[45] We are currently investigating on a more "cognitive approach", which is based on reasoning about the robots' intentions.

From a technical viewpoint, the proposed protocol is based on the explicit exchange of data about the status of the environment, as reconstructed by each player, and is based on simple forms of negotiations. Simplicity in the protocol stems from the need to make rather weak assumptions on each robot's capability. An increase of such capabilities would lead to more complex protocols. However, we believe that a major issue in coordination is to find a suitable balance between the robot individual capabilities and the form of cooperation realized.

The distributed coordination method has been successfully employed by all the members of the ART team during the 1999 RoboCup competition.[45] The effectiveness of the method has been proved by the fact that we were always ready to substitute any robot with another one, without endangering the coordinated behavior of the overall team.

ACKNOWLEDGMENTS

Our research on Cognitive Robotics has been supported on ASI, CNR, and MURST grants.

The work reported here has been carried on with the invaluable contribution of the students of the School of Engineering of the Università di Roma "La Sapienza". In particular, we are grateful to: Claudio Castelpietra, Giuseppe De Giacomo, Alberto Finzi, Luca Iocchi, Gianfranco Piazzolla, Riccardo Rosati, and Milko Vaccaro.

Finally, we thank the (human) members of the ART team.

REFERENCES

1. D. Kortenkamp, R.P. Bonasso, and R. Murphy eds., *Artificial Intelligence and Mobile Robots: Case Studies of Sucessful Robot Systems*, AAAI Press/The MIT Press (1998).
2. R. Fikes and N. Nilsson, STRIPS: A New Approach to the Application of Theorem Proving to Problem Solving. *Artificial Intelligence Journal*, 2(3-4):189-208, (1971).
3. P. Hart, N. Nilsson, and B. Raphael, A Formal Basis for the Heuristic Determination of Minimun Cost Paths, *IEEE Transactions on Systems Science and Cybernetics*, SSC-4(2):100-107, (1968).
4. R.A. Brooks, A Robust Layered Control System for a Mobile Robot, *IEEE Transactions on Robotics and Automation*, 2(1):14-23, (1986).
5. R.A. Brooks, Intelligence without Representation, *Artificial Intelligence Journal*, 47:139-159, (1991).
6. R.A. Brooks, Intelligence without Reason, *IJCAI 91*, pages 569-595, (1991).
7. R. Reiter, The Frame Problem in the Situation Calculus: A Simple Solution (sometimes) and a Completeness Result for Goal Regression, in V. Lifschitz ed., *Artificial Intelligence and Mathematical Theory of Computation: Papers in Honor of John McCarthy*, pages 359-380, Academic Press, San Diego, CA, (1991).
8. R. Reiter, Proving Properties of States in the Situation Calculus, *Artificial Intelligence Journal*, 64:337-351, (1993).
9. H. Levesque, What is Planning in the Presence of Sensing? *AAAI 96*, (1996).
10. Y. Lesperance, An Approach to the Synthesis of Plans with Perception Acts and Conditionals, Technical report, Department of Computer Science, University of Toronto, (1996).
11. H. Levesque, R. Reiter, Y. Lespérance, F. Lin, and R. Scherl, GOLOG: a Logic Programming Language for Dynamic Domains, *J. of Logic Programming, Special Issue on Actions*, 31(1-3):59-83, (1997).
12. H. Levesque, F. Pirri, R. Reiter, Foundations for the Situation Calculus, *ETAI*, 3(18), (1998).
13. G. De Giacomo and H. Levesque, An Incremental Interpreter for High-level Programs with Sensing, in H.

Levesque and F. Pirri eds., *Logical Foundations for Cognitive Agents*, Springer Verlag, AI series, pages 86-102, (1999).
14. F. Pirri and R. Reiter, Some contributions to the metatheory of the Situation Calculus, *Journal of the ACM*, 46(3): 325-362, (1999).
15. R. Reiter, *Knowledge in Action: Logical Foundations for Describing and Implementing Dynamical Systems*, Unpublished draft, 2001.
16. W. Burgard, A.B. Cremer, D. Fox, D. Hähnel, G. Lakemeyer, D. Schulz, W. Steiner, and S. Thrun, The Interactive Museum Tour-Guide Robot, in *AAAI 98*, (1998).
17. G. Lakemeyer, On Sensing and Off-line Interpreting in GOLOG, in H. Levesque and F. Pirri eds., *Logical Foundations for Cognitive Agents*, Springer Verlag, AI series, pages 173-189, (1999).
18. E. Sandewall, *Features and Fuents*, Oxford University Press, (1994).
19. M. Shanahan, *Solving the Frame Problem: A Mathematical Investigation of the Common Sense Law of Inertia*, MIT Press, (1997).
20. G. De Giacomo, L. Iocchi, D. Nardi, and R. Rosati, A theory and implementation for mobile robots, *Journal of Logic and Computation*, (1999).
21. L. Iocchi, D. Nardi, and R. Rosati, Planning with sensing, concurrency, and exogenous events: logical framework and implementation, in *Proceedings of KR 2000 (Principles of Knowledge Representation and Reasoning)*, Breckenridge, Colorado, USA, Morgan Kaufmann, pages 678-689, April 2000.
22. A. Finzi, F. Pirri and R. Reiter, Open World Planning in the Situation Calculus, in *AAAI 00*, (2000).
23. F. Pirri and A. Finzi, An approach to perception in theory of actions: Part I, *ETAI*, 4, (1999).
24. F. Pirri and R. Reiter, Planning with natural actions in the situation calculus, in J. Minker ed., *Logic-Based Artificial Intelligence*, Kluwer, (2000).
25. D.S. Weld, Recent Advances in AI Planning, *Artificial Intelligence Magazine*, Summer 1999: 93-123.
26. J. McCarthy and P. Hayes, Some Philosophical Problems from the Standpoint of Artificial Intelligence, in B. Meltzer and D. Michie eds., *Machine Intelligence*, volume 4, pages 463-502, Edinburgh University Press, Edinburgh, Scotland, (1969).
27. C. Green, Theorem Proving by Resolution as a Basis for Question Answering Systems, in B. Meltzer and D. Michie eds., *Machine Intelligence*, volume 4, pages 183-205, Edinburgh University Press, Edinburgh, Scotland, (1969).
28. C. Green, Application of Theorem Proving to Problem Solving, *IJCAI 69*, (1969).
29. G. Piazzolla, M. Vaccaro, A. Finzi and F. Pirri, ArmHand0: a mobile manipulator for the blocks world, ECAI Workshop on Cognitive Robotics, Bonn, (2000).
30. A. Finzi, F. Pirri and M. Schaerf, A System Integrating high and low level planning of complex tasks with a 3 dimensional visualizer, *SAIRAS 99*, pages 521-528, (1999).
31. S. Cameron, Dealing with geometric complexity in motion planning, *IEEE Transactions on Robotics & Automation*, April 1996.
32. D. Nardi, Artificial Intelligence in Robocup, in *ECAI 00*, Springer-Verlag, (2000).
33. RoboCup. The Robot World Cup Initiative, Robocup officiale site, http://www.robocup.org.
34. H. Kitano, M. Asada, Y. Kuniyoshi, I. Noda, E. Osawa, and H. Matsubara, Robocup: A challenge problem for AI, *Artificial Intelligence Magazine*, 18:73-85, (1997).
35. I. Noda, S. Suzuki, H. Matsubara, M. Asada, and H. Kitano, Robocup-97 the first robot world cup soccer games and conferences, *Artificial Intelligence Magazine*, 19(3):49-59, (1998).
36. M. Asada, M. Veloso, M. Tambe, I. Noda, H. Kitano, and G. Kraetzschmar, Overview of robocup-98, in M. Asada and H. Kitano eds., *RoboCup-98: The Robot World Cup II*, volume 1604 of Lecture Notes in Computer Science, pages 1-21, Springer-Verlag, (1999).
37. S. Coradeschi, L. Karlsson, P. Stone, T. Balch, G. Kraetzschmar, and M. Asada, Overview of RoboCup-99, *Artificial Intelligence Magazine*, (2000).
38. S. Coradeschi and L. Karlsson, Robocup-99 final report, Technical report, RoboCup, The Robot World Cup Initiative, November 1999.
39. M. Asada, H. Kitano, I. Noda, and M. Veloso, Robocup: Today and tomorrow - What we have have learned, *Artificial Intelligence Journal*, 110:193-214, (1999).
40. L. Iocchi, *Design and Development of Cognitive Robots*, PhD thesis, DIS, Univ. "La Sapienza", Roma, Italy, (1999).
41. K. Konolige, K.L. Myers, E.H. Ruspini, and A. Saffiotti, The Saphira architecture: A design for autonomy, *Journal of Experimental and Theoretical Artificial Intelligence*, 9(1):215-235, (1997).
42. D. Nardi, G. Clemente, and E. Pagello, ART: Azzurra Robot Team, in *RoboCup-98: The Robot World Cup II*, Springer-Verlag, (1999).
43. D. Nardi, G. Adorni, A. Bonarini, A. Chella, G. Clemente, E. Pagello, and M. Piaggio, ART-99: Azzurra Robot Team, in *Proceedings of the RoboCup-99 Workshop*, Springer-Verlag, Stockholm, Sweden, 1999.
44. P. Stone and M. Veloso, Task decomposition, dynamic role assignment, and low-bandwidth communication for real-time strategic teamwork, *Artificial Intelligence Journal*, 110(2):241-273, (1999).
45. C. Castelpietra, L. Iocchi, D. Nardi, M. Piaggio, A. Scalzo, and A. Sgorbissa, Coordination among heterogeneous robotic soccer players, in *Proceedings of IEEE IROS (International Conference on Intelligent Robots and Systems)*, Takamatsu, Japan, IEEE/RJS, November 2000.

PANEL SUMMARY: **RATIONAL AND NOT RATIONAL BEHAVIORS IN DECISION-MAKING**

Panelists: Roberto Bordogna[1] *(chairperson)*, Vito Di Gesù[2], Bertrand Zavidovique[3]
Reactions: Virginio Cantoni

[1]MST
Pavia University, Italy
e-mail: bordogna@vision.unipv.it
[2]Dip. Matematica ed Applicazioni
Palermo University, Italy
e-mail: digesu@math.unipa.it
[3]I.E.F., Paris 11 University
91405 ORSAY, France
e-mail: zavido@ief.u-psud.fr

FOREWORD *(R. Bordogna)*

The aim of this panel is to address the complexity of deciding as it is stated in the workshop agenda which says, as it is known:

Decision Theory provides a methodology for dealing with such circumstances. Moreover, is decision a simple consequence of reasoning on facts? This simple question put several drawbacks concerning the decision level (hierarchical view), its complexity and representation.

To introduce the discussion I have proposed the following preliminary statement.

If we identify actor's personal and common ambit as a contingency in a dynamic natural-artificial/social-cultural continuum, deciding may be described as a continuos field established in an ambit that may originate discrete results (decisions).

In order to explain this statement and to prepare the discussion of concepts such as "rational" and "non rational" decisions making, perhaps it may be useful to shortly summarize some known results providing also a pictorial image of the preliminary statement as an example of a possible syncretic conceptualization. Improved understanding of human learning and decision making is considered here (among other possible considerations) a fundamental social development factor and an engineering / marketing problem for artificial agents (as "intelligent" machines and other human-centered systems).[1]

A PICTORIAL DESCRIPTION OF DECISION MAKING

As it is known the act of writing is commonly considered a "rationalization" process, especially using a western language where the linearity of writing is commonly designed for the process of reading (left to right – top-down), including assumptions about readers' knowledge and culture. All this affects concepts (object) identification and relationship (generally made *ex-ante* as subject-object / attributes relationships). While the act of writing may be generally considered a rationalization process nothing can be said about the actual content, which may be considered "rational" or "not rational", as it is known, accordingly to a specific community of readers. But there is more, a poetry may express something that is not even mentioned in the text, typically because it is difficult to explain with plain words or common concepts – such as a particular "feeling", that can be often understood by reading the text as a whole (with a syncretic operation made by the reader leveraging on his knowledge and experience). In this case we are facing not only a problem of *bounded rationality,*[2] because of lack of information, or a behavior related to a contingent actor agenda (for instance *agency goals* as opposed to *well-being* goals[3]), but a scenario that can be described (among many possible conceptualizations) as a generation of a "meaning" for a reader that may influence his future decisions (for instance to buy or not another book of poetry of that kind for a given price).

This is the reason because in the preliminary statement I have stressed the need to analyze the decision process as a contingency in a continuum made by the actor and a common (natural, artificial, cultural) ambit. In fact using a very simple description (out of many possible taxonomies) of the world we may see that in every circumstance an human actor (with a given experience and a personal agenda) is in direct or mediated relation with several visible (as a pen or a car) or observable "objects" (that can be "seen" with instrument such as a microscope as a bacteria, or a camera, a computer and so on), and with many "historical" objects (such as laws or institutions) or invisible "objects" (such as religions or culture) each of them with a reference community which defines the nature of truth,[4] and what is rational and what is not. In fact the various subject-object-context relationship suggests a continuum[5] approach about personal-common ambit, particularly if we think at individual-group/ individual-context (natural, artificial, cultural ambit)

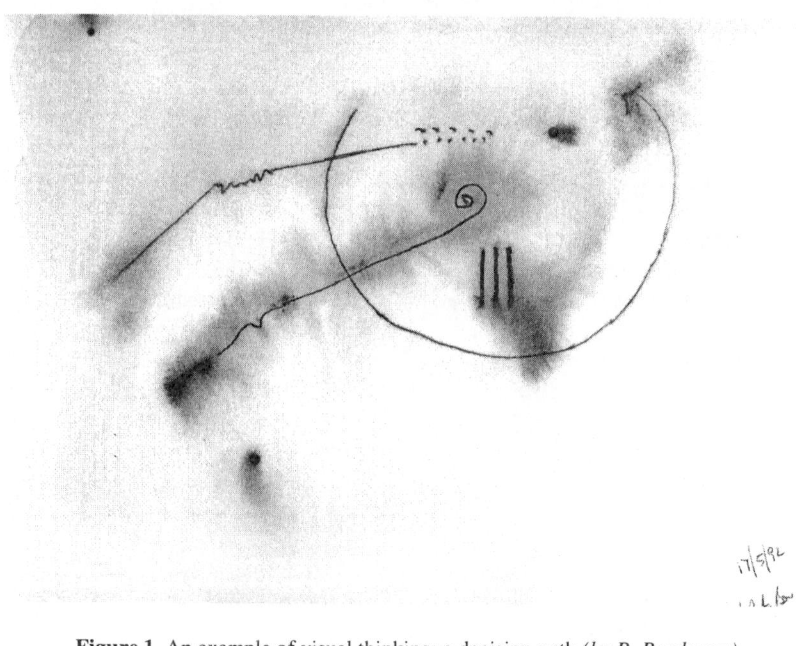

Figure 1. An example of visual thinking: a decision path *(by R. Bordogna).*

dynamics. But for what we have just said, learning and decision making, appear to be an ambit biased process.

In Figure 1 I propose a watercolor of mine; a communication and conceptualization mean that usually is not associated with rationality. It is an attempt (honestly I do not know how effective) to express the preliminary statement visually, in one "shot". This is done using a Kandisky[6] inspired visual language (who did studies on children spontaneous drawings and that desired to introduce the concept of field of force in visual arts). In the picture points and lines define paths and artifacts in a field (the background colors) where a process (the line with an arrow) take place to generate a decision (represented by the dark - blue in color - area). That watercolor is also a (poor) tentative to emulate Rudolf Arnheim experiments of what he calls *visual thinking.*[7] Arnheim's researches on perceptions as cognition processes (that he considers as continuum) in visual arts[8] are interesting particularly if we think that the civilization process is based on artificial and cultural ambit (that is to say on "objects" such as institutions, laws and other instances that can not be "seen"). In fact quite often, and in almost every practice, it is possible to note that people tends to use metaphor, symbols and other physical substitute to deal with abstractions, (as happen with paper money). These metaphors may be considered a kind of frameworks for meaning creation and therefore for behaviors. If this cognition complexity may appear out of reach for artificial agents, on the other hand human sometimes are encouraged to (or do) act as automatons. Means like music may affect behavior and decisions. Macchiavelli noted "*as a man that is dancing and keeps time with the music, cannot make a false step; so an army that properly observes the best of its drums cannot easily be disordered*", as stressed by A.B Crosby,[9] in a work that emphasize *pantometry* as a fundamental driver of western civilization, and a source of a superior natural and artificial/social phenomena knowledge. Drums may be useful when one mind must command thousands of arms, but with automatons (hopefully) are not needed any more.

STANDARDIZATION AND TECHNOLOGY FOR DECISIONS TAKING AND ANYTIME ANYWHERE DISTRIBUTED CONTINUOUS LEARNING

Moderns instrument of standardization to create a common background for learning, meaning creation and perhaps "rational" (right) decision making are metadata standards. The need of common description about "objects" increases in fact with society complexity and interdependence for many reasons. Science and new economy as well as the society as a whole is increasingly intertwined and need worldwide common information: something that requires a common global background (a common global ambit). Strategic development programs as the ADL Advanced Distributed Learning sponsored by the Us Department of Defence (http://www.adlnet.org),[10] the Prometeus European project ISO/IEC-JTC/SC36 Design of Electronic Learning Environment 2000 (http://www.igd.fgh.de)[11] or IEEE Learning Technology Standards Committee, LTSC IEEE P1484,[12] http://www.manta.ieee.org/p14184, are good examples of efforts aimed to create *anytime, anywhere distributed, continuos learning ambits,* empowered by multimedia (wired and wireless communication infrastructure) and a common culture. Technology trends on the other hand support this vision: wireless network for personal/home environment, advanced multimedia standard infrastructure and human-center devices are assets for empowered living and working/acting ambits.

GENERAL TRENDS AND HOT POINTS

Metadata standards are not related to a specific technology, but they generally need information and communication technology to be effectively implemented. It is more and more evident that the knowledge is increasingly socially distributed and intertwined as well as learning and decision processes. Therefore it seems increasingly difficult to define ex-ante what is "rational" (and what is a right decision) and what is not. This is another reason for which markets and freedom appear to be strategic social development driver. If social development more and more appears related on distributed proper learning and decision making, for what we have said, there is the need for both coordination and standardization and freedom /creativity to operate in contingent and specific ambits or domains of action. A possible solution is a standardization of *common objects that are part of the domain of action*. In this way common standard data descriptions may facilitate proper action ("rational") and (for example more "participated") decision making. This standardization, as a drum for the ancient times army, may facilitates (and is aimed to facilitate) the interactions of humans and artificial agents. It seems reasonable to say that human-centered artificial systems should be capable to take in some account also human psychological forces to better "understand" their counterparts. But on the other hand there is too the need to define metadata suitable to human thinking (learning – decision taking and so on) and creativity, as the proposed visual thinking example was intended to suggest. There are in fact several evidences that natural and artificial/cultural entities (as persons/groups or systems process) evolution dynamics are increasingly intertwined and that the identification of the contingent field/system is a dynamic decision as well. Technology diffusion (particularly multimedia communications wired and wireless) in working and living objects and ambits, and powerful personal digital assistants, offer the opportunity for a new generation of more human-centered devices, intelligent agents and infrastructures, and for new enriched "human and machine" interactions.

Therefore the hotspot appears to be on artificial supported learning / communicating / meaning building instruments as metadata and visual languages standard for better human artificial agent interaction and improved deciding dynamics.

All that said, I propose to this distinguished audience to farther discuss the proposed hotpoints.

How physiology and the environment influence the person, the community and learning and decision dynamic? How emotion, myths and rationality influence perceptions and decisions? Is the concept of bounded rationality comprehensive?

How could be modified a perception and decision ambit dynamics in presence of intelligent artificial systems agents?

RATIONAL AND NOT RATIONAL BEHAVIORS IN DECISION-MAKING

(*Vito Di Gesù*)

I would like to make few comments on the meaning of the terms "rational" and "irrational" in relation to their use in daily life. We said that a behavior is "rational" if it produces some thing that has a common sense and "usefulness". For example, "I am in the desert, the temperature is 45 degree Celsius and the humidity is 10% there are two bottles one filled with vodka a second one with water. If I drink water everybody will say that my behavior is rational. The evaluation could be different if I was in a party in Moscow". However, I could be against alcohol and affected of a form of diabetes for which it is necessary to drink often water. In this case the common sense judgment could be misleading. This example may suggest that the concept of "rationality" ("irrationality") can't fully exploit if both the "context" where the action is performed and the internal status of the acting system are not fully known. I suppose that in "economy" can be found similar and more concrete examples of "apparently irrational" behavior, due to dummy information.

A second comment regards the "consciousness" of the system. Can I do "rational unconscious" decision? Again the problem is of definition, natural systems do a lot of actions that are unconscious and very useful for their survival (our heart works pumping blood). The genetic program does the job and it is "rational" but unconscious. From this point of view, artificial and natural systems have analogies; so far we can attribute "rational" behavior to an artificial system too. The analogy can be carried out further. This is the case of a natural system fault (a mutation generate a wrong instruction) as well as an artificial system in which the "software code" has been corrupted by a hardware conflict.

A third comment is related to the dynamical change of a context. Cultural background, costumes and habitudes may change due to a war, to economic revolution, to a climate evolution. Rational behavior and the related decisions should adapt (if possible) to the new situations to avoid the extinction. The same problem may occur in the case of artificial systems that must be programmed in order to consider the dynamic of the environment where they act in order to be robust enough and act properly on it.

RATIONAL AND IRRATIONAL DECISION MAKING: DEFINITIONS

(Bertrand Zavidovique)

According to the examples given, we can propose the following definitions:

Def. 1: ***Rational:*** obeying some premises according to some rules (both commonly agreed on by a sufficiently strong* corpus of "systems")

Def. 2: ***Irrational***: any production of the weak part of the population that does not follow the couple premises/rules as previously stated

Comment/remark: there is *recursivity* (still some hope remains!) in that rules. Do evolve *dynamically*, yet following Def. 1. This is one way, among many, *time* gets into the picture. So rationality is bound to consensus, and then a matter of communication and even of conviction (in some conditions any mean will do). That makes language one (among many) vector of irrationality. Another consequence is that if a goal is defined (made explicit, agreed on ...) then deciding between rational or not would become trivial, since a metric is then available, were it not for noise and dynamics i.e. observation. And again complexity becomes one among other parameters of rationality (common sense says that any process not to succeed on time does not exist), would it be only because an absolute intrinsic perception does not mean anything. If the burden and the reward are unclear, then minimizing one while maximizing the other is utilitaristic for sure, except utility gets uncertain and risks to be biased (where economy and politics overcome logic!). Eventually we are left with: rationally deciding, as any automation process, depends on observation and aims, but Descartes said already that all depends on the couple. And a last example, ultimate in irrationality: "There is this guy who tries to get a thread through the eye of a needle; he is drunk, he is on a boat in stormy weather, and he intends to sew a lost button on his fly." (Irrational, isn't it?). "Yes but the boat is sinking and they won't let him get in the life boat if his fly is undone!!!" (What about that?).*

Virginio Cantoni questions:

I have two questions:

1. It could be of interest to know what are the analogies and the divergencies between the two large projects of Advanced Distributed Learning of the USA and the 'Prometeus' project of the European Community.

2. I found very interesting the picture that you presented and I want to show to the author my enthusiastic congratulations for the impression on the audience. Nevertheless, following the quoted Rudolf Arnheim which inspired the picture, I would like to have explicit from the author the methods and the discoveries of the science of representation (or depicting following one of the terms of the workshop) which overwhelm "Gestalt" and that were used in this composition.

---

* numerous is just an example of strength i.e. dominance according to democracy

Roberto Bordogna answers:

About your first question, mentioned standard metadata based development programs as USA Advanced Distributed Learning Program and other initiative, including the European Prometeus Design seems that might find a common ground on standard matter in IEEE P1484 proposal. On the other hand the Advanced Distributed Learning program, accordingly to the information provided, involves a major number of educational institutions and of major corporations, recognizing that information superiority is a strategic asset (particularly for security). Similar initiative to improve learning by means of technology in Europe are at the moment mainly a research project as Prometeus, and the European Commission program "eEurope - An Information Society for All" aimed to improve innovation and knowledge in Europe, started very recently (without a particular emphasis on reinforcing both democracy and security/defense).

About the second questions relative to the watercolor shown I tried to use Arnheim's idea of images (symbols) to express functions. So I have used a line with an arrow to express the idea of a process (a movement) producing a result (symbolized by a dark – blue in color - area) in a given context (or ambit) of psychological "forces" represented by dots (attraction points), lines (obstacles / path) and colors. This to propose an example of so called, visual thinking.

REFERENCES

1. R Bordogna, *Technology learning driven general social economic growth dynamic*, Preprints, 7Th IFAC Symposium on automated Systems Based on Human Skill, Aachen, Germany 2000.
2. Herbert A. Simon, *Administrative Behaviour (1947)* in Italian Edition *Causalità, Razionalità, Organizzazione*, Il Mulino Bologna, (1985).
3. Amartia Sen, *Inequality Reexamined*, Oxford University Press, Oxford, (1992) Italian edition La diseguaglianza, Il Mulino, Bologna, (1994).
4. Kzysztof Poniam, *Filosofia/Filosofie*, in Enciclopedia Einaudi, (1975).
5. R. Bordogna, *Leveraging on multimedia technologies to modernize social economic ambit development paths*, 6 th IFAC Symposium Automated Systems Based on Human Skill, Kranjska Gora, Slovenia, (1997).
6. W. Kandisky, *Punt und Linie zu Flaeche* (1926), Italian Edition Punto linea superficie, Adelphi, Milan, (1991).
7. R.Arnheim, *Visual Thinking*, The Regents of the University of California, Los Angeles (1969) Italian edition *Il pensiero visivo*, G. Einaudi, Turin, (1974).
8. R Arnheim, *Arte e percezione visiva*,. Italian Edition, Feltrinelli, Milan, (1997).
9. W. Crosby, *The Measure of Reality* Cambridge University Press, UK (1997).
10. ADL (DoD Advanced Distributed Learning - 1996), Us Department of Defence, http://www.adlnet.org
11. ISO/IEC-JTC/SC36 Design of Electronic Learning Environment 2000, http://www.igd.fgh.de
12. IEEE Learning Technology Standards Committee, LTSC IEEE P1484, http://www.manta.ieee.org/p14184

BIOLOGICAL BRAIN AND MIND

Edoardo Boncinelli

Università Vita-Salute, Scientific Institute San Raffaele
Milan, Italy
e-mail: boncinelli.edoardo@hsr.it

INTRODUCTION

It is particularly appropriate to speak in these days about brain, that is biological brain, and its relationship with mind. A lot has been learnt about this topic in the past 20 years or so. This is due primarily to the convergence of three powerful research lines: experimental psychology (nowadays often termed cognitive science), biology (especially molecular biology) and brain imaging.

Brain is an organ, admittedly very complicated, but an organ, composed of cells. Every cell contains the entire genome characteristic of the corresponding individual and consults it continually, in order to survive, operate and duplicate. Every biological organism has this unique property: it contains multiple copies of its design and realization plan, one in each cell. In humans there are more than 10^{13} cells. The genome present in every cell carries the genetic information typical of that species and originated through millions years of biological evolution.

Having said that, it is clear that neural cells, or neurons, are very special cells. They have many particular features. Among them stands out their large surface, or better their big surface-to-volume ratio. They may have a surface even three orders of magnitude higher than that of other types of cells of comparable volume, that is containing a comparable organic mass. A cell characterized by such a large surface is especially fitted for communication. In fact, the higher the surface, the larger the number of possible exchanges of matter or energy per time unit. Neurons also possess axons, capable of reaching distant cells and exchange information with them. It is rather difficult to imagine how higher organisms could achieve neural co-ordination and activity using cells of a different type. But the question remains as such, as it happens for most biological "why" questions. We only know of a single type of life, evolved in a certain way, through endless more or less random events accumulated through time. Only if we could compare this one with different types of life, we might answer question about how necessary a specific biological feature actually is.

Communication between neural cells takes place through specific microstructures present at the end of axons and termed synapses. Impressive is the number of synapses present in the central nervous system. In the human cerebral cortex, for example, there are not less than 10^{11} neural cells, each one possessing an average of 10^4 synapses. This makes a

Human and Machine Perception 3: Thinking, Deciding, and Acting, edited by Cantoni *et al.*
Kluwer Academic/Plenum Publishers, New York, 2001.

grand total of approximately 10^{15} synapses. Common wisdom claims at present that the essence of our cerebral cortex resides in these 10^{15} synapses. They mediate and embody the complex of cerebral functions, from memory to reasoning, thinking and artistic creation. They are necessary to be a human being and uniquely characterize every single individual, through their common features and individual differences, respectively.

BRAIN EVOLUTION

A fundamental question, in this light, is how all these synapses are formed. It is relatively well established that the biological instructions encoded in the genome of every individual are not sufficient to dictate and specify all of them. A considerable portion of them, but by no means all of them, are in fact established through the implementation of our biological instructions. Some of them are certainly established on the basis of experience, the collection of major or minor events we all go through in our life, especially in the early months and years of our life. The concept of experience must contain everything: what we ate, what we did not eat, what we saw, what we listened to, which diseases we had, which diseases we had not, what satisfaction and what frustration we experienced, how much we were cared of and by whom, and subsequently what we were taught and what we learned. It is widely believed that all this shapes the wiring of our cerebral circuits, especially by setting, erasing, reshuffling, confirming and establishing most of our synaptic connections.

A general consideration is in order in this respect. Human new-borns are particularly immature and unable to cope with the challenges of the external world. A general trend in vertebrate and especially mammalian evolution actually implies that animals are born more and more immature, and this progression reaches its peak in our species. This phenomenon is mainly due to the immaturity of our brain. At birth, in fact, most of the cerebral cells are in place, but some gross connections and a remarkable number of synapses are not yet established. They will be established in the first months and years of our life under the influence of our sensations and perceptions. Our brain completes its own development while eyes are seeing and ears hearing. Many events leave a permanent trace in our cerebral circuitry and microwiring, so that part of them become an integral portion of our brain. Using a metaphor drawn from the computer world, we may say that part of the hardware of our brain derives from the internalization of the software we run on it early in our life.

Nevertheless, detailed inspection of cerebral circuits in both lower and higher organisms reveals that all this is still not sufficient to determine synaptic connection and sometimes also to establish brain circuitry. Actually, we are forced to admit a significant stochastic component. Since it is necessary that some connections, both major and minor, are established at some fixed time-windows, if the input of experience is not sufficient to reach this goal, chance and only chance will do that. Though we do not know exactly what percentage of all synaptic connections are set-up by chance, we expect this fraction not to be negligible. Microanatomical features have been reported to be often different even in the nervous system of kin crustaceans. This is even more the case in mammals and humans.

In conclusion, three are the sources of instructions for cerebral microwiring: genome, life experience and chance. The combination of these three actions ensures that no brain is identical to any other. This is important in connections with at least two hot issues: human cloning and consciousness. Firstly, even if somebody will embark in the future in the insane project of cloning human beings, it is essentially impossible that two brains, not to speak of two minds, will be identical, even in closely related individuals. Secondly, if consciousness is at least in part a product of brain, the irreducible individuality of every brain should reflect in the uniqueness of individual consciousness and in the ensuing difficulty of grasping it, not to say understanding it. Science primarily deals with reproducible events and is in a particularly bad position in coping with the problem of individual consciousness.

A particularly interesting subject is the evolutionary origin of brain. A sort of brain is present in essentially all higher animals, but cerebral cortex is exclusively apparent in mammals. This cerebral structure increases in size and complexity during mammalian evolution and shows dramatic changes in higher primates and in the transition from apes to man. The past two decades of molecular biology have much contributed to improve our understanding of evolutionary events. It is certainly not possible to summarize here all this and I will confine myself to a few issues.

First of all, conservation. Even if we are inclined to underline biological diversity, it is increasingly clear that an incredible number of features are conserved throughout natural kingdoms. The genetic code, the structure of the cell and cell membrane, the machinery required for protein synthesis, most metabolic pathways and many other processes are remarkably conserved in evolution. In the past fifteen years we also learned that most of relevant regulatory genes are exactly the same in all animals.

Developmental events are genetically controlled by a number of genes. Among them stand out some high level regulatory genes often termed master control genes. They take decisions of high generality and hierarchical level like establishing where the head must be, where the thorax and abdomen; where upper limbs and where lower limbs, where eyes and where the nose; whether a pair of limbs must be legs or wings and how many vertebrae have to be present in a given body region. These genes take rather general decisions and shape the overall body plan. In addition to this, there are batteries of downstream genes controlled by master control genes and deputed to the implementation of their orders. The master control genes are the same throughout higher animals, if not throughout animals. In particular, genes establishing head versus trunk, the subdivision of the trunk and the shaping of the brain are the same in all Bilateria, animals exhibiting bilateral symmetry.

A second major issue in evolution is discontinuity. Although we are accustomed to think of evolution as a continuous process implying a number of small changes, the most important events in evolution actually occurred suddenly and altogether, as a result of genetic rearrangements and/or major events in the external world. If we follow the evolution of a specific animal species living today, e.g. the horse, we observe the accumulation of an infinity of small changes. But the appearance of this and all other species is a sudden, discontinuous event. Discontinuous events are at the basis of true evolutionary innovation and often imply several types of different anatomical and functional changes at a time. There are also evolutionary periods where a clustering of major changes is observed. The most important of these clusters of events is probably the so-called Cambrian explosion: in a relatively short period of some 20 millions years, around 600 millions years ago, suddenly appeared all existing animal phyla, from Chordates to Arthropods, even if we are hardly able to imagine what really took place.

Evolution is also a largely random and opportunistic process. In our current view, evolution results from two components. One is the random, nondirectional and aimless appearance of genetic mutations. Most of them are neutral or deleterious. From time to time, individuals bearing new forms of some genes survive and outcompete individuals bearing the old and established forms of the same genes. In the long run, a new equilibrium is reached and the genetic composition of a given population has changed. There is a selection, termed natural selection, of some types of individuals of a given population at the expense of all others. It is the environment in its entirety that selects the characteristics of some individuals rather than others. Natural selection on populations is operated by the specific conditions of their environment and is strictly dependent on it. A change in the local environment usually changes the criteria followed in the selective process. Natural selection is in a position to choose among given individuals bearing various forms of some genes, but cannot create or mould new gene forms. As a consequence of all that, every physical space on our planet is inhabited by individuals belonging to as many different species as possible, each one of them exploiting a particular ecological niche in this space.

Such an opportunistic behavior of selection is only bound by the demand of survival of at least some individuals and their capability to reproduce themselves. Living organisms have genomes that result from the stratification of many different genetic changes. There is little room for adjustments and rearrangements of previous assets on route and no *de novo* design. This is specially true for the primates brain, since this organ evolved fairly rapidly.

The discovery of master control genes contributed a lot to help us explain discontinuity and innovation in evolution. In fact, it is clear that a single mutation in some of them may originate big phenotypic changes, as it involves changes in the action of many downstream genes. It is not even necessary to have a structural mutation in these genes. It may be sufficient that the new mutation changes the timing and/or the place of action of someone of the master control genes. (The two phenomena are termed heterochrony and ectopic expression, respectively.) Heterochrony in particular may explain a variety of phenomena. Let us suppose that a regulatory gene acts for some time during development and promotes the growth of a specific body structure. A delay in the time at which the gene is switched off may cause an abnormal growth of this organ, transforming a nose into a proboscis or a thin cerebral cortex into a thick one. Furthermore, many regulatory genes and most master control genes regulate a number of different body structures. A change in their expression causes variation in many anatomical structures at the same time. In the past we studied, for example, a regulatory gene, OTX1, that controls among other things the shape of the cerebral cortex, of the larynx and kidneys. A contemporary change in brain and larynx might explain exciting evolutionary innovations in the evolution of primates.

INFORMATION REPRESENTATION

Brain function essentially consists in the elaboration of information. The information must be first acquired and this happens through our sensory organs. The single most relevant aspect of the action of sense organs is that they do not observe passively the world around us, but rather poses to it specific, stereotyped questions. The answers to these questions belong to a discrete set of possibilities: is there a vertical bar or not? is the visual scenery bright or dark? how much of salty and how much of bitter is there in this food? Both in peripheral sensors and in intermediate nervous centers deputed to the elaboration of sensations the rule is uniformly that of assigning sensory response to a preformed array of possibilities. The keywords in this context are discreteness and coding. All this aims at reducing signal-to-noise ratio and belongs to a general strategy of living beings to reduce information loss.

This is most important for the very persistence of a living cell or organism. In physical sciences one usually deals with sets of billions and billions of molecules and even if they are individually subjected to random fluctuations, their statistical properties are relatively stable and predictable. In biology, we are often faced with a very reduced number of molecules of a given species present in every cell and they turn out to behave in a predictable quasi-determined way. This is particularly striking considering that some protein molecules may be present in a few hundreds of copies and DNA itself as only one or two copies per cell. Life seems to escape most pitfalls deriving from the second principle of thermodynamics, without violating it, through the judicious use of free energy. This, I maintain, is achieved by means of coding discrete, possibly Yes/No, choices at every level.

First of all, it is known that living objects have a double level of reality: their exterior apparent aspect, as everything else, and a hidden blueprint capable of maintaining and eventually reproducing their exterior aspect. These two realities are called phenotype and genotype, respectively. There is no living object without a genotype or genetic complement. This genotype is guarded in the innermost corner of every cell and is organized as an encoded message. Its material support is DNA (or sometimes RNA), a long macromolecule consisting of a string of elementary constituents termed nucleotides. This message is used for the regulated synthesis of all the proteins necessary for life to exist.

A message is a sequence of symbols. In abstract terms, symbols are alternative entities selected from a closed inventory or alphabet. These symbols have to be arranged in a sequence where the order matters considerably, as AB is generally different from BA. The existence of symbols implies in turn the existence of a code as a correspondence, generally unambiguous, between members of two inventories. This correspondence is fixed but usually arbitrary, a concept expressed in Linguistics as *l'arbitraire du signe*. Symbols may in turn exist at many different levels of coding. In natural languages it is customary to speak about articulations: at least two, at the level of alphabet letters and at the level of words.

A short sequence of DNA, that may represent a portion of a gene, consists of a chain of nucleotides. What matters in this sequence is not only the actual identity of the various nucleotides – there are only four choices: A or G or C or T - but, more importantly, their order. It is well known that this message is read in triplets; every group of three nucleotides encodes a specific amino acid. The sequence of the various triplets present in a gene unambiguously specifies the sequence of amino acids constituting the corresponding protein. For example, the triplet CCA encodes a proline residue. The C nucleotide present in the middle of this triplet may change as a consequence of a biological error, but it may only change into three possible directions: it may become a G or a A or a T. It cannot become anything else and, most importantly, it may never be something like 2/3 of C and 1/3 of G or 2.546 C. The choice is among discrete alternatives. This prevents a large portion of mistakes deriving from random fluctuations. From the point of view of biochemistry, it is not impossible that a given nucleotide shares some features of, let us say, a C and some features of a T. The point is that the resulting nucleotide will be read either as C or as a T at the moment of assembling the corresponding protein. Thus, it is not the intrinsic nature of nucleotides that prevents random fluctuations but their status of coding entities.

This phenomenon is not confined to DNA sequences and their coding capacity. Even a gene can be either active (on) or inactive (off) and every cellular state is determined by the on/off state of some of its genes. Even considering only a set of few genes, they may be either all on, or some of them on and others off. There are a high number of possible combinations but also this inventory is discrete and countable. Every cell may be in one of these states characterized by the combination of genes which are on and genes which are off, in a conceptual scheme frequently called combinatorial. This is also true at the level of tissues and even body regions. Early embryonic epidermis may turn into mature epidermis or into neuroepithelium. A given embryonic body segment of a developing insect may turn into one of the 15 or so body segments, but very rarely turns into something intermediate between two body segments.

Another example is the firing of neurons. Neurons are well known to fire or not to fire. Many different signals converge on a single neuron that integrates them according to rules, yet unknown, but the eventual output is of the all-or-none type. Finally, modern neuroscience and cognitive science tell us that even at the highest level of mentation there are predetermined schemes - perceptive, representational or behavioral - and that we learn, evaluate and make decisions on the basis of these schemes. In every instance, life implies choice among a discrete set of alternative possibilities, most of them dictated by the generative project encoded in the genetic complement present in every cell.

The identity of some of the alternative possibilities or internal biological states just mentioned appears to be tightly controlled. For example the exact pathway of a given nerve fibre or the shape of a spot present on the wing of a butterfly is often controlled by more than one process, to be traced back in turn to the action of more than one gene. Accidental silencing of one of these genes may not result in an alteration of these structures. These phenomena are usually ascribed to a sort of redundancy of the genetic information. In fact, it is often stated that biological information is redundant. Even if this cannot be completely excluded, there are good evolutionary arguments against the maintenance of genetic redundancy per se. The point is, to me, that most biological events are overdetermined. Overdetermination of biological internal states should rather be seen in terms of intersecting regulatory networks.

A consideration is here in order on the general meaning of experience. Biological processes are obviously influenced by the environment and in general by external events. We may call this interaction experience in broad sense. Following this definition, experience may apply to elementary biological entities like genes or cells, up to learning and behavior in everyday life. In the light of what we have said so far, modern biology and neuroscience suggest that we may envisage experience as an influence of external inputs on the actual selection among alternative biological internal states belonging to a closed inventory.

According to this view, experience is selection among pre-existing possible states representing discrete alternatives. As a consequence of this process, the repertoire of internal states available to a cell, an organ or a system is momentarily restricted, possibly to a single choice only. Whether this repertoire will remain restricted, evolve toward a different configuration implying new possible choices or bounce back to the previous situation, is a matter of successive events which may or may not be confirmatory. It is also clear that inventories themselves may change along with biological evolution - biological evolution is actually the evolution of these inventories - but the time-scale implied is quite different. The very concept of experience has broad implications as it bears on the general theme of innate vs. acquired. In particular it bears on the epistemological problem, because knowledge acquisition is a particular form of experience.

Information on external world gained through senses must be subsequently elaborated in a variety of different ways and may end up in eliciting an action, a response to a stimulus. In unicellular and lower multicellular organisms a response almost invariably follows a stimulus. This response is often very quick and selected within a very limited inventory of possible responses. In moving from lower to higher organisms and humans, the time elapsing from the stimulus to its response progressively increases and may become remarkably long. In humans many stimuli do not appear to generate a response and many of our actions do not appear to be responses to any stimulus. Elaboration of the information on the status of the external world may require minutes, hours or days and takes many different forms. Operationally, we may call thinking all this, i.e. what happens in our brain between the reception of a stimulus and the enacting of the corresponding response. All animals have some degree of information elaboration, that is thinking, but we certainly have more possibilities of elaboration and, eventually, response. In this light, we may define freedom the relatively large extension of the inventory of possible responses to the same stimulus. We are free because we can react in many different ways to a given stimulus or series of stimuli and more free than a dog since many more different responses are available for us.

Storing and Forgetting

Memory is another issue on which we learned a lot, even if we would like to know much more. We now know for example that we have at least two types of memories, belonging to either a short term or a long term memory compartment. We store for fractions of a second or for a few seconds many issues, like digits of a telephone number which we are dialing, figures of mathematical operations, words of a sentence we are reading and so on. After some time, none of these memories is present in our minds, unless some of them are transferred to the long term memory compartment. This is both useful and unavoidable. It is hard to imagine what our life might be, if we had to remember every single issue we memorized at one moment or another of our existence. The importance of forgetting is clear, even if it would be better to speak of a selective forgetting. Some memories are in fact transferred to the long term compartment where they last for months and years. We do not know yet where memories are stored and in which physical form, but we know that for the passage of notions from the short term to the long term compartment, the integrity of our hippocampus is required. There is also preliminary evidence that a broad short term (or working) memory is a component of superior intelligence.

We also do not know what is the basis for the decision about which short memories have to be erased and which one promoted to long term memories. There is no doubt, however, that this process is heavily influenced by emotional aspects of individual issues and their relationship with the interests and the attentional focus of the subject. We may say that the decision is taken on the basis of the relevance of the various issues for the subject. Relevance may in turn be considered as the higher mental correlate of selectivity, an intrinsic property of living matter present at every step of life. We may envisage a progression, a sort of red thread, of unique abstract features throughout the world of life: discretization -> coding -> selection -> relevance -> meaning. Meaning, that is making sense of something, may be seen as the symbolic correlate of relevance and of all that lies behind it. This brings me to the issue of comparison between our mind and computer activity.

Information Processing Models

Comparison of modes of action of computers and brains has been terribly helpful in understanding how mind works and how mind does not work. We could mention many issues in this connection, but I will confine myself to one aspect. Computers are able to process, one after the other, chains of several logical steps. The length of the chain is only limited by the capacity of their memory and the rate of the processing by the quality of the central processing unit. For a computer there is no relevant difference between step 11 and step 26 of any running program. Our mind is different. We have some difficulty at processing a chain of several logical steps and from time to time we need to stop reasoning and try to summarize what happened up to that point. In doing that, we attach, more or less consciously, tags of interpretation and meaning to the various steps. We need to interpret, according to anyone of many possible interpretation schemes, and make sense of practically every step in a reasoning chain. This necessity has many consequences, both on the operational level and for our own attitude toward life and its interpretation. We tend for example to interpret and overinterpret everything, arriving at posing basic universal questions that are intrinsically impossible to answer, like the meaning of our life.

Another unique feature of our mind is its limited rationality. We are perfectly aware of the limited role played by reason and rationality in our everyday life, but we tend to assume that whenever we make an effort to use our mind in a rational way we may achieve a perfect rationality. It has recently been shown that this is not always the case. Everybody is inclined to provide wrong estimates and evaluations for some specific questions and this is largely independent of the degree of intelligence and education of the person tested. These questions mostly deal with evaluation of probability and risk. These phenomena are surprising and their study revealing, but it is not impossible to make sense of them, if not to account for them. Our brain is essentially that of at least 200 000 years ago, when life conditions were very different. Our brain evolved to take rapid relevant decision to ensure survival and reproduction to our ancestors. We cannot understand the functioning of our mind if we do not take into consideration these requirements. Genetically, we are not born to be insurance agents or chess players. Our central nervous system is perfectly fitted for allowing us to escape a danger or to chase a prey. We need not to be perfectly logical, but rather to take rapid decisions on relevant issues. How we make judgements of relevance is not known and represents a formidable task for future studies, whereas rapidity of decision making is understandable in terms of the generally parallel nature of information processing in our central nervous system.

Let us now consider briefly the issue of *consciousness*. This is probably the most difficult subject to study. It is possibly too late for philosophy and too early for science. Nevertheless, many books have recently been published on this topic and many ideas have been aired. I think that all this should be better approached if one makes some highlights. In fact, the word consciousness is used with at least three different meanings.

Awareness. Many higher animals are more or less aware of what happens and what they are about to do. This awareness is common to several animal species and small children and shows many degrees. It can be best studied observing their behavior, with the possible help of our theory of mind, a mental function we all share. There is preliminary but suggestive evidence that awareness may imply synchronization of neural signals in some cortical areas.

Consciousness. There is a form of shared consciousness common to all of us. Humans are able to communicate through language. Therefore their consciousness can be studied not only observing their behavior, as for animals and infants, but also through speech and conversation. This renders human consciousness somewhat different from simple awareness. Consciousness requires awareness but the contrary is not true.

Phenomenal consciousness. Humans have also private sensations, feelings and emotions, not always easy to communicate. From a rigorous point of view, they are accessible exclusively through introspection. Every individual is aware of his/her own ideas and private feelings and mental states. Let´ us call this mental property phenomenal consciousness, as opposed to a general concept of shared consciousness. It is a property of one single subject: me. On these grounds, some thinkers claim that it is operationally impossible to penetrate the phenomenal consciousness of individuals and this private mental state will remain for ever outside the realm of observation and henceforth of scientific investigation. Many illustrations of this concept have been put forward and it looks to me apparent that I can never penetrate completely the inner phenomenal consciousness of a person different from me. It is, by the way, not even clear whether phenomenal consciousness is necessary for our mind to work.

With this distinction in mind, we may hope to reach some understanding of consciousness in its first and second meaning and this needs to be vigorously pursued. As far as the third meaning is concerned, there is probably nothing to do, but who cares? The impossibility to study private, phenomenal consciousness, not to be confused with consciousness in general, should not be an obstacle to the scientific study of shared aspects of consciousness. This is a field of great interest and difficulty, where only a limited number of successes have been so far reported. Nevertheless, there is a metaphor I like most for consciousness. According to this view, consciousness is a sort of funnel transforming a fraction of the parallel processes taking place in our central nervous system into something intrinsically serial and sequential. In this process a collection of neurostates is transformed into a psychostate. This happens continually and most of the time we do not even realize that this is actually an operation and has precise requirements. Only when something goes wrong, due to the assumption of drugs, tiredness or age, we realize the existence of this great mechanism and the related problems. Linguistic utterances are serial by definition and this is even more true for written statements. In this light, it is conceivable that language, whenever it first appeared, contributed over time to shape our own form of consciousness and that literate people develop in their life unique features shaping and characterizing some details of their own consciousness. It is evident that there is a great deal to investigate and presumably a great deal of fun around the corner!

FINAL THOUGHTS

Finally, I want to address the audience with three personal messages, general messages from a biologist who has been listening to the presentations of this very specialized meeting.

1. People attempting to devise thinking machines and robots should first consider a fundamental choice: try to imitate biological brains and bodies or design their artifacts from scratch. After all, cars and airplanes have been designed with little or no reference to living objects. None of the latter has wheels or helices.

2. It is not clear to me whether learning and education can be achieved through association only. This is certainly not the way our brain works. Apart from associative processes, a relatively easy mechanism to simulate, our mind has some degree of internal organization and works via a set of predetermined schemes and potential circuits waiting only to become operative. This is actually a restricted or guided association.

3. Individual animals and humans are very good at performing many tasks, but occasionally make mistakes and are not totally reliable. It is conceivable that machine designers ask too much to their creatures. They should be as efficient as living agents, but error-free and reliable. It is possible that this may not be at all an attribute of single individuals but only of collective agents.

VISUAL ATTENTION SYSTEMS

Virginio Cantoni and Massimo Cellario

Dipartimento di Informatica e Sistemistica
Università di Pavia
Via Ferrata 1, 27100 Pavia, Italy
e-mail: {virginio.cantoni, massimo.cellario}@unipv.it

ABSTRACT

The attention processes of vision systems are thought-out outlining some analogies between biological vision and the solutions given by computer scientists in artificial vision. The studies on human visual attention identify two phases: a pre-attentive one in which the visual system detects regions of interest within the full field of view, also through simple and fast alerting mechanisms, and an attentive one in which each detected region is analyzed in detail. The same computational paradigm is considered for artificial vision systems in order to successfully deal with the enormous quantity of raw data transduced by standard vision sensors. Such a paradigm exploits the variable resolution grids, according to the image detail required for the task at hand, so fully utilizing the capabilities of multi-resolution and pyramid computer vision systems.

PRE-ATTENTIVE AND ATTENTIVE MECHANISMS

The technology evolution of these last decades leads to *active* visual sensors:[1] the camera evolved towards a visual front-end in which single or multiple points of view are on-line controlled, giving in this way support for an effective implementation of the 'perception and action' framework of biological systems.[2] These new equipments supported the development of new areas of research such as the exploratory vision, term introduced by Tsotsos,[3] in which the environment is analyzed through a controllable camera exploiting investigations analogue to the one supported by eye/head/body movements in human vision. At a different control level, are the sub cases in which acts only the control system for camera orientation. In biological and in particular in human vision, the research at this level started several decades ago. In fact, well known are the *scan-paths* recorded by Yarbus.[4] In this work, sequences of observer's eye fixation points were analyzed (the mean time between two fixation points is around 250 ms). At a third level of the hierarchy,

Human and Machine Perception 3: Thinking, Deciding, and Acting, edited by Cantoni *et al.*
Kluwer Academic/Plenum Publishers, New York, 2001.

characterized by more subtle and detailed scans, there is the element-by-element scrutiny done by the fast moving aperture of a *search-light of attention* (with steps lasting 15÷50 ms each). In this case no "mechanical" movements are required, eyes are directed to some "gross center of gravity"[5] of points of interest, and then step by step a search-light scrutinizes each single point of interest.

Moreover, from the retinal neurons to the central cortical ones, the level of data abstraction gets higher, while precision in pattern location decreases: there is a semantic shift from 'where' towards 'what' in the scene.[2,6] In fact, as information passes through the visual pathway, the increasing receptive field size of neurons produces a translation invariance. The large receptive fields of neurons at late stages lead to the mis-localization and the consequent mis-combination of features. Moreover, the simultaneous presence of multiple objects increases the difficulties in determining which features belong to which objects. This difficulty has been termed *binding problem.*[7] The precise solution may be provided by an *attention mechanism* that allows only the information from a selected spatial location to act at the higher stages.[8,9]

From the computational view point, early visual information processing is characterized by two functionally distinct modes:[10,11]

• *pre-attentive mode* - the information is processed in a spatially parallel manner, without the need for attention resources. The parallel execution is characterized by:

i) a complete independence from the number of elements;
ii) an almost instantaneous execution;
iii) the lack of sophisticated scrutiny;
iv) a large visual field.

• *attentive mode* - the allocation of attention resources to specific locations or objects is needed for more complex analysis. A sequence of regions of interest is scrutinized by a small aperture *focal attention.*[11] The object can be consciously identified only in this attention aperture.

Pre-attentive processes perform primitive analysis in order to segment or isolate areas of the visual field, for detecting potential regions of interest to which attention should subsequently be allocated for more sophisticated, resource-demanding processing. Attentional dispatching can be voluntary, i.e. guided by strategic goals, or involuntary, i.e. reacting to particular stimuli. These two spatial attention modes have been named *endogenous* and *exogenous*, respectively.[12] Nevertheless, there are actions which directly and only reflect pre-attentive processing.

DISCONTINUITIES IN SPACE AND TIME

A number of different models have been proposed to describe the pre-attentive behavior. A first family of models includes the set of early-vision primitives of the *reconstructionist approach.*[13] A second family includes the reacting models acting to capture the attention by detecting discontinuities and anomalies in space and/or in time.

Early vision in the sense of Marr, is a data-driven process in which sets of *intrinsic images* are derived from the visual scene in order to support a 3-D reconstruction of the scene. The early processing is considered to be organized in a hierarchy of increasingly complex feature detectors.[14] The set of pre-attentive primitives is potentially wide and can vary substantially with the context. In particular they usually include geometrical and physical entities such as color, morphology, and motion. The reconstructionism framework is based on these physical property representations which support quantitative feature analysis (such as slant, tilt, reflectance, depth, and motion) as a primal step for qualitative judgments and actions. In this connection, the most discussed paradigms in psychology are the pop-out theory[8] and the individuation stage introduced by Pylyshyn[15] in which a small number of salient items in the visual field are indexed and thereby made readily accessible for a variety of visual tasks.

Interrupt theories, in the sense of Green,[16] justify how the observers can detect targets by a *difference signal* arising from a discontinuity which automatically attracts the attention towards the location of 'substantial' differences, without identifying the features. These processes can still be classified as data-driven triggered by a discontinuity in space[11] or in time:[17,18]

- *discontinuity in space* - According to B. Julesz,[11] the visual scene is pre-attentively represented as a map of differences in textons. That is, local areas that substantially differ from their neighbors on certain features are detected, even if their feature identities are not recognized.
- *discontinuity in time* - A new event, like the abrupt appearance of an object in the visual field draws the attention toward the ex novo object. Yantis and Jonides[17,18] proposed the existence of a mechanism of attention capture that is tuned to abrupt onset and directing the visual attention to the onset location for the efficient identification of the newcomer.

SELECTIVE AND SPATIAL ATTENTION

The processing capabilities of the visual system are resource-limited and it is not possible to deal in exhaustive mode with the huge amount of information that visual sensors gather. For a long time attention has been conceived as a way to contend, with limited processing resources, the enormous quantity of data-information.

These capacity models of attention[19,20] assume that the limited resources can be distributed across a spatial region, that varies in size by a order of magnitude, depending on the task. Some capacity models postulate a strictly serial processing;[8,21] others postulate a parallel stage with limited resources.[22] All these models justify the decrease in performance due to an increase in the amount of items to process, by an information bottleneck in the serial case, and by the necessity to subdivide the attention resources among the foveated items.

Two kinds of attention are usually considered.

- *Selective attention.* This is the *goal-driven*[23] case of search for a known target and regards how appropriate an instance is selected from the scene. Like in all resource limited systems the discrimination of the peculiarities of the case at hand is critical. In most vision problems the effective implementation of a goal is strongly dependent on the capability of selecting the proper features for the recognition, location, or interpretation processes.

In this connection selective attention exploits a trade-off between resolution and spatial extent: the same channel can be activated to attend to a large visual area with low resolution or to a small visual area with fine resolution. This trade-off is needed because of the limited amount of information that attentional mechanism for high level processing can select. Nakayama[24] suggested that the visual system is composed of both a pyramidal analyzer and an iconic visual memory. The visual field is analyzed in parallel at different levels with different degrees of resolution. The memory contains a set of icons linked by associative connections. Each icon consists of a very small amount of information because of the limited attentional bandwidth.

In searching for a target that is defined as the odd element among distractors, two mechanisms can be used to direct attention:

i) a top-down mechanism, which is exploited when the target is known;

ii) a bottom-up mechanism, activated when the target is unknown and the search is for a possible odd object.

These two modes of processing are known as *feature search mode* and *singleton detection mode*, respectively. Note that, when the target feature is known, both these strategies are available, however, only the latter strategy is possible for the odd target.

A mechanism for directing attention by a bottom-up process has been suggested:[25] the local (lateral) inhibition within feature maps is responsible for the detection of target conspicuousness and its subsequent status of focus of attention. The activity at each location is summed across the feature maps, and attention is directed to the most stimulated location. Since the inhibitory connections between detectors in each feature map are thought to be local, it is reasonable to suppose that this process is most effective when the distractors are close together.

Summarizing, two factors determine search performance: the resolution required to make the judgment and a priori knowledge about the target. Bravo and Nakayama[26] compared different search tasks that required different resolutions. A low resolution task is used to detect or identificate the target's unique color; this can be done at low resolution. In these cases the response time (RT) is not affected by the number of distractors in the scene (*flat search function*). Here no effect due to target knowledge resulted, because the detection of an odd target can be achieved without focusing attention. On the contrary, if a sophisticated discrimination is required, a small aperture of attention is necessary. However, a critical aspect concerns a priori knowledge of the target's feature. In fact, this knowledge allows to direct attention, by a top-down mechanism, to the location of the highest activity for that feature. The effectiveness of this mechanism turns out to be independent of the number of distractors, as it has a characteristic feature not belonging to the target (flat search function). Bottom-up directing of attention is needed when the unique target feature is not known. In accordance to the lateral inhibition model this is a local process. In this case, it was seen that increasing the number of distractors the RTs decreases (*negative search function*).[26] Finally, in the case described by Treisman and Gelade[8] mentioned above, in which a serial search through all items in the scene is required (discriminating between target and distractors which share some features), a *positive search function* ensues.

A similar multi-resolution approach has been widely investigated by the computer vision community.[27]

• *Spatial attention*. This is the mechanism of orienting the attention to a particular location in the scene. Orienting is usually defined in terms of foveation towards a stimulus (*overt orienting*). Stimulus foveation should improve the efficiency of target processing by exploiting acuity. This process can be voluntary or involuntary; in this last case it is *data-driven* and involves an odd target. The abrupt appearance of a cue produces the automatic orienting of the focus of attention on the cue location (*stimulus-driven*[28]). It is worthwhile pointing out the relationship between voluntary and automatic orienting of attention (for example due to an abrupt onset). It has been demonstrated[18] that when the location of the target is known exactly, the subject can override the attentional capture produced by abrupt onsets in other locations.

Different models of spatial attention have been proposed:

i) the spot-light model[29] in which attention is conceived as a beam;

ii) the zoom-lens model[30] in which the focus of attention can be differently sized, and the more narrow the focus the greater the benefit in processing;

iii) the gradient model[31] in which attention corresponds to a distribution of resources across the scene with suitable shape and dispersion.

In the two former models the beam is moved from one location to another corresponding to the shift of attention, whereas in the gradient model the attention is not moved per se, rather a resource gathering emerges at the cued location.

Note that, when attention is moved elsewhere following an exogenous cue the processing of stimuli at the previously cued location is inhibited. This important conduct is called *inhibition of return*. On the other hand, no inhibition occurs following an endogenous

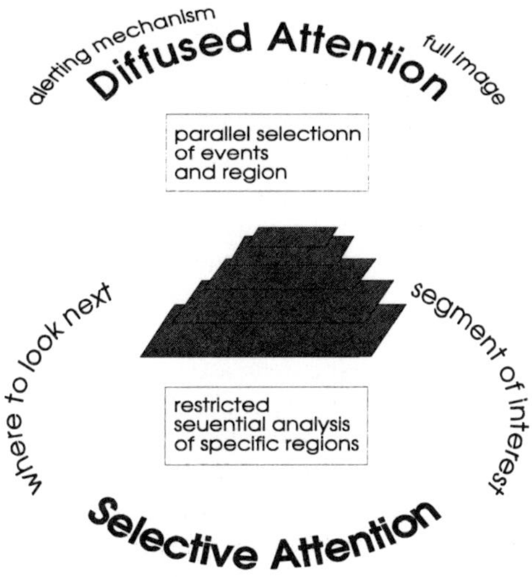

Figure 1. The 8 cycle of biological and artificial attention systems.

cue. The inhibition of return is particularly important in avoiding further exploration of locations and items that have already been inspected. The eye movements, which take a long execution time and are followed by a refractory period, require a particular effective control mechanism.

ATTENTION MECHANISMS IN ARTIFICIAL SYSTEMS

Among the different proposals of artificial systems that are inspired to the mentioned bio-approaches the most promising ones are based on the logmap transformations of the spatial domain,[32,33] and on multi-resolution processing.[34-36,27] Both these approaches consist of analyzing the images through a data structure which supplies the scene content at variable/different resolution respectively, so allowing the exploitation of the most appropriate data-details for the location, the operation, the task, and the image at hand.[37]

A framework for these techniques, in which a multi-resolution environment based on the pyramid theory is exploited,[38] is given in Figure 1. The approach illustrated for the artificial case is as well applicable to describe the bio-solution. A peripheral behavior, aiming to the selection of the relevant areas, consists of a parallel implementation of simple operators, on the complete field of view, at a low resolution. This pre-attentive like phase is often implemented through specialized or massively parallel hardware. Then the full capabilities of the system, at the most sophisticated levels of detail, are required to analyze sequentially selected image segments to interpret the scene reality and follow its evolution.

The a priori knowledge about the scene guides the interpretation phase, and, when a result is not achieved, a new selection on the salient spatial feature map is performed. It is worthwhile stressing how multi-resolution hardware systems can noticeably raise the process performance by supplying features at different scales at low cost, supplying locality measures, and allowing correlation along the scales in both coarse-to-fine and fine-to-coarse modalities. In particular, these features facilitate the implementation of the sequence of hypothesis-test steps for the recognition of complex objects in presence of similar 'distractors'.

As a general strategy, large components can be analyzed at low resolution, while small components are tackled at full resolution. Nevertheless, independently of the targets size, a considerable reduction in the amount of data to be dealt with is gained every time that the property involved is scale-independent.

Two artificial systems which represent implementations that more closely follow the receptors distribution in the retina and the biological solution are the proposal of Bendenson et al.[39] for the Cortex I and the one of Mc Gill University for the FOVEA.[40,41]

In this chapter more details are given for the second approach that achieved the first commercial available multi-resolution systems, exploiting the emulation of the focus of attention typical of biological systems through a multi-resolution representation. In this connection, different proposals of representation have been proposed; among these the following three are particularly relevant:

- *pattern tree*[37] where the sub-patterns are represented as nodes of a tree; each descendant node corresponds to a detail of the sub-pattern of the parent node;
- *models feature graph*[36] based on an acyclic graph; in this representation too, nodes correspond to sub-patterns or features with a hierarchical organization. The difference with respect to the pattern tree is that nodes may be reached via different paths from the roots;
- *refinements grammar*[42,43] in which a linguistic approach to describe 2-D objects at different resolution levels by means of a context-sensitive grammar is used. Production rules codify the image variations from a coarser to a finer level.

These representations allow the guidance of the focus of attention mechanisms. An object recognition process may be considered as a node-to-node sequence of partial matches. At each stage, the successive step is determined according to the partial results so far achieved. This strategy is composed of two basic steps:

i) the selection of a low level scale to hypothesize the presence of what we are looking for;

ii) the zooming on the exact position at a higher scale to achieve evidence of the presence/absence of the target.

The selection of the optimum level at which the search has to start is a particularly critical task that depends on the characteristics of the object-target at hand. In general, for reasonable constraints (like the original spatial resolution, the maximum shift allowed to the target in the search space, etc.) the number of used levels turns out to be less or equal to 3 or 4. These cases correspond to data reduction factors of 64 or 256.

The second phase of the approach is generally called the *refinement* and is effectively implemented by proceding top-down through iterative re-arrangements. More precisely, starting from the results of the first screening of the region of interest, the refinement is accomplished step-by-step at every intermediate level, until the maximum resolution level is reached; at each level the confidence of the selection is verified, and the spatial position of scrutiny is refined by checking (e.g., by thresholding a suitable figure of merit) the near neighbors corresponding to the position of the previous level.

To implement the quoted computational platform two main approaches have been proposed. The former is based on an 'ad hoc' hardware for multi-resolution construction, the latter is based on a pyramidal structure. Two different implementations of the latter class have been pursued: the former with bin- and quad-pyramid topology, and the second adopting a logical pyramid made up of a flat array. In the following sections these basic architectures are discussed.

THE MULTIRESOLUTION CONSTRUCTOR MODULE

A multiresolution processing environment can be achieved by a pipeline architecture specialized for decimation and expansion. This approach was conceived almost ten years

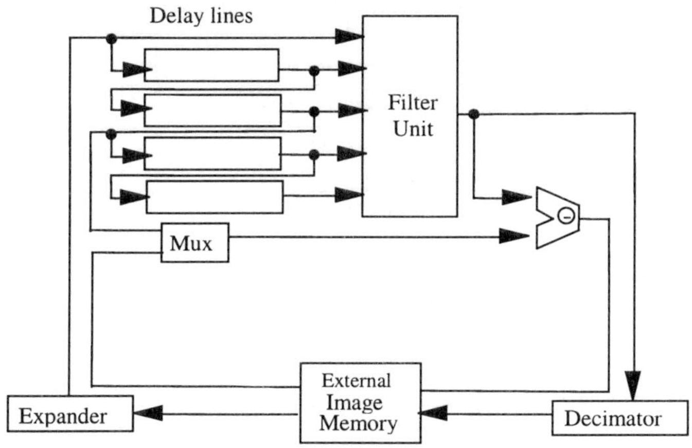

Figure 2. Simplified main loop of the pipeline multiresolution constructor.

ago at the D. Sarnoff Center (Princeton, USA - SENSAR)[44] and it has been continuously upgraded with new hardware and software facilities. The multi-resolution constructor constitutes the minimum hardware for implementing a 'smart sensor' that not only supplies the multi-resolution environment, but which is also suitable, being a convolver, for implementing low level operators.

The main loop of the multiresolution constructor is shown in Figure 2. It contains:

i) a filter unit which is fed by the sub-array elements obtained by suitable one-line-long delay units;

ii) a decimator unit which extracts the subset of the output pixels of the filter unit that form the next reduced resolution image;

iii) an expander unit which doubles the image size by padding with zeros the input pixels and, through the filter unit (with a suitable interpolating kernel), forms the next expanded resolution image;

iv) the image memory that stores all versions of the decimated image;

v) multiplexing and subtractor units complete the circuit supplying the capability of building on-the-fly simple arithmetic operations between different resolution images.

The loop is specialized to implement particular pyramidal transforms (Gaussian and Laplacian) on the original image. With this hardware, all the resolution versions are obtained in 4/3 of the scanning time of the original image. In fact, the scanning of the first level, receiving only 1/4 of the image data, requires only 1/4 of the base scanning time, the second level requires $1/4^2$, the third $1/4^3$, and so on, for a total amount of 1/3. The host system, by adopting this hardware, can exploit the multi-resolution environment with a delay of 1/3 of the frame time. Moreover, the multi-resolution constructor, integrated with simple off-the-shelf hardware, can be seen as a specialized co-processor capable of supplying selected windows of variable sizes and resolutions to the higher level or host system thus realizing a cheap and powerful Visual Front End (VFE) (ACADIA).[45]

COMPACT PYRAMIDAL ARCHITECTURES

The pyramidal architectures combine the operational capabilities of the mesh arrays in each layer with those of the trees for inter layer operations. The most popular configuration

Figure 3. The quad-pyramid interconnection PE network.

for pyramidal architectures is the quad-pyramid shown in Figure 3. Some authors proposed alternative solutions adopting bin-pyramids, flip-trees, multi-trees, hypernets, etc..[46]

Two basic approaches have been followed in physically building a pyramidal network of processing elements (PE): one with fine granularity, where, in the limit, there is one processor per image pixel of the base and another with coarse granularity, where one powerful processing unit (PU) is associated to an image block.[47] The former group belongs to the family of "compact" pyramids and corresponds to SIMD machines in which several layers of processors are organized in a tightly orchestrated mode. The approach is to set up systems using the largest number of processing elements (compatible with economical and technological constraints) executing in each layer the same operation in parallel, in spite of the single processing unit power. For this reason the word length is one bit and serial arithmetic is adopted. Itemizing, the arrays are characterized by:

i) simple single-bit custom PEs working with serial arithmetic;

ii) fine granularity, i.e. one PE per pixel on the base;

iii) concurrent access to both the lateral and hierarchical near neighbors (often implemented through a broadcasting/gating modality allowing to compute in one clock cycle a boolean function of the whole neighborhood);

iv) near neighbor recursive instructions;

v) small amount of local memory per PE;

vi) at most, one controller for each layer of the pyramid to implement a Multi-SIMD control modality;

vii) image loading/unloading, usually overlapped with the program execution on the PEs.

Moreover, the pyramid structure can be considered also as the three-dimensional extension of the two-dimensional binary tree, then two more features can be useful in computer vision. The first is related to the well-known order-logarithmic path-length followed in the interprocessor communication; the second depends on the possibility of working on different images flowing in a pipeline mode towards the apex. These systems are suitable for the implementation of planning strategies, multi-grid solutions, and low level image processing, but the implementation of high level image processing is cumbersome and less effective.

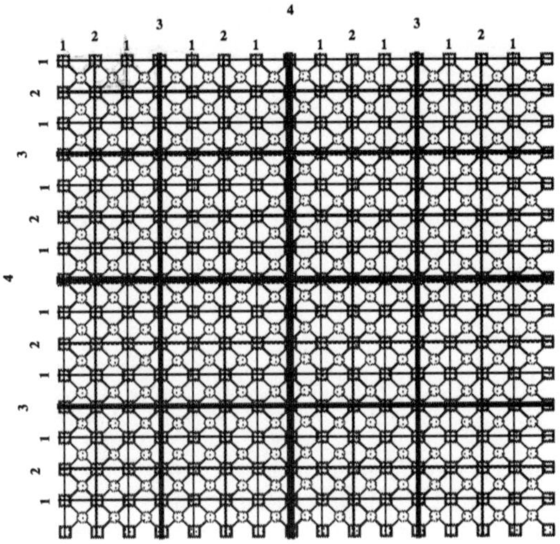

Figure 4. A simplified scheme of the PAPIA II system. Square boxes represent PEs, circles represent switches. Only six levels are included.

Some multi-resolution approaches have been implemented on standard flat array. An effective solution is shown in Figure 4.[48] Five logical planes are represented: each element is a PE of the base of the logical pyramid; PEs of the levels above the base are indicated by a number equal to the level they belong to. Each processor at most belongs to two different planes: the full resolution plane and a higher level one. To maintain the full connectivity of the pyramid topology each PE needs bi-directional connections towards its eight near neighbors plus the capability to short-port its connections both in diagonal and orthogonal directions. Since this pyramid is a virtual one, and its physical structure is instead a mesh, according to the position of the region of interest one may consequently place the apex of the logical pyramid so that a sub pyramid of suitable size can be placed wherever necessary on the image. Furthermore, the fault tolerance mechanisms are those usually employed on regular and homogeneous PEs arranged on a bi-dimensional array. For these reasons, and for the simplification obtained in the chip and board interconnections, this 'flat' configuration is the most suitable for massive implementations. The design of a chip containing 8x16 PEs has been carried out with a 1 μ, two-metal CMOS technology; the periphery of the chip consists of 232 pins.[49]

CONCLUSIONS

Computer vision systems are tackling the problems of real-time vision. This challenging goal can be pursued also through solutions that are inspired to the biological ones. In this connection, different strategies for the selection of region of interest and the re-allocation of available resources lead to different architectural solutions. Some of these approaches, still under development, have been analyzed, to offer the broadest view on this promising field.

REFERENCES

1. M. Savini, Active vision, in V. Cantoni ed., *Human and Machine Vision: Analogies and Divergencies*, New York, Plenum Press, pp. 159-170, (1994).

2. D. Sagi, Where and What for - Perception Made Ready for Action, in V.Cantoni, M. Marinaro, and A. Petrosino eds, *Visual Attention Mechanisms*, New York, Kluwer Academic/Plenum Publishers, in press.
3. J.K. Tsotsos, There is no one way to look at vision, *CVGIM: Image Understanding*, Vol. 60, N. 1, pp. 95-97, (1994).
4. L. Yarbus, *Eye movements and vision*, New York, Plenum Press, (1967).
5. J. M. Findlay, Global processing for saccadic eye movements, *Vision Research*, Vol. 22, pp. 1033-1045, 1982.
6. P.J. Burt, C.H. Anderson, J.O. Sinniger, and G. van der Wal, A Pipilined Pyramid Machine, in V. Cantoni and S. Levialdi eds., *Pyramidal Systems for Computer Vision*, Berlin, Heidelberg, Springer-Verlag, NATO ASI Series F, Vol 25, pp. 133-152 (1986).
7. C. von der Malsburg, and W. Schneider, A neural cocktail-party processor, *Biological Cybernetics,* 54 29-40 (1986)
8. A. Treisman and G. Gelade, A feature-integration theory of attention, *Cognitive Psychology*, 12, pp. 97-136 (1980).
9. S.J. Luck and S.A. Hillyard, Spatial filtering during visual search: evidence from human neurophysiology, *Journal of Experimental Psychology: Human Perception and Performance*, 20 (5), pp. 1000-1014 (1994).
10. U. Neisser, *Cognitive psychology*, New York, Appleton-Century-Crofts (1967).
11. B. Julesz, Texton gradients: the texton theory revisited, *Biological Cybernetics*, 54, pp. 245-251 (1986).
12. J. Jonides, Voluntary versus automatic control over the mind's eye, in J. Long & A. Baddeley eds., *Attention and performance IX*. Hillsdale, N.J.: Lawrence Erlbaum Associates (1981)
13. D. Marr, *Vision*, Freeman, New York, 1982.
14. D.H. Hubel and T.N. Wiesel, Receptive fields, binocular interaction and functional architecture in the cat's visual cortex, *Journal of Physiology*, 160, pp. 106-154 (1962).
15. Z. Pylyshyn, The role of location indexes in spatial perception: a sketch of the FINST spatial-index model, *Cognition*, 32, pp. 65-97 (1989).
16. M. Green, Visual search, visual streams, and visual architectures, *Perception and Psychophysics*, 50 (4), pp. 388-403 (1991).
17. S. Yantis and J. Jonides, Abrupt visual onsets and selective attention: evidence from visual search, *Journal of Experimental Psychology: Human Perception and Performance*, Vol. 10, pp. 601-621, (1984).
18. S. Yantis and J. Jonides, Abrupt visual onsets and selective attention: voluntary versus automatic allocation, *Journal of Experimental Psychology: Human Perception and Performance*, 16 (1), pp. 121-134 (1990).
19. D.E. Broadbent, *Perception and communication*, London, Pergamon Press, (1958).
20. P. Verghese and D.G. Pelli, The information capacity of visual attention, *Vision Research*, 32 (5), pp. 983-995 (1992).
21. J.M. Wolfe, K.R. Cave, and S.L. Franzel, Guided search: an alternative to the feature integration model for visual search, *Journal of Experimental Psychology: Human Perception and Performance*, 15 (3), pp. 419-433 (1989).
22. J. Duncan and G.W. Humpheys, Visual search and stimulus similarity, *Psychological Review*, 96 (3), pp. 433-458 (1989).
23. S. Yantis,. Goal-directed and stimulus-driven determinants of attentional control. In S. Monsell & J. Driver eds., *Attention and Performance XVIII*, Cambridge, MA: MIT Press, pp. 73-103, (2000).
24. K. Nakayama, The iconic bottleneck and the tenuous link between early visual processing and perception, in C. Blakemore ed., *Vision: Coding and efficiency*, Cambridge: Cambridge University Press, pp. 411-422, (1990).
25. C.Koch, and S. Ullman, Shifts in selective visual attention: towards the underlying neural circuitry, *Human Neurobiology*, Vol. 4, pp. 219-227, (1985).
26. M. J. Bravo, and K. Nakayama, The role of attention in different visual-search tasks, *Perception and Psychophysics*, Vol. 51, pp. 465-472, (1992).
27. V. Cantoni and M. Ferretti, *Pyramidal Architectures for Computer Vision*, New York: Plenum Press, (1994).
28. S. Yantis, Stimulus-Driven and Goal Directed Capture Attentional Control, in V. Cantoni, M. Marinaro, and A. Petrosino eds, *Visual Attention Mechanisms*, New York, Kluwer Academic/Plenum Publishers, in press.
29. M.I. Posner, Orienting of attention, *Journal of Experimental Psychology*, 32, pp. 3-25 (1980).
30. C.W. Eriksen and St. J.D. James, Visual attention within and around the field of focal attention: a zoom lens model, *Perception and Psychophysics*, 40 (4), pp. 225-240 (1986).
31. D. LaBerge and V. Brown, Theory of attentional operations in shape identification, *Psychological Review*, 96 (1), pp. 101-124 (1989).
32. E.L. Schwartz, Computational Anatomy and Functional Architecture of Striate Cortex: A Spatial Mapping Approach to Perceptual Coding, *Vision Research*, Vol. 20, pp. 645, (1980).

33. Y. Yeshurum and E.L. Schwartz, Shape Description with a Space Variant Sensor: Algorithms for Scan-path, Fusion, and Convergence over Multiple Scans, *IEEE Trans. on PAMI*, Vol. 11, pp. 1217-1222 (1989).
34. S.L. Tanimoto and A. Klinger eds., *Structured Computer Vision: Machine Perception Through Hierarchical Computation Structures*, New York, Academic Press, (1980).
35. A. Rosenfeld ed., *Multiresolution Image Processing*, Berlin, FRG, Springer Verlag, (1984).
36. C.R. Dyer, Multiscale Image Understanding, in *Parallel Computer Vision*, L. Uhr ed., Orlando, FL, Academic Press, pp. 171-213 (1987).
37. P. J. Burt, Attention Mechanisms for Vision in a Dynamic World, *Proc. 9th Int. Conf. on Pattern Recognition,* pp. 977-987, (1988).
38. V. Cantoni and M. Mosconi, Approaches for Allocation of Attention in Computer Vision Systems, *WOPPLOT*, Springer-Verlag, in press.
39. B.B. Bendenson, R.S. Wallace, and E.L. Schwartz, A miniaturized active vision system, *Proc. 11th Int. Conf. on Pattern Recognition,* The Hague, NL, pp. 58-61, (1992).
40. T. Baron, M.D Levine, and Y. Yeshurum, Exploring with a Foveated Robot Eye System, *Proc. of 12th Int. Conf. on Pattern Recognition,* Vol. d, pp. 377-380, (1994).
41. M. Bolduc, G. Sela, and M.D. Levine, Fast Computation of Multiscalar Symmetry in Foveated Images, *Proceedings CAMP'95*, IEEE Computer Society Press, pp. 2-11, (1995).
42. V. Cantoni, L. Cinque, C. Guerra, S. Levialdi, and L. Lombardi, Describing Object by a Multi-resolution Syntactic Approach, in A Nakamura, M. Nivat, A. Saudi, P. S. P. Wang, and K. Inoue, eds., *Parallel Image Analysys,* Heidelberg, FRG: Springer Verlag, pp. 54-68, (1993).
43. V. Cantoni, L. Cinque, C. Guerra, S. Levialdi, and L. Lombardi, Describing Object by a Multi-resolution Syntactic Approach, *Pattern Recognition*, pp. 54-68, (1998).
44. G. van der Wal, and P. Burt, A VLSI Pyramid Chip for Multiresolution Image Analysis, *Int. Journal of Computer Vision,* 8:3, pp. 177-189, (1992).
45. G. van der Wal, M. Hanse, and M. Piacentino, The Acadia Vision Processor, *Proc. CAMP 2000,* Padua, Italy, Sept. 11-13, pp. 31-40, (2000).
46. V. Cantoni, M. Ferretti, and L. Lombardi, A Comparison of Homogeneous Hierarchical Interconnection Structures, *Proc. IEEE,* Vol. 79, N. 4, pp. 416-428 (1991).
47. P. E. Danielsson, Vices and Virtues of Image Parallel Machines, in S. Levialdi, *Image Analysis and Processing*, London: Pitman Book, (1983).
48. M. G. Albanesi, V. Cantoni, U. Cei, M. Ferretti, and M. Mosconi: Embedding Pyramids into Mesh Arrays, in: H. Li, and Q. F. Stout eds.: *Reconfigurable Massively Parallel Computers*, Englewood Cliffs: Prentice Hall, pp. 123-140, (1990).
49. M. G. Albanesi, V. Cantoni, M. Ferretti, and F. Maineri, A VLSI 128-Processor Chip for Multiresolution Image Processing, *Proc. Int. Conf. on Massively Parallel Computer*, IEEE CS Press, pp. 296-307, (1994).

PANEL SUMMARY: **PLANNING AND PLASTICITY IN ARTIFICIAL AND NATURAL SYSTEMS**

Panelists: Vito Di Gesù[1] and Settimo Termini[1]
Reactions: Virginio Cantoni[2] and Bertrand Zavidovique[3]

[1]DMA - University of Palermo
Via Archirafi 34, 90123 Palermo, ITALY
e-mail: (digesu,termini)@math.unipa.it
[2]DIS – University of Pavia
Via Ferrata 1, 27100 Pavia, ITALY
e-mail: cantoni@unipv.it
[3]Université Paris-XI
91405 ORSAY Cedex, FRANCE
e-mail: zavido@ief.u-psud.fr

INTRODUCTION TO HOT POINTS *(V. Di Gesù)*

Boncinelli underlined often, during our workshop, that simplest organisms are the more robust and they survived longer than more evolved organisms (what about the quality of their life?). This depends on the fact that complex systems either natural or artificial need for more complex controls. Moreover, planning strategies, one of the steps for finding optimal solutions, must be carefully designed and it is complex too.

In the following I will restrict my discussion to artificial systems (for example robotic systems). In this case serial and parallel *actions* are assigned to proper *agents/actors* in a given *(dis)order*, depending on the *structure* and *complexity* of the problem to be solved. Here, disorder means that agents can execute their actions with a certain degree of freedom.

Distributed and interacting planners drive complex artificial systems. For example, two planners may send in a concurrent way a command to the same agent; and this may cause the needs for conflicting solutions (see Fig.1). Moreover, the same *actuator* can be used by more that one agent. An actuator can be considered as the simpler element of the system: it does simple and exactly coded actions.

From this sketch it follows that a planner module is not a simple scheduler; in fact, *agent/actors* can either co-operate or compete, they can both co-operate and compete.

The picture can be more complex if we allow agent/actors to be endowed with autonomous behaviors. This case must be considered whenever the system is engaged with a hard problem. In this case planners take care of the general strategy leaving the details of the solution of a particular problem to the agent/actor. For example, in the case of an artificial

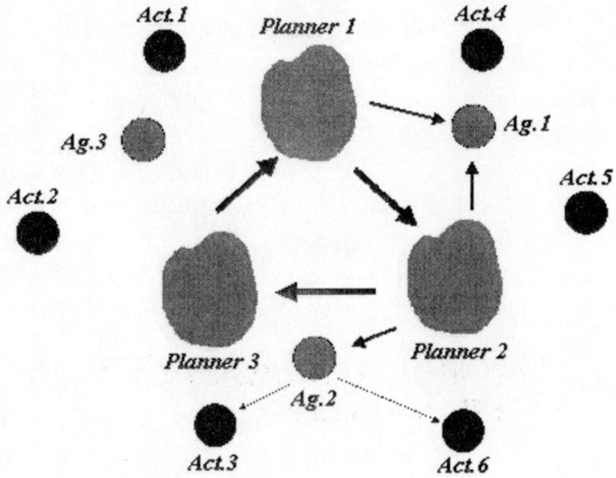

Figure 1. The general architecture of a complex system planner.

visual system, active detectors may take case of preliminary computation (detect the region of interest, filter the image…).

Let me introduce another point: *the environmental conditions may change dynamically;* i.e. the system must be ready to react against new conditions in order to solve its problem. The conditions may change with an uncertainty that makes it harder to find stable solutions. This means that planners must be designed in order to be able to adapt themselves to these changes.

After these considerations let me list some of the hot points I would like to discuss:

• How systems may adapt their strategies to the environment changes? orHow can we include *plasticity* in artificial planners? I use the term plasticity in the context of artificial systems: i.e. plasticity is *the planner-capability to change its strategy and its interaction with other planners.*

• Is the internal "time" (both biological and quartz clock) effected by the nature of the problem to be handled? This is a big problem whenever we must solve problem at different level of complexity. In this case we need to switch the computation among different computers, depending on their computation power that is related to the computer clock.

• How will planning strategies change with the dramatic growth of distributed information systems? Internet is now connecting the entire world and data (*potential information*) are available if needed be. The future will be, as it is conventionally said, the distributed computing. Millions of interconnected computers will cooperate in solving problems. This point will have a dramatic impact in the design of planning strategies. The planner will become himself a *very* complex system.

HISTORIC REMARKS *(S. Termini)*

When Vito invited me to join him in thinking about this panel, at first sight, I thought that its title (*planning and plasticity in artificial and natural systems*) was very far from my interests and, subsequently, it was difficult for me to think about arguments and considerations that could aim at stimulating discussion and provoking comments. However, after pondering a little about what there is (or could be) behind the names involved ("planning" and "plasticity") - especially if theory were seen against the background of the connections between "the natural" and "the artificial", I realized that a whole new brave world could open before our eyes.

Both planning and plasticity - besides the specific technical reference they have to present-day research in AI and similar fields - point to a very huge number of interesting problems, unitary affordable in the setting of research paradigms with specific features. Many of these have been already at the core of the original scientific program of Cybernetics and the association in the title of the panel of the natural and the artificial cannot reinforce the same flux of thoughts.

Present day attitude towards the roots of our discipline(s) is a little bit schizophrenic in the sense that the spectacular results (mainly of technological type) induce us to forget that any new achievement has always been *counterbalanced* by the arising - in a sort of Dantean "law of retaliation" - of a host of new (theoretical and conceptual) problems. It has been always so in our new disciplines in these last fifty years (and perhaps it has been always so in the first decades of emerging fields of investigation.

Let me observe that this has happened not only for accomplishments on the technological side but also for those scientific advances of purely theoretical flavor, as well as for the conceptual clarifications which have been obtained. We must explicitly recognize that still unsolved problems of conceptual type are present, and a good attitude should be the one of taking into account the lessons of the history of this same field. Having from one side the humility of considering that many important but scattered results cannot be in their final form but only steps towards the construction of a more general and comprehensive theory. The attempts at unifications have been - no doubt - the most critical pages of the brief history of Information Sciences, but we must come to terms with these problems and not hide ourselves behind reduced presentations of the same problems.

If we look at the history of the past fifty years, we realize that we have seen again and again on the stage things, concepts (and also pseudo-concepts) which had been considered dead years or decades before and which reappeared again under new names.

Both planning and plasticity can be certainly related to old notions like feedback, adaptivity, learning and the fact of taking them into account in artificial AND natural systems cannot but remind the unifying ambitions of Cybernetics as well as its problems.

One of the more devastating criticism which - periodically - is arisen against the most general ambitions of scientific enterprises as Cybernetics in the forty/fifties/sixties, AI in the subsequent decades, Cognitive sciences or Consciousness Studies now, is to present again and again essentially the same general questions and problems providing only new names.

In fact it seems that there is an insurmountable gap between the ambitions (to look for unified theories which would encompass -meaningful- aspects of both the *natural* and the *artificial*) and what, periodically, has been actually realized.

One could, of course, certainly escape out of this dilemma and others related big problems by saying that one is only interested in making some qualitative comparison and in borrowing some general ideas from one field to the other. This means that the actual, real scientific work will be done - in fact - only inside one specific field.

Of course, this is possible; however, the fact that from the early fifties, people has been inclined to look for some unifying framework and is still using a language which makes non trivial references to the possibility of passing from one level to the other, shows - in my view - that the problem of looking at the possibility of a working unification is a question which must not be exorcized but one to which one must try to provide answers at least for what regards the two following points:

1) an historical assessment of what has happened in the past 50 years;
2) the real import of the difficulties which have been faced and which we must face.

PLASTICITY AND RECONFIGURABILITY *(V. Cantoni)*

I cannot avoid quoting at least two comments of the first workshop of this series (or second, considering also the "Human and Machine Vision" of Pavia, 27-30 September 1993),[1] and precisely the panel session on: 'Plasticity and reconfigurability in sensory systems', chaired by S.Vallerga:[2]

1) (S. Vallerga, biologist): "Natural neural networks are highly plastic in structure and function and may adapt to long lasting environment changes. Plasticity and adaptation are not confined to lower level of processing ...[same session] I should say to my surprise the memorable amount of recent concepts acquired by neurophysiology, that have already been absorbed by artificial intelligent studies".

2) (V. Roberto, computer vision scientist) ... "'plasticity' is a key property of a machine: on general grounds, it denotes the capability to modify its configuration in response to a stream of stimuli".

I strongly believe that plasticity and reconfigurability are two terms describing the same functionality, in the realms of natural and artificial systems, and that was the reason for which we, Vito, introduced that very exiting panel. Now my question is: considering the biological solution (plasticity) still more effective than the artificial equivalent what you consider the basic weak aspects that must be more developed in artificial system to reach the quality of the other realm? May be the capability of evolving? The adaptation to dynamic environments? The control effectiveness? etc.

SYSTEM COMPLEXITY *(B. Zavidovique)*

A system appears complex for several reasons: there are lots of active units and intricate exchanges, or it solves complex problems, or we do not understand how it works, etc...Most planners for real systems are meant to be complex in some of these senses. Like Vito Di Gesù said, the available technology in computing and storing likely leads to distributed versions: several actual entities communicate and compete to order a given set of moves. The multiplicity is due to the number of potential actions and situations with respect to reaction times allowed in a realistic world. Moreover, external conditions to the planning for instance related to energy (power supply), to the exo-system (aggressiveness), or to the system itself (aging, getting tired or used or overwhelmed...) make that time is neither solely nor fully dependent on the problem or the system status.

All the above suggests that being able to (self-)reconfiguring the machine to carry planning out would help. So would the capability to put, on demand, more and different resources at work or to rearrange those already run : that is plasticity, at least if the result lasts! Note that a first step toward plasticity then is for a system (or part of a system) to adopt and follow a plan (or part of a plan) that it did not produce, integrating it in its own plan. That poses problems of plan compression to start. More generally, representations that systems manipulate and reason on do vary with time and according to space (point of view from sensors, data unavailable or getting obsolete etc...): optimizing planner variations with the representations would then be a second step to plasticity.

It has been a permanent question in HMP, how to make sure about a given property of a complex system when it is observed from outside. It is not clear where plasticity shows in human planning. Richard Walker suggested that an expert (chess great master, music concert performer...) proves plasticity in the brain. Actually, they might well suggest the contrary: in such cases of extreme skills, the architecture stops adapting to the complexity of planning, whether permanently or not. Instead, when data have been understood and stored from enough previous repeated experience where such plans occurred necessary, our access to data is made more efficient and starts predominating. Improvement can be on the storage itself or on the retrieval. In all case when it comes to ultimate efficiency recognition (situation assessment with direct link to action) seems to overcome planning and make it simpler. It brings a more static then less combinatorial, possibly acquired-reflex and definitely executive process in strategic analysis (only few good moves are considered) as well as in actuator/sensor coordination (the right gesture is anticipated). The same phenomenon is trivially true in automation: compilation, pre-stored control etc. can rarely be avoided in case of increased time or situational difficulty. But does it tell anything about planner plasticity for less constraining cases?

REFERENCES

1. V. Cantoni ed., *Human and Machine Vision: Analogies and Divergencies*, Plenum Press, New York, (1994).
2. S. Vallerga, G. Grisanti, and V. Roberto, Panel Summary: Plasticity and Reconfigurability in Sensory Systems, in V. Cantoni, V. Di Gesù, A. Setti, and D. Tegolo eds., *Human and Machine Perception: Information Fusion*, Plenum Press, New York, pp.139-145, (1997).

NOTION FORMATION IN MACHINE LEARNING

Lorenza Saitta[1] and Roberto Esposito[2]

[1]Dipartimento di Scienze e Tecnologie Avanzate
Università del Piemonte Orientale "Amedeo Avogadro"
Corso Borsalino 54 - 15100 Alessandria, Italy
e-mail: Lorenza.Saitta@unipmn.it
[2]Dipartimento di Informatica
Università di Torino
Corso Svizzera 154 - 10149 Torino, Italy
e-mail: esposito@di.unito.it

INTRODUCTION

In Artificial Intelligence, and, specifically in Machine Learning, to form a new notion usually means to build up a "concept" or a "category". The simplest way to consider a category is extensional: a category is a set of "equivalent" objects,[1] i.e., objects that share properties.[2] Categories are organized into *taxonomies,* linked through the set inclusion relation.[1] A concept is sometime considered as the intensional representation of a category, i.e., a description of the objects in the category,[2] called also the *instances* of the concept. Actually, most frequently, the denotations "concept" and "category" are used as synonyms.

Intension and extension are dual properties of concept hierarchies. For exemplification sake, let us consider the simple concept hierarchy reported in Figure 1. Visiting the network *bottom-up*, a set inclusion relation holds between concepts extensions; visiting the network *top-down*, a set inclusion relation holds for concept intensions. Any node in the hierarchy inherits *instances* from its descendants and *properties* from its ancestors.

The intensional approach to the definition of a concept includes at least three different views. In the *classical* view, which goes back to Aristotle, a concept is a "name" corresponding to a set of necessary and sufficient conditions.[4] The instances are not given explicitly: they are all those objects that satisfy the conditions. This definition, well suited for mathematical concepts, is inadequate for everyday-life domains;[5] then, weaker definitions have been sought. The *heuristic* view of a concept only keeps sufficient conditions, whereas the *exemplar* view considers a concept as a *prototype,* i.e., a real or virtual example with the "typical" properties occurring in the instances.[6]

To form and use categories is a fundamental activity in human reasoning, one that is so natural that we may be tempted to consider the underlying cognitive process easy to understand and reproduce. On the contrary, category formation appears to be quite a sophisticated process, still far from being understood in its completeness.[7]

Human and Machine Perception 3: Thinking, Deciding, and Acting, edited by Cantoni *et al.*
Kluwer Academic/Plenum Publishers, New York, 2001.

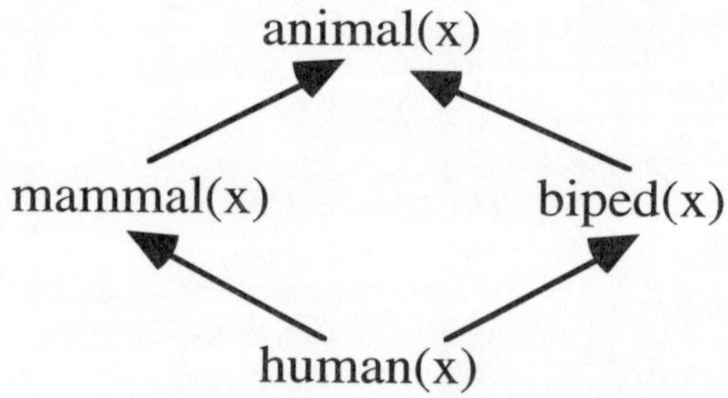

Figure 1. Simple concept hierarchy.[3]

In artificial systems, the heuristic and prototypical views of concepts are the dominant ones, especially in Machine Learning, where, in particular, concepts are most often equated with their description in a chosen language.[8,9] In reasoning, the prototypical view is embedded in the Case-Based approach.[10,11,12] The definition of a concept through a prototype has the advantage of being able to model the gradual nature of membership, clearly emerging from several psychological studies on semantic memory.[1,13]

Rosch[1] claims that categorization in humans is guided by two principle: the first is the *Cognitive Economy* principle, which asserts that the categorical structure should allow the maximum number of properties to be predicted for an object, knowing which category it belongs to. Given a goal, properties irrelevant to the goal should be ignored. The second principle states that the category structure should reflect the discontinuities and the correlations among attributes occurring in the world. Then, Rosch proposes a two-dimensional category structure, with a vertical dimension based on set inclusion (superordinate, basic, and subordinate levels), and a horizontal one based on prototypes. Membership in a category is based on parameters such as *cue validity*, *family resemblance*[14] and others.

Rosch's theory of predominance of the basic level in learning and reasoning has been criticized under various aspects.[15,16] In particular, several researchers have pointed out that categorization is a process that may be strongly influenced by background knowledge. Specifically, Murphy and Medin have pointed out that similarity is not sufficient to explain conceptual coherence;[2] in fact, similarity measures are too much dependent on feature weighting and selection. Also, there are usually too many correlations among attributes, and it is difficult to identify the relevant ones. On the contrary, categorization coherence can be explained by assuming that it is based on the theories of the world that people form. Concepts seem to be organized in the context of naïve theories that are based on everyday experience. This background knowledge constrains the type of attributes considered for categorisation.[17] From this perspective, the *functional* properties of objects are also important.[18] In fact, access to conceptual knowledge is needed to determine which concepts are involved in a given problem, and to infer their properties and relations with other concepts. There are substantial findings that suggest that *categorization* is a flexible process in which background knowledge and task demands determine which attributes and dimensions should be considered.[19]

A more complex and dynamic view of concepts is proposed by Barsalou, who points out that concepts show a degree of *instability:* ranking of membership degrees in a category varies across cultures and individuals, even across psychological states or contexts for the same individual.[19] Actually, people do not seem to use an *invariant* categorical knowledge for different situations, but they build up different context-dependent concepts from the same category. At least two are the reasons for the instability: *accessibility* and *context-dependency*. The hypothesis is that concepts are dynamically formed by moving part of the large quantity of information permanently associated to a category in the long-term memory into the short-term memory, according to the current needs. The information associated to a category is of three types: Context-independent Information (CI), Context-dependent

Information (CD), and Recent Context-dependent Information (CD_{rec}). Then, the most recently formed concepts bias the new ones.

CONCEPT FORMATION IN ARTIFICIAL SYSTEMS

Given a concept and a set of objects, we can classify the objects into *examples* (positive instances) of the concept, and *counterexamples* (negative instances) of the concept. The most frequent setting in Machine Learning is the *learning concept from examples,* which consists in the task of acquiring an intensional description of an unknown concepts from positive and negative instances. An important distinction in learning settings is between *supervised* and *unsupervised* learning. Supervised learning assumes that a teacher is able to supply a set of labeled data, from which the learner must derive a general concept description. Unsupervised learning does not exploit pre-labeled examples, but searches for spontaneous emergence of patterns in the data.

CONCEPT FORMATION

Unsupervised learning encompasses three basic tasks: concept formation through *conceptual clustering,* theory formation and quantitative discovery. As opposed to learning from examples, conceptual clustering is a bottom-up method which forms sets of instances and possibly arranges them into a hierarchy. As an example, in Figure 2 a clustering algorithm should discover the three emerging groups of objects.

Unsupervised learning is the most ambitious and long term form of concept learning. Michalski's first approach to conceptual clustering, which led to the implementation of the system CLUSTER/2,[20] evolved toward the introduction of background knowledge and classification goals, encoded in Goal Dependency Networks,[21] allowing more meaningful clusters to be constructed. Lebowitz describes the system UNIMEM,[22] which incrementally groups exemplars into overlapping conceptual categories, based on predictive features. DISCON[23] is devoted to the formation of a hierarchical clustering, given a list of interesting attributes. Fisher presents COBWEB,[24] a conceptual clustering system that organizes data in order to maximize inference ability. Whereas these systems all produce tree-like hierarchies, KLUSTER produces a lattice in the representation formalism of KL-ONE[25].

Conceptual clustering is a problem similar to categorization in Cognitive Science. For instance, the system COBWEB [24] has been strongly influenced by Rosch's work[1].

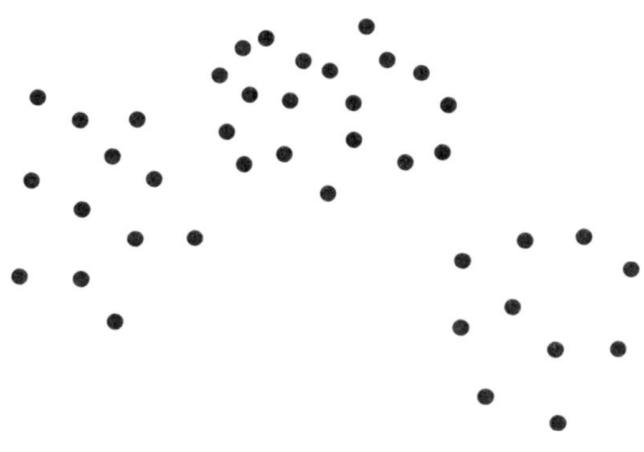

Figure 2. Three groups of objects (concepts) that a clustering algorithm should identify.

ABSTRACTION

A recent proposal by Goldstone and Barsalou to human conceptualization advocates the importance of perception even in abstract domains that seem far from the physical world.[26] The authors also suggest a number of cognitive operations that transform perceptions. This perception-based approach to conceptualization suggests that concepts need more flexible and articulated representations, each one better suited to a particular task and goal.

One dimension along which concept flexibility and goal-dependency can be achieved is the *abstraction level* of the representation. In this chapter we briefly describe a computational model of abstraction, which supports Barsalou and Goldstone's conceptual operations. Our approach originates from the observation that the conceptualization of a domain involves at least four different levels: Perception, Structure, Language, and Theory.

The main point of this analysis is that, at least in representing and manipulating concepts, abstraction should be grounded on perception, and that modifications of the structure and the language should follow the changes at the level of the perceived world.

More precisely, underlying any source of experience there is the *world* W. However, what is important for an observer is not the world *per se,* but the perception P that s/he has of it. P specifies the nature of the elements that constitutes the results of perception.

An actual world perception is obtained through a process $\mathcal{P}$ of signal/information acquisition from/about the world:

$$P = \mathcal{P}\,(W) \tag{1}$$

$\mathcal{P}$ detects objects, properties and relations specified by P through mechanisms that we leave implicit. At the perception level, the percepts "exist" only for the observer, and only during the act of perceiving. In order to let the stimuli become available over time, for retrieval and further reasoning, they must be memorized and organized into a *structure* S. This structure is an *extensional* representation of the perceived world, in which stimuli perceptually related one to another are stored together.

We denote by $\mathcal{M}$ the memorization process:

$$S = \mathcal{M}\,(P) \tag{2}$$

In order to describe in a symbolic way the perceived world, and to communicate with other agents, a *language* L is needed. L allows the perceived world to be described *intensionally*. Assigning "names" to the tables in the structure S and to their entries is a process of description:

$$L = \mathcal{D}(S) \tag{3}$$

Finally, a theory allows reasoning about the world. We call *formalization* the process of expressing the theory in the language L (possibly enriched to accommodate domain-independent background knowledge):

$$T = \mathcal{F}\,(L) \tag{4}$$

The four levels are ordered as in Figure 3. The meaning of an arrow $X \rightarrow Y$ and, consequently, of the functional denotation $Y = f(X)$, indicates that the syntactic and semantic definition of level Y must take into account the content of level X.

As no world is totally isolated, a body of background knowledge provides, in principle, input to each level, especially to the theory, where general laws and domain-independent facts may be needed. The primary role of perception (level P in Figure 3) in biasing the upper levels is strongly advocated by Goldstone & Barsalou. This dependency, however,

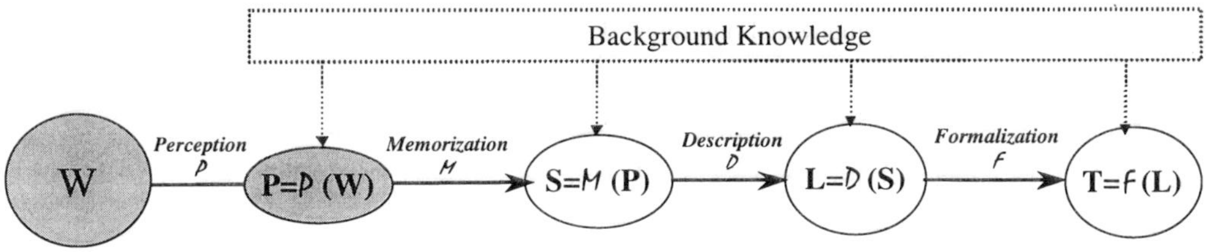

Figure 3. Levels involved in representing and reasoning about a world W. P denotes a perception of objects and their physical links in W. S is a set of tables, each one grouping objects sharing some property. L is a formal language, whose semantics is evaluated on the tables of S. Finally, T is a theory formulated using L, in which properties of the world and general knowledge are embedded. General background knowledge may provide inputs to any level.

does not mean that the content of the levels is generated strictly bottom-up: a complex bi-directional interplay may exist, and conceptualization or task information may even affect perception.[26]

Given the previous definitions, we call a description framework over a world W any quadruple $\mathcal{D}(W) = (P, S, L, T)$. We will now define abstraction as a mapping between description frameworks over the same world W. In the following definition it is worth noting that the abstraction process operates on the perception level P, and that the other levels are automatically influenced by P.

Definition 1 – Given a world W, let $\mathcal{D}_g(W) = (P_g, S_g, L_g, T_g)$ and $\mathcal{D}_a(W) = (P_a, S_a, L_a, T_a)$ be two description frameworks over the same world W, which we conventionally label as *ground* and *abstract,* respectively. An *Abstraction* is a mapping

$$\mathcal{A} : \mathcal{D}_g(W) \rightarrow \mathcal{D}_a(W) \tag{5}$$

such that P_a is "simpler" than P_g.

To state formally what we mean by abstraction we still have to formalize the meaning of the "simpler" relation between P_g and P_a. To this aim we define the set of possible configurations detectable by a perception process, and then we show how Kolmogorov's algorithmic complexity can be used to introduce the notion of "simplicity" that we need for the definition.

More precisely, given a world W, let P be a perception of W resulting from a process $\mathcal{P}$ that uses a set of sensors. A set of values provided by the sensors in $\mathcal{P}$ is called a *signal pattern* or a *configuration*; we denote by Γ the set of possible configurations detectable by $\mathcal{P}$. If, for example, the world is the real world and we perceive it through a digital camera $\mathcal{P}$, then Γ is the set of all the possible bitmaps that the camera can acquire, given its resolution and its color depth.

In order to capture the intuitive notion that a more abstract configuration should be described in a simpler way, and that its simplicity should not depend upon the tool used for the description, but should be intrinsic to the configuration itself, we can exploit the notion of *algorithmic complexity* introduced by Kolmogorov.[27]

Given $\mathcal{P}$, and, hence, Γ, any configuration $\gamma \in \Gamma$ can be described by a program π on a universal computer $\mathcal{U}$. Then, a complexity measure can be associated to Γ according to the following definition:

Definition 2 – Given a perception system $\mathcal{P}$ and the associated set Γ, let $\gamma \in \Gamma$ be any configuration. Given a universal machine $\mathcal{U}$ let

$$\mathbb{K}(\gamma) = \min_{U(\pi)=\gamma} \ell(\pi) + C \tag{6}$$

be the Kolmogorov complexity of γ, defined as the length $\ell(\pi)$ of the shortest program π that output γ on the universal machine. The complexity of Γ can be defined as:

$$\mathcal{K}(\Gamma) = \underset{\gamma \in \Gamma}{\mathrm{Max}}\ \mathcal{K}(\gamma) + C \tag{7}$$

The above definition has the advantage to be an "absolute" measure of the complexity of Γ, because it is machine-independent up to an additive constant. The additive constant C can be interpreted as the length of the program that describes the common structure of the configurations, i.e., the length of the program necessary for the machine $\mathcal{U}$ to interpret the descriptions π. Kolmogorov complexity is not computable. Nevertheless, we can use computable approximations of $\mathcal{K}$, sufficient for practical purposes, provided that the same approximation is used uniformly over all the Γ's.

Kolmogorov's algorithmic complexity can be thought of as a measure of the simplicity of the configuration detected by a perception process. Then, we can say that a perceived world P_2 is simpler than the perceived world P_1 when the algorithmic complexity of the first is smaller then the one of the second. More formally:

Definition 3 – Given a world W, let $\mathcal{P}_1$ and $\mathcal{P}_2$ be two perception processes generating P_1 and P_2, respectively. Let γ_1 and γ_2 be the corresponding configuration sets. The perceived world P_2 will be said *simpler* than P_1 iff:

$$\mathcal{K}(\gamma_2) \leq \mathcal{K}(\gamma_1) \tag{8}$$

Assuming that a ground representation framework $\mathcal{D}_g(W) = (P_g, S_g, L_g, T_g)$ has been defined over a world W, we will now investigate the relations between $\mathcal{D}_g$ and a more abstract representation framework $\mathcal{D}_a(W) = (P_a, S_a, L_a, T_a)$. These relations are schematically illustrated in Figure 4.

The symbols ω, σ, λ and τ in Figure 4 denote *abstraction operators,* belonging to four sets Ω_W, Ω_S, Ω_L and Ω_T, respectively, acting at the four considered levels. The operators belonging to Ω_W, Ω_S, Ω_L, and Ω_T will be called P-operators, S-operators, L-operators, and T-operators, respectively.

In Figure 4 two dimensions appear: the vertical one, along which the *nature* of the representation changes, and the horizontal one, along which *abstraction,* for each type of description, occurs. As abstraction is, in the proposed model, grounded on perception, the P-operators are the ones that have to be defined first, whereas the other operators must reflect, if possible, the perceptual transformations. If a perceptual transformation linking $\mathcal{P}$ and $\mathcal{P}'$ exists, no problems of consistency should appear;[26] instead, it is possible that the language (and/or the theory) becomes unable to describe the new perceived world.

Definition 4 – An abstraction P-*operator* ω denotes a procedure that takes as input a perception $P_g(W)$ of a world W and outputs a *simpler* perception $P_a(W)$ of the same world. The *domain of application* of a P-operator represents the subset of the world perceptions to which it applies.

Given the abstraction P-operator ω and, hence, the abstracted perception of the world $P_a(W)$, the domain conceptualization can be performed using the previously defined processes of memorization, description and formalization. However it is not always obvious how to memorize the abstracted perception of the world; also, the subsequent steps may be even more difficult. On the other hand, it has been found that most of the time it is much simpler to translate directly S_g in S_a, L_g in L_a and T_g and T_a (and this explains the presence of the other abstraction operators in Figure 4). While it seems to be useful to map each ground level into the abstract counterpart, it is also simple to create mappings that doesn't reflect the actual abstraction process. In order to avoid this problem we require that the S-,

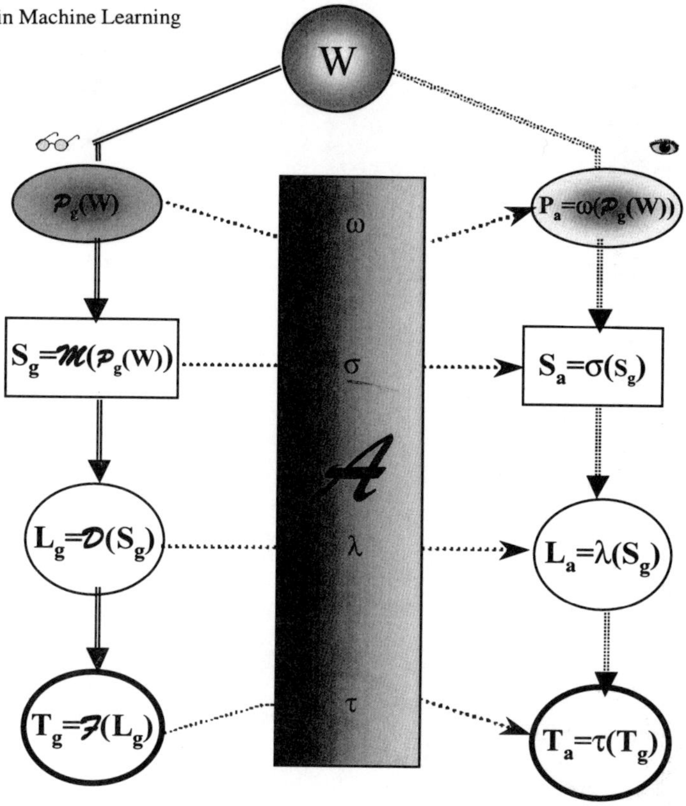

Figure 4. Abstraction mapping between a ground representation framework $\mathcal{D}_g(W) = (P_g, S_g, L_g, T_g)$ and a more abstract one $\mathcal{D}_a(W) = (P_a, S_a, L_a, T_a)$.

L- and T-operators are *compatible operators*, where this definition means that the result of the application of the abstraction operator must equate the result that would be obtained starting from the abstract perception directly.

AN EXAMPLE OF APPLICATIONS

In this section we provide an example of a domain, where the representation/abstraction framework introduced in the previous section finds a natural application. The domain is that of Cartographic Generalization.[28-31] Cartographic generalization is the process of producing maps; it involves focussing on the relevant features of the objects to be represented, applying to them transformations that highlight relevant pieces of information and hide others. In Figure 5 a mapping between the levels identified in Figure 3 and those used in cartography is described.

The topographic map production process closely parallels the described model for representation (conceptualization) and abstraction, because it can be analyzed according to the same two dimensions. Let us first consider the scheme of Figure 3 applied to cartography. The first step of cartography is to collect data from the geographic world or part of it (W). Data collection is usually done through aerial photographs, satellite the perceived world P(W). Objects contained in these photographs are located and labeled to create a geographic database (GDB). This GDB is the set of data organized in a structure (S). The GDB is then displayed by means of cartographic symbols applied to objects stored in it. This procedure consists of the creation of a map, involving the definition of an iconic language (L). Finally, maps are created for specific tasks (e.g. space analysis, search for itineraries, geographic theory construction). The theory (T) is the result of map analysis and

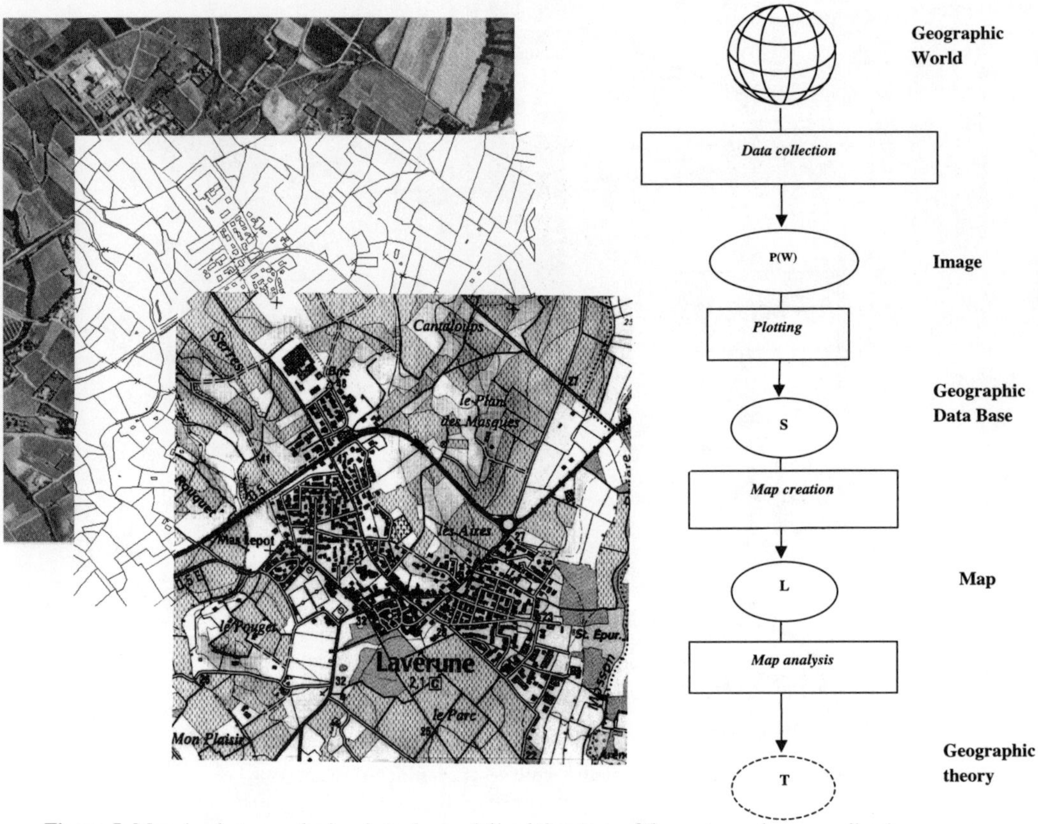

Figure 5. Mapping between the levels in the model and the steps of the cartography generalization process.

is guided by all the background facts and laws allowing one to reason about geographic configuration.

But cartography is not just knowledge representation. All steps of map creation also involve knowledge abstraction, because each step retains only part of the available information (e.g., a GDB does not contain the complete information seen on an image). In particular, map creation (which includes the generalization process) is both a knowledge representation process, when objects are symbolized, and a knowledge abstraction process, when objects relevant to the theory construction are identified. So described, map creation is represented as a diagonal process in Figure 6.

The representation and abstraction model we have introduced exhibits key properties for cartography. It allows the process of representation (change of language) to be distinguished from the process of abstraction (change of level of detail). These two processes are usually very much entangled in cartography. This distinction provides the basis for automating cartographic knowledge acquisition, as a combined acquisition of specific knowledge for abstraction and knowledge for changing representation.

The process we briefly sketch requires two steps: knowledge abstraction and knowledge representation. Knowledge abstraction input is a geographic object, and its output is an abstracted qualitative description of it. Knowledge representation inputs are the abstracted description (to help to determine how to represent) and the geographic object itself (to perform the geometric transformation). Knowledge representation output is the cartographic object representing the geographic object. Nevertheless knowledge acquisition from experts for this architecture is problematic, as shown by researches on cartographic generalization.[32] Machine Learning is also made difficult because the description space of inputs and outputs of the system are huge.

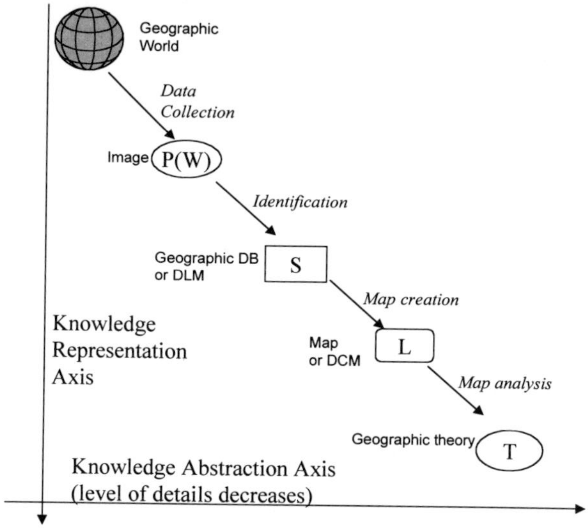

Figure 6. Cartography in the abstraction / representation space.

The knowledge abstraction is done in two steps: the first one is the description of an object as a set of measures (this abstraction is a way to supply the system with characteristics the cartographer may focus on). The measurements to be used are acquired from an expert of the domain. The description of the considered object as a set of measures is therefore greatly simplified over the initial vector representation. Then, the second abstraction step goes from this set of measures (surface, elongation...) to a qualitative description of the object. That means that a set of abstracted qualitative descriptors (size, shape...) are determined to enhance the most important characteristics of the object to deal with (this is the description the cartographer will use to reason about).

The separation between the steps of knowledge representation and abstraction allowed the automatic learning of transformation rules to be improved and simplified.[33]

CONCLUSIONS

In this chapter we have described two aspects of automatic concept acquisition: the emergence of concepts as sets of objects sharing common features, and the need to adapt the level of details in their representation in order to fit the requirements of specific goals.

The organization of knowledge in hierarchical levels allows the relationships among different representations to be clarified, whereas the mapping between different abstraction levels allows irrelevant information to be hidden.

REFERENCES

1. E. Rosch, Principles of categorization, in E. Rosch & B. Lloyd eds., *Cognition and Categorization*, Erlbaum, Hillsdale, NJ (1978).
2. L.G. Murphy and D.L. Medin, The role of theories in conceptual coherence, *Psychological Review, 92*, pp. 289-316 (1985).
3. J.P. Desclés, Implication entre concepts: la notion de typicalité, in M. Riegel & I. Tamba eds., *L'implication dans les langues naturelles et dans les langages artificiels,* Klincksieck, Paris, France, pp. 179-202 (1987).
4. G. Frege, *Grundgesetze der Aritmetik, Begriffsschriftlich Abgeleitet, Vol. I.* Jena, Germany (1893).
5. L. Wittgenstein, *Philosophical Investigations*, Blackwell, Oxford, UK (1953).
6. E. Rosch and C.B. Mervis, Family resemblance studies in the internal structure of categories, *Cognitive Psychology, 7*, pp. 573-605 (1975).
7. L. Barsalou, *Cognitive Psychology*, Erlbaum, Hillsdale, NJ (1992).

8. R. Michalski, A Theory and methodology of inductive learning. In R. Michalski, T. Mitchell & J. Carbonell eds., *Machine Learning, Vol. I*, Tioga Publ. Co., Palo Alto, CA, pp. 83-134 (1983).
9. T. Mitchell, Generalization as search, *Artificial Intelligence,18*, pp. 203-226 (1982).
10. A. Aamodt, Knowledge-intensive Case-based Reasoning and Sustained Learning, in *Proc. of the ECAI Conference*, Stockholm, Sweden, pp. 1-6 (1990).
11. J. Kolodner, Judging which is the best case for a Case-Based reasoner, in *Proc. of the Case-Based Reasoning Workshop*, Pensacola, FL, pp. 77-81 (1989).
12. R. C. Schank and D.B. Leake, Creativity and learning in a Case-Based explainer, *Artificial Intelligence, 40*, pp. 353-385 (1989).
13. C. Smith, J. Snir, and L. Grosslight, Using conceptual models to facilitate conceptual change: the case of weight-density differentiation, *Cognition and Instruction, 9*, pp. 221-283 (1992).
14. A. Tversky and I. Gati, Studies of similarity, in E. Rosch & B. Lloyd eds., *Cognition and Categorization*, Erlbaum, Hillsdale, NJ (1978).
15. D. L. Medin, W. D. Wattenmaker, and S. Hamson, Family resemblance, conceptual cohesiveness and category construction, *Cognitive Psychology, 19*, pp. 242-279 (1987).
16. L.J. Rips, Similarity, typicality, and categorization, in S. Vosniadou & A. Ortony eds., *Similarity and Analogical Reasoning,* Cambridge University Press, New York, NY, pp. 23-59 (1989).
17. D.L. Medin, R.L. Goldstone, and D. Gentner, Respects for similarity, *Psychological Review, 100*, pp. 254-238 (1993).
18. J. Hoffman and M. Ziessler, The integration of visual and functional classifications in concept formation, *Psychological Research, 48*, pp. 69-78 (1986).
19. L.W. Barsalou, Deriving categories to achieve goals, in G.H. Bower ed., *The Psychology of Learning and Motivation: Advances in Research and Theory,* Academic Press, New York, NY, pp. 1-64 (1991).
20. R. S. Michalski and R. Stepp, Learning from observation: conceptual clustering, in R. S. Michalski, J. G. Carbonell & T. Mitchell eds., *Machine Learning: An Artificial Intelligence Approach, Vol. I,* Tioga Publishing Co., Palo Alto, CA, pp. 331-363 (1983).
21. R. Stepp and R.S. Michalski, Conceptual clustering, inventing goal-oriented classifications of structured objects, in Michalski R., Carbonell J. & Mitchell T. eds., *Machine Learning: An AI Approach,* Vol. *II,* Morgan Kaufmann, Los Altos, CA, pp. 471-498.
22. M. Lebowitz, Experiments with incremental concept formation: UNIMEM, *Machine Learning, 2*, pp. 103-138 (1987).
23. P. Langley and J. M. Zytkow, Data-driven approaches to empirical discovery, *Artificial Intelligence, 40,* pp. 283-312 (1989).
24. D. H. Fisher, Knowldge Acquisition via Incremental Conceptual Clustering, *Machine Learning, 2*, pp. 139-172 (1987).
25. J-U. Kietz and K. Morik, A polynomial approach to the constructive induction of structural knowledge, *Machine Learning, 14*, pp. 193-218 (1994).
26. R. Goldstone and L. Barsalou Reuniting perception and conception, *Cognition, 65*, pp. 231–262 (1998).
27. M. Li, and P. Vitányi, *An Introduction to Kolmogorov Complexity and Its Applications,* Springer-Verlag, New York, NY (1997).
28. K. Brassel, and R. Weibel, A review and conceptual framework of automated map generalization, *Int. J. of Geographical Information Systems, 2*, pp. 229-244 (1988).
29. M.P. Armstrong, Knowledge classification and organisation in map generalization: Making rules for knowledge representation, in McMaster & Buttendfield ed., Longman Scientific & Technical, pp. 86-102.
30. J-F. Hangouët, *Approaches et Méthodes pour l'Automatisation de la Généralisation Cartographique; Application en Bord de Ville*, Ph.D. Thesis, Univ. of Marne-la-Vallée, France (1998).
31. S. Mustière, GALBE: Adaptive Generalisation. The need for an adaptive process for automated generalisation, an example on roads, in *Proc. GIS'Planet-1*, Lisbon, Portugal (1998).
32. R. Weibel, S. Keller, and T. Reichenbacher, Overcoming the knowledge acquisition bottleneck in map generalization: the role of interactive systems and computional intelligence, in *Proceedings of COSIT'95*, pp. 139-156 (1995).
33. S. Mustière, J-D. Zucker, and L. Saitta, An abstraction-based machine learning approach to cartographic generalization, in *Proc. Int Conf. on Spatial Data Handling*, Bejing, China, in press.

TOWARDS ARTIFICIAL ACTION: TEACHING BY SHOWING

T. Drummond and R. Cipolla

Department of Engineering
University of Cambridge
Cambridge CB2 1PZ, UK
e-mail: {twd20|cipolla}@eng.cam.ac.uk

ABSTRACT

This paper describes an approach to using machine vision to provide sensor feedback to robotic manipulators and shows how tasks may be taught, represented and repeated. This work was undertaken in order to increase the flexibility within a plant in which robotic manipulators are used to weld ship parts. This task has several features which are characteristic of a broad range of tasks in flexible manufacturing; the robot is required to accurately position a tool relative to a target structure (the workpiece) which may be inaccurately located with respect to the robot. This requirement imposes the constraint that the robot must be able to sense the structures in its environment in some way and machine vision provides a flexible way to achieve this.

INTRODUCTION

Visually guided robotics is a key application domain which draws together sensing and acting in order to provide hand-eye coordination for performing an activity. It is also an important enabling technology for many manufacturing tasks such as automotive or shipbuilding.

Here we present an approach to visually guided robotics based on teaching by showing. This paradigm is a natural one to use in the context of robotics and is currently used to specify tasks for a large number of industrial flexible automation activities. The principle is to specify the task to be performed by manually taking the robot through the activity.

Typically the task is broken down into a series of steps and information about the pose of the robot for each step is recorded and the supplemented with transition information (such as motion velocity, welding parameters, etc.) Traditionally this stage is performed blind in robotic applications and the robot simply records the positions of each of its joints using kinaesthetic sensors. For visually guided robotics, however, the aim is to design a system which can cope with misplaced parts and adjust the task execution accordingly. This capability places two requirements on the system.

Human and Machine Perception 3: Thinking, Deciding, and Acting, edited by Cantoni *et al.*
Kluwer Academic/Plenum Publishers, New York, 2001.

• Hand-eye coordination must be available at run-time in order to supply information that allows the system to correct for misplaced parts.

• When the task is taught to the robot, information describing the task must be represented in such as way that the robot performs the correct actions on the target workpiece(s).

The approach presented here uses information extracted from video images during both the teaching and execution phases to address both of these requirements. The system employs a camera mounted on the robot's end-effector in a position which is fixed relative to the tool which will perform the task (for example a welding tool). In this configuration the video image encodes information about the position of the tool relative to the workpiece. In particular, the image of the workpiece seen by the camera when the tool is in the correct position to perform the task is the same regardless of where the part has been placed in the world.

Thus a system constructed in this way could operate by simply recording the image as seen from the target position during teaching. At runtime, the robot can then be controlled until the view seen from the camera matches that stored during teaching. In practise, however, it is not necessary to store the entire image as seen from the target position; it is only important to store enough information that the specified viewpoint can be recovered. The approach presented here makes use of real-time model-based visual tracking technology to extract this critical information from the images. This visual tracking system uses a CAD model of the workpiece and uses the video images to generate accurate estimates of the pose (position and orientation) of the workpiece relative to the camera.

During the teaching by showing phase, the robot is manually placed in a series of positions at various stages through the task. The visual tracking system provides the pose of the workpiece relative to the camera in the robot end-effector in each of the positions and this information is recorded. The aim at runtime is to return the robot to these positions.

MODEL-BASED VISUAL TRACKING

Because a video feed contains a very large amount of data, it is important to extract only a small amount of perceptually important information, if real-time frame (or field) rate performance is to be achieved.[1] This observation leads to the notion of *feature based* tracking[2] in which processing is restricted to locating strong image features such as contours.[3,4]

A number of successful systems have been based on tracking the image contours of a known model. Lowe[5] used the Marr-Hildreth edge detector to extract edges from the image which were then chained together to form lines. These lines were matched and fitted to those in the model. A similar approach using the Hough transform has also been used.[6] The use of dense two-dimensional image processing incurs a significant computational cost and both of these systems make use of special purpose hardware in order to achieve frame rate processing.

An alternative approach is to render the model first and then use sparse one-dimensional search to find and measure the distance to matching (nearby) edges in the image. This approach has been used in RAPID,[7] CONDENSATION[8] and other systems.[9-11] The efficiency yielded by this approach allows all these systems to run in real-time on standard workstations. The approach is also used here and discussed in more detail in the section on tracking edges.

Using either of these approaches, most systems (except CONDENSATION) then compute the pose parameters by linearising with respect to image motion. This process is reformulated here in terms of the Lie group SE(3) and its Lie algebra. This formulation is a natural one to use since the group SE(3) exactly represents the space of poses that form the output of the system. The Lie algebra of a group is the tangent space to the group at the identity and is therefore the natural space in which to represent differential quantities such as velocities and small motions in the group. Thus the representation provides a canonical method for linearising the relationship between image motion and pose parameters. This approach can also be generalised to other transformation groups and has been successfully applied to deformations of a planar contour using the groups GA(2) and P(2).[12] Further, this can also be extended to provide a means of on-line calibration of internal camera parameters.

Outliers are a key problem that must be addressed by systems which measure and fit edges. They frequently occur in the measurement process since additional edges may be present in the scene in close proximity to the model edges. These may be caused by shadows, for example, or strong background scene elements. Such outliers are a particular problem for the traditional least-squares fitting method used by many of the algorithms. Methods of improving robustness to these sorts of outliers include the use of RANSAC,[13] factored sampling[8] or regularisation, for example the Levenberg-Marquadt scheme used in Lowe.[5] The approach used here employs iterative re-weighted least squares (a robust M-estimator) which is then extended to incorporate a number of additional saliency measures.

There is a trade-off to be made between robustness and precision. The CONDENSATION system, for example, obtains a high degree of robustness by taking a large number of sample hypotheses of the position of the tracked structure with a comparatively small number of edge measurements per sample. By contrast, the system presented here uses a large number of measurements for a single position hypothesis and is thus able to obtain very high precision in its positional estimates. This is particularly relevant in tasks such as visual servoing since the dynamics and environmental conditions can be controlled so as to constrain the robustness problems, while high precision is needed in real-time in order for the system to be useful.

Occlusion is also a significant cause of instabilities and may occur when the object occludes parts of itself (self occlusion) or where another object lies between the camera and the target (external occlusion). RAPID handles the first of these problems by use of a pre-computed table of visible features indexed by what is essentially a view-sphere. By contrast, the system presented here uses graphical rendering techniques to dynamically determine the visible features and is thus able to handle more complex situations (such as objects with holes) than can be tabulated on a view-sphere.

External occlusion can be treated by using outlier rejection, for example in Armstrong and Zisserman[13] which discards primitives for which insufficient support is found, or by modifying statistical descriptions of the observation model (as in MacCormick and Blake[14]). If a model is available for the intervening object, then it is possible to use this to re-estimate the visible features.[15,6] Both of these methods are used within the system presented here.

An important application domain for real-time visual tracking is that of visual servoing in which the output from such tracking systems is used for control of robotic manipulators. A distinction is often made in visual servoing[16] between an *image-based* approach in which the control loop is closed in the image by driving image features home to desired target positions[17,18] and a *position-based* approach in which the control loop operates in an explicitly three-dimensional space.[19,20] The approach presented here is position-based but closes the control loop by projecting the action of three-dimensional camera motion into the image where it is fitted to image measurements in a manner similar to Wunsch and Hirzinger.[6] Since the eye-in-hand approach is used, this generates a motion-to-image Jacobian (also known as the interaction screw[17]) which can be used to generate robot control commands to minimise the image error.

RIGID BODY TRACKING SYSTEM

Theoretical Framework

The approach used here for tracking known 3-dimensional structures is based upon maintaining an estimate of the camera projection matrix, P, in the co-ordinate system of the structure. This projection matrix is represented as the product of a matrix of internal camera parameters:

$$K = \begin{bmatrix} f_u & s & u_0 \\ 0 & f_v & v_0 \\ 0 & 0 & 1 \end{bmatrix} \tag{1}$$

and a Euclidean projection matrix representing the position and orientation of the camera relative to the target structure:

$$E = [\,R\,t] \qquad \text{with } RR^T = I \text{ and } |R| = 1 \tag{2}$$

The projective co-ordinates of an image feature are then given by

$$\begin{pmatrix} u \\ v \\ w \end{pmatrix} = P \begin{pmatrix} x \\ y \\ z \\ 1 \end{pmatrix} \tag{3}$$

with the actual image co-ordinates given by

$$\begin{pmatrix} \tilde{u} \\ \tilde{v} \end{pmatrix} = \begin{pmatrix} u/w \\ v/w \end{pmatrix} \tag{4}$$

Rigid motions of the camera relative to the target structure between consecutive video frames can then be represented by right multiplication of the projection matrix by a Euclidean transformation of the form:

$$M = \begin{bmatrix} R & t \\ 000 & 1 \end{bmatrix} \tag{5}$$

These M, form a 4×4 matrix representation of the group SE(3) of rigid body motions in 3-dimensional space, which is a 6-dimensional Lie Group. The generators of this group are typically taken to be translations in the x, y and z directions and rotations about the x, y and z axes, represented by the following matrices:

$$G_1 = \begin{bmatrix} 0&0&0&1 \\ 0&0&0&0 \\ 0&0&0&0 \\ 0&0&0&0 \end{bmatrix}, G_2 = \begin{bmatrix} 0&0&0&0 \\ 0&0&0&1 \\ 0&0&0&0 \\ 0&0&0&0 \end{bmatrix}, G_3 = \begin{bmatrix} 0&0&0&0 \\ 0&0&0&0 \\ 0&0&0&1 \\ 0&0&0&0 \end{bmatrix}$$
$$G_4 = \begin{bmatrix} 0&0&0&0 \\ 0&0&-1&0 \\ 0&1&0&0 \\ 0&0&0&0 \end{bmatrix} G_5 = \begin{bmatrix} 0&0&1&0 \\ 0&0&0&0 \\ -1&0&0&0 \\ 0&0&0&0 \end{bmatrix}, G_6 = \begin{bmatrix} 0&-1&0&0 \\ 1&0&0&0 \\ 0&0&0&0 \\ 0&0&0&0 \end{bmatrix} \tag{6}$$

These generators form a basis for the vector space (the Lie algebra) of derivatives of SE(3) at the identity. Group elements can be obtained from the generators via the exponential map:

$$M = \exp(\alpha_i G_i) \tag{7}$$

(with Einstein summation convention over Latin indices used throughout this paper). Thus, if M represents the transformation of the structure between two adjacent video frames, then the task of the tracking system becomes that of finding the α_i that describe the inter-frame transformation. Since the motion will be small, M can be approximated by the linear term:

$$M \approx I + \alpha_i G_i \tag{8}$$

Consequently, the motion is approximately a linear sum of that produced by each of the generators. The partial derivative of projective image co-ordinates with respect the ith generating motion can be computed as:

$$\begin{pmatrix} u' \\ v' \\ w' \end{pmatrix} = PG_i \begin{pmatrix} x \\ y \\ z \\ 1 \end{pmatrix} \tag{9}$$

with

$$L_i = \begin{pmatrix} u' \\ v' \end{pmatrix} = \begin{pmatrix} \frac{u'}{w} + \frac{uw'}{w^2} \\ \frac{v'}{w} + \frac{vw'}{w^2} \end{pmatrix} \tag{10}$$

giving the motion in true image co-ordinates. A least-squares approach can then be used to fit the observed motion of image features between adjacent frames. This process is detailed in the "Computing the Motion" section.

Tracking Edges

The features used in this work for tracking are the edges that are present in the model. These are strong features that can be reliably found in the image because they have a

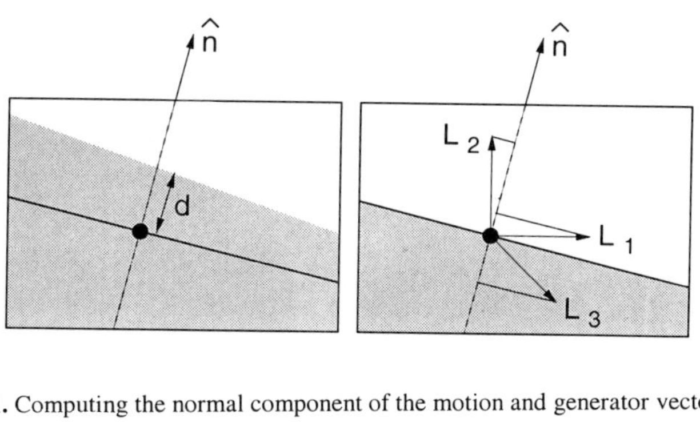

Figure 1. Computing the normal component of the motion and generator vector fields.

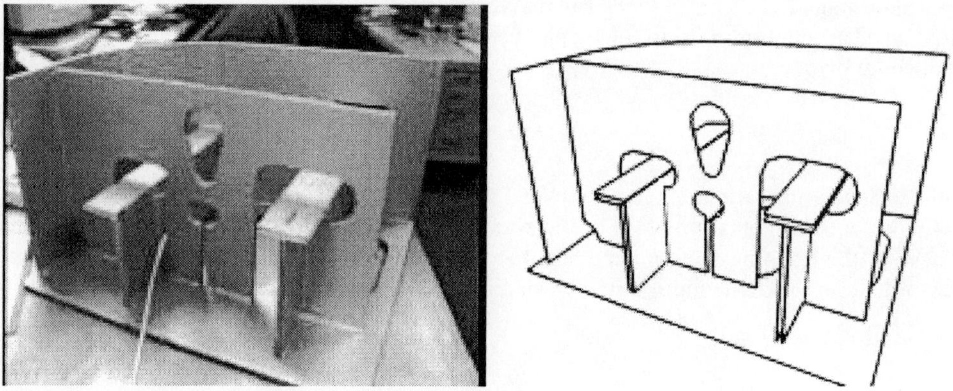

Figure 2. Image and CAD model of ship part.

significant spatial extent. Furthermore, this means that a number of measurements can be made along each edge, and thus they may be accurately localised within an image. This approach also takes advantage of the aperture problem (that the component of motion of an edge, tangent to itself, is not observable locally). This yields a significant benefit since the search for intensity discontinuities in the video image can be limited to a one dimensional path that lies along the edge normal, $\hat{n}$ (see Figure 1) and thus has linear complexity in the search range, rather than quadratic.[21] This reduction in complexity makes it possible to track complex structures in real time on a standard workstation without special hardware. The normal component of the motion fields, L_i are then also computed (as $f_i = L_i \hat{n}$) and d can be fitted as a linear combination of the f_i to give a linearised estimate of the 3D motion.

In order to track the edges of the model as lines in the image, it is necessary to determine which (parts of) lines are visible at each frame and where they are located relative to the camera. This work uses binary space partition trees[22] to dynamically determine the visible features of the model in real-time. This technique allows accurate frame rate tracking of complex structures such as the ship part shown in Figure 2. As rendering takes place, a stencil buffer is used to locate the visible parts of each edge by querying the buffer at a series of points along the edge prior to drawing the edge. Where the line is visible, tracking nodes are assigned to search for the nearest intensity discontinuity in the video feed along the edge normal (see Figure 3).

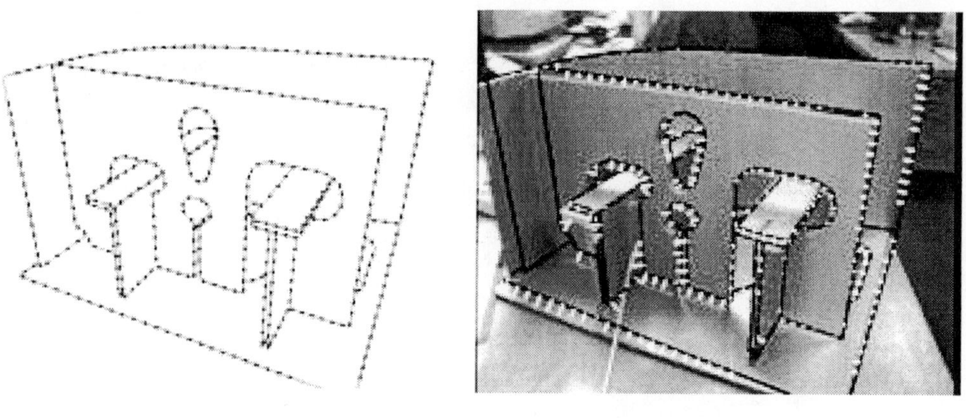

Figure 3. Sample points are assigned and distances measured.

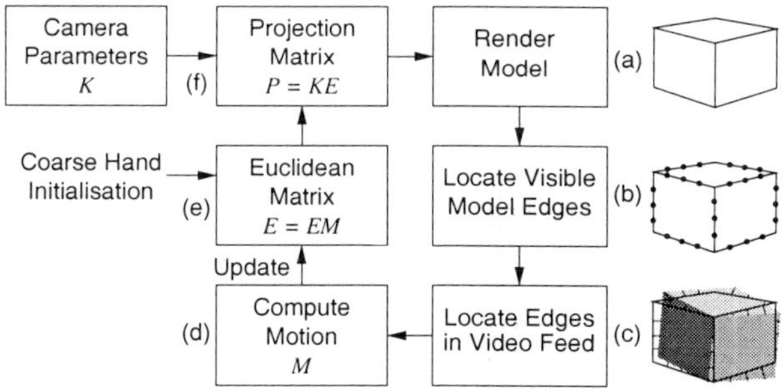

Figure 4: Tracking system operation

Figure 4 shows system operation. At each cycle, the system renders the expected view of the object (a) using its current estimate of the projection matrix, P. The visible edges are identified and sample points are assigned at regular intervals in image co-ordinates along these edges (b). The edge normal is then searched in the video image for a nearby edge (c). Typically $m \approx 400$ samples are assigned and measurements made in this way. The system then projects this m-dimensional measurement vector onto the 6-dimensional subspace corresponding to Euclidean transformations (d) giving the least squares estimate of the motion, M. The Euclidean part of the projection matrix, E is then updated by right multiplication with this transformation (e). Finally, the new projection matrix P is obtained by multiplying the camera parameters K with the updated Euclidean matrix to give a new current estimate of the local position (f). The system then loops back to step (a).

Computing the Motion

Step (d) in the process involves the projection of the measurement vector onto the subspace defined by the Euclidean transformation group. This subspace is given by the f_i^ξ which describe the magnitude of the edge normal motion that would be observed in the image at the ξth sample point for the ith group generator. These can be considered as a set of m-dimensional vectors which describe the motion in the image for each of the six modes of Euclidean transformation. The system then projects the m-vector corresponding to the measured distances (to the observed edges) onto the six dimensional subspace spanned by the transformation vectors. This corresponds to the geometric transformation of the part which best fits the observed edge positions and is found by minimising the square error between the transformed edge position and the actual edge position (in pixels). This process is performed using the standard least squares algorithm as follows:

$$v_i = \sum_{\xi} d^\xi f_i^\xi \tag{11}$$

$$C_{ij} = \sum_{\xi} f_i^\xi f_j^\xi \tag{12}$$

$$\alpha_i = C_{ij}^{-1} v_j \tag{13}$$

Giving α_i which are the six coefficients of the projected vector. It can be seen that setting $\beta_i=\alpha_i$ gives the minimum (least-squares) solution to

$$S = \sum_{\xi} (d^{\xi} - \beta_i f_i^{\xi})^2 \tag{14}$$

since

$$\frac{\partial S}{\partial \beta_i} = -2\sum_{\xi} f_i^{\xi} (d^{\xi} - \beta_j f_j^{\xi}) \tag{15}$$

and setting $\beta_i=\alpha_i$ and substituting Equation (13) gives

$$\frac{\partial S}{\partial \beta_i} = -2\sum_{\xi} f_i^{\xi} d^{\xi} - f_i^{\xi} f_j^{\xi} C_{jk}^{-1} \sum_{\xi'} f_k^{\xi'} d^{\xi'} \tag{16}$$

$$= -2\sum_{\xi} (f_i^{\xi} d^{\xi}) + 2C_{ij} C_{jk}^{-1} \sum_{\xi'} f_k^{\xi'} d^{\xi'} \tag{17}$$

$$= 0 \tag{18}$$

The α_i thus define a linear approximation to the Euclidean motion which minimises the sum squared error between the model and the observed lines. When more complex configurations are examined, it becomes important to consider how the sum squared error varies when $\beta_i \neq \alpha_i$. Setting $\beta_i=\alpha_i+\varepsilon_i$, Equation 15 gives

$$\frac{\partial S}{\partial \beta_i} = 0 + 2\sum_{\xi} f_i^{\xi} f_j^{\xi} \varepsilon_j \tag{19}$$

$$2C_{ij}\varepsilon_j \tag{20}$$

and integrating gives

$$S = S_0 + \varepsilon_i C_{ij} \varepsilon_j \qquad \text{where} \quad S_0 = S\big|_{\varepsilon=0} \tag{21}$$

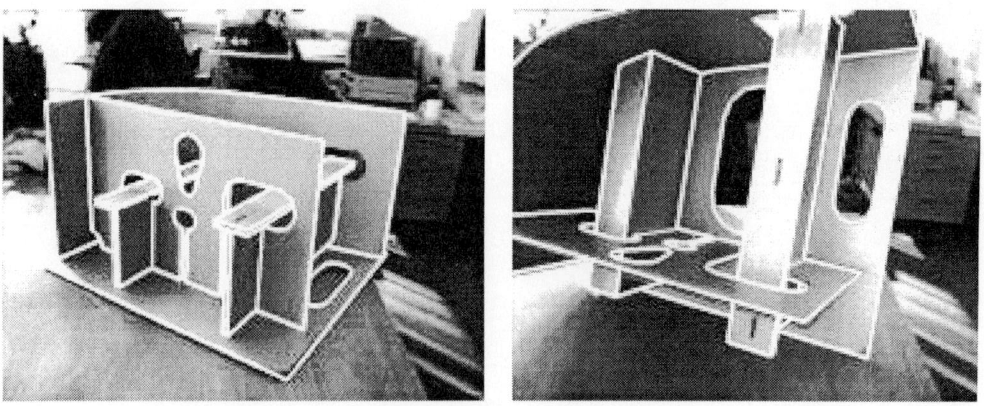

Figure 5. Frames from video of tracking sequence. The CAD model of the ship part is superimposed on the video image using the estimate of the projection matrix.

All that remains for the rigid body tracker is to compute the matrix for the motion of the model represented by the α_i and apply it to the matrix E in Equation (2) which is done by using the exponential map.

$$E_{t+1} = E_t \exp(\textstyle\sum_i \alpha_i G_i) \tag{22}$$

The system is therefore able to maintain an estimate of E (and hence P) by continually computing the coefficients α_i of inter-frame motions (see Fig. 5). One key advantage of this approach is that it is possible to extend it to include more complex situations such as tracking camera parameters in addition to object motion.

APPLICATION TO VISUAL SERVOING SYSTEM

A visual servoing system has been developed using the visual tracking system described in the previous sections. This system (shown in Figure 6) takes the Euclidean matrix, E, output from the tracking system and uses this within a non-linear control law to provide feedback to servo the robot to a stored target pose. These poses are learned using the principle of teaching-by-showing in which the robot is placed into the target pose by the supervisor and records the observed pose given by the tracker, E_t. The inverse of this target matrix, E_t^{-1}, is easily computed and the product of this with the current position matrix yields the transformation from the target position to the current position (Figure 6(A)).

$$T = EE_t^{-1} \tag{23}$$

The translation and rotation vectors that must be applied to the robot are then easily extracted from this representation (Figure 6(B)). (here i,j,k = 1,2,3):

$$t_i = T_{i4} \tag{24}$$

$$r_i' = \frac{1}{2}\varepsilon_{ijk} T_{jk}$$
$$r_i = \frac{r_i' \sin^{-1}(|r'|)}{|r'|} \tag{25}$$

The vectors t and r are then multiplied by a gain factor and sent to the robot as end effector translation and rotation velocities (Figure 6(C)). The gain is dependent on the

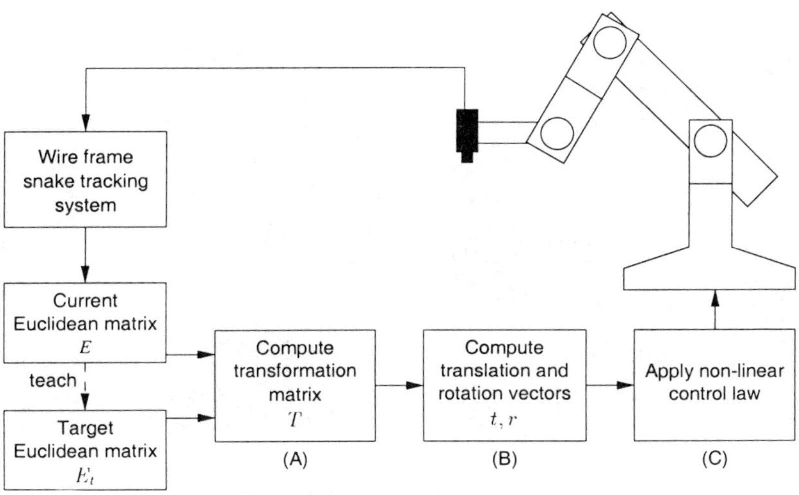

Figure 6. Visual servoing system operation.

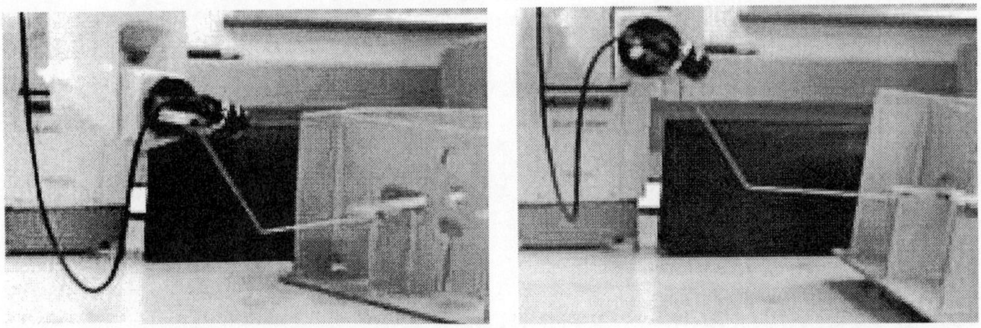

Figure 7. Closed-loop visual servoing. The task is to maintain a fixed spatial position relative to the workpiece.

magnitudes of t and r so that small velocities are damped to obtain higher precision, while large errors in position may be responded to quickly. A maximum velocity clamp is also applied for safety reasons and to prevent possible instabilities due to latency. Figures 7 and 8 show the visual servoing system in action, performing closed loop control tracking a moving part and tracing a path between recorded waypoints.

Experimental Results

The tracking system and visual servoing system have been tested in a number of experiments to assess their performance both quantitatively and qualitatively. These experiments were conducted with an SGI O2 workstation (225 MHz) controlling a Mitsubishi RV-E2 robot.

Stability of the tracker with respect to image noise. The stability of the tracker with a stationary structure was measured to assess the effect of image noise on the tracker. The standard deviation of position and rotation as measured from the Euclidean matrix were measured over a run of 100 frames. From a viewing distance of 30cm, the apparent r.m.s. translational motion was found to be 0.03mm with the r.m.s. rotation being 0.015 degrees.

Accuracy of positioning. The accuracy of positioning the robot was measured with two experiments. Firstly, the ship part was held fixed and the robot asked to home to a given position from a number of different starting points. When the robot had ceased to move, the program was terminated and the robots position queried. The standard deviation of these positions was computed and the r.m.s. translational motion was 0.08mm with the r.m.s. rotation being 0.06 degrees.

Figure 8. Visual servoing. The task is to trace out a trajectory relative to the workpiece.

The second accuracy experiment was performed by positioning the ship part on an accurate turntable. The part was turned through fifteen degrees in one degree rotations and the robot asked to return to the target position each time. Again, the position of the robot was queried and a circle was fitted to the data. The residual error was computed and found to give an r.m.s. positional error of 0.12mm per measurement (allowing for the three degrees of freedom absorbed into fitting the circle).

Artificial actions. Recently we have been able to test the complete Teach by Showing system on a real task involving the welding of part of a ship at the Odense shipyard in Denmark. The robot was shown the key positions for approaching the object and welding and then was able to successfully return to these positions after the ship part and robot had been moved.

REFERENCES

1. C. Harris, Geometry from Visual Motion, in A. Blake and A. Yuille eds., *Active Vision*, MIT Press, chapter 16, pp. 263-284, (1992).
2. G. Hager, G. Grunwald and K. Toyama, Feature-Based Visual Servoing and its Application to Telerobotics, in V. Graefe ed., *Intelligent Robotic Systems*, Elsevier, (1995).
3. D. Terzopoulos and R. Szeliski, Tracking with Kalman snakes, in A. Blake and A. Yuille eds., *Active Vision*, MIT Press, chapter 1, pp. 3-20, (1992).
4. R. Cipolla and A. Blake, Image Divergence and Deformation from Closed Curves, *International Journal of Robotics Research*, 16(1), pp.77-96, (1997).
5. D. G. Lowe, Robust model-based motion tracking through the integration of search and estimation, *International Journal of Computer Vision*, 8(2), pp.113-122, (1992).
6. P. Wunsch and G. Hirzinger, Real-time visual tracking of 3-Dobjects with dynamic handling of occlusion, in *Proceedings of the 1997 International Conference on Robotics and Automation*, pp.2868-2873, (1997).
7. C. Harris, Tracking with rigid models, in A. Blake and A. Yuille eds., *Active Vision*, MIT Press, chapter 4, pp.59-73, (1992).
8. M. Isard and A. Blake, CONDENSATION - Conditional Density Propagation for Visual Tracking, *International Journal of Computer Vision*, 29(1), pp.5-28, (1998).
9. N. Daucher, M. Dhome, J. T. Lapresté and G. Rives, Modelled object pose estimation and tracking by monocular vision, in *Proceedings of the British Machine Vision Conference*, pp.249-258, (1993).
10. A. D. Worrall, G. D. Sullivan, and K. D. Baker, Pose refinement of active models using forces in 3D, in J. Eklundh ed., *Proceedings of the Third European Conference on Computer vision (ECCV'94)*, (2), pp.341-352, May 1994.
11. E. Marchand, P. Bouthemy, F. Chaumette, and V. Moreau, Robust real-time visual tracking using a 2D-3D model-based approach, in *Proceedings of ICCV'99*, (1), pp. 262-268, Kerkyra, Greece, 20-27 September 1999.
12. T. Drummond and R.Cipolla, Application of Lie Algebras to Visual Servoing, *International Journal of Computer Vision*, 37(1), pp.21-41, (2000).
13. M. Armstrong and A. Zisserman, Robust object tracking, in *Proceedings of Second Asian Conference on Computer Vision*, pp.58-62, (1995).
14. J. MacCormick and A. Blake, Spatial Dependence in the observation of visual contours, in *Proceedings of the Fifth European Conference on Computer vision (ECCV'98)*, pp. 765-781, (1998).
15. M. Haag and H-H. Nagel, Tracking of complex driving manoeuvres in traffic image sequences, *Image and Vision Computing*, (16), pp.517-527, (1998).
16. S. Hutchinson, G.D. Hager, and P.I. Corke, A Tutorial on Visual Servo Control, *IEEE T-Robotics and Automation*, 12(5), pp.651-670, (1996).
17. B. Espiau, F. Chaumette, and P. Rives, A New Approach to Visual Servoing in Robotics, *IEEE T-Robotics and Automation*, 8(3), pp.313-326, (1992).
18. A.C. Sanderson, L.E. Weiss, and C.P. Neumann, Dynamic Sensor Based Control of Robots with Visual Feedback, *IEEE Journal of Robotics and Automation*, (3), pp.404-417, (1987).
19. R. Basri, E. Rivlin, E. and I. Shimshoni, Visual Homing: Surfing on the Epipoles, in *Proceedings of International Conference on Computer Vision (ICCV '98)*, pp.863-869, (1998).
20. W.J. Wilson, C.C. Williams Hulls, and G.S. Bell, Relative End-Effector Control Using Cartesian Position Based Visual Servoing, *IEEE T-Robotics and Automation*, 12(5), pp.684-696, (1996).
21. R. Cipolla and A. Blake, The dynamic analysis of apparent contours, in *Proceedings of IEEE 3rd International Conference on Computer Vision*, pp.616-623, Osaka, December 1990.

22. M. Paterson and F. Yao, Efficient binary space partitions for hidden surface removal and solid modeling, *Discrete and Computational Geometry*, 5(5), pp.485-503, (1990).

PANEL SUMMARY: **FRONTIERS OF HUMAN-MACHINE INTERACTION**

Panelists: Gianni Rigamonti[1], Nathan Intrator[2], and Hezy Yeshurun[3]

[1]DMA, University of Palermo, Italy
e-mail: rigamont@math.unipa.it
[2]Brown University and
Computer Science Department, Tel Aviv University, Israel
e-mail: nin@math.tau.ac.il
[3]Computer Science Department, Tel Aviv University, Israel
e-mail: hezy@math.tau.ac.il

HOT POINTS AND INTRODUCTION (*H. Yeshurun*)

The hot points presented to the panel were the following:

• What are the broader definitions of Human/Machine interaction? For example:
- non direct connection (using currently available computer GUI);
- first phase direct connection (linking sensors to the sensory system);
- second phase (direct connections into the Central Nervous System).

• What are the technologies that should be developed in order to enable each of the previous phases?

• What are the scientific research issues that are related to such phases?

• What are the possible implications of a direct human-computer link on the society? (compared to the Internet revolution?)

There are several phases in Human-Machine interaction. The first, and most trivial one, involves using virtual reality and augmented reality devices, that allow even today manipulation of objects by interactively superpositioning artificial objects and data on real images. The next interesting phases, however, involves direct data transfer from artificial sensors to the nervous system. In that regard, we can show various implementations of artificial auditory and visual sensors. Hearing implants are being coupled into the auditory system, and a direct link to the auditory nerve does exist. Regarding visual data, the task is more complicated, and current technologies are being developed in order to implant a sensor into the retina and connect it to the optic nerve.

Another interaction between computers and humans could be envisaged by making digital data directly retrievable for humans. This could be implemented in the future by using controlled brain signals picked up by external sensors to navigate through the data.

Human and Machine Perception 3: Thinking, Deciding, and Acting, edited by Cantoni *et al.*
Kluwer Academic/Plenum Publishers, New York, 2001.

However, the most interesting phase might be a direct link between artificial devices and the central nervous system. This involves external or internal electrodes that can pick neural information and analyze it, and on the other hand, inject direct signals into the nervous system. Currently, we lack the technology as well as the fundamental theory regarding the operating of the CNS that is required for such ambitious task. It might be possible, however, to design such systems in the future. In that regard, we can present even today experimental data that show that a direct correlation could be found between cortical activity and complex, high level cognitive tasks. These results are obtained mainly by functional brain imaging techniques, that correlate observed activity in certain brain areas to well defined cognitive activity.

MAN MACHINE INTERACTION (*G. Rigamonti*)

At present human-computer interaction uses mostly GUI.I don't think GUI will ever become obsolete as an interaction form, for its flexibility is unlimited and there are an infinity of different things it can do; but there also are, of course, things it cannot do. That is where direct connection of a computer to the nervous system, either peripheral or central, comes into play. Anyway, those connections that are possible today are, as far as I can see, NOT interactive: the human being whose sensory organs or cerebral cortex are connected to some computer either sends inputs to the latter or receives them, but does not do both things. In these areas, then, interaction is at present merely speculative, while it is a matter of fact with GUI. However, I am going to try some speculation myself.

Take vision, and imagine a direct link between somebody's eyes and a computer. The sort of link I have in mind cannot be realized by means of an ordinary video monitor, for to interact with such a device one must use a keyboard and a mouse that is, GUI.

Imagine instead goggles so big, so opaque and fitting the person's face so tightly that no light reaches his or her eyes from outside; but inside there are two computer-controlled video monitors, one for each eye, and the images they display are seen by the person, while nothing else is.

Now, if that is all, we cannot call this interaction: the human's eyes are simply fed by the computer, and there is no feedback. This is not essentially different from vision through a microscope, or a telescope.

But we can also imagine there are buttons on the outside of the goggles: these buttons correspond to such commands as "up", "down", "left", "right", "zoom forward", "zoom back", "darken", "brighten" and so on, and the person can operate them in a blind fashion. Or (which would be far more sophisticated and difficult to achieve) the goggles might have inner sensors responding to the movements of the person's eyes, head or body so as to shift the TV image in a way matching these movements as closely as an ordinary retinal image would do. I don't know if goggles or the first or the second sort will ever be made; anyway, I hope I have given, with this example, an idea of what I mean by human-computer INTERACTION via sensory organs.

Turning now to links with the central nervous system, I have quite apart from their interactive or non-interactive nature, two doubts: one is about feasibility, the other is ethical.

Concerning feasibility, let us suppose a microelectrode is implanted into a single neuron (this seems to be already possible today, or in a very near future). Then we can observe neural activity on a very fine-grained scale; however, there still is a long way between this and "reading" thoughts directly in the brain. In fact, we cannot be sure this sort of thought-reading will ever be possible, for there is no evidence that thoughts, at least in the sense of mental propositions, are localized in the brain. Can there be, for example, an "I love you" neuron? Does it make sense to suppose there is one?

But if thought-reading were possible (here I come to my second doubt), it would represent an intolerable violation of privacy. And there might be something even worse than that: for, supposing a computer could be employed to ready somebody's thoughts, then it would not be so strange to try and use it to IMPLANT alien thoughts into a person. At present this possibility is just science-fiction, but I believe a warning does make sense even today.

MAN MACHINE COOPERATION (*N. Intrator*)

Researchers have begun studying the idea of having a computer attached to a person at all times. This computer is expected to perform various tasks which are supposed to ease every-day activity in many respects.

It is likely that such a computer can be thought of as a co-pilot which takes care of many routine tasks and possibly gets involved in more advanced tasks. While the technology is already mature enough to enable such an idea, there are several fundamental issues that have to be addressed before such technology can be used in practice. It is likely that such a computer should have sensory input that is similar, or even more advanced, than our sensory system. This is needed in order not to put such a computer in an inferior position to humans, when it is about to make decisions. Are we ready for analyzing realistic sensory input?

We shall explore some aspects of this question in this brief summary on the subject by focusing on the difference between conventional machine learning tasks and the tasks that a co-pilot computer may have to deal with. We shall refer to the co-pilot computer as CPC in short.

Variability of the Input Space. In a conventional machine learning task, the input/output relations are well defined, and it is assumed that the data available for training the expert is a good representative of the environment where the expert is expected to perform its tasks. In the CPC task however, the I/O relations are not well defined, as there are several modalities that can affect the CPC in an unexpected manner. Dealing with all possible effects of combinations of these modalities on the output may be a lifetime task. Thus, the realistic data available for training is marginally sufficient, and the rule extraction and generalization abilities required from the CPC are much more demanding than in a regular machine learning task. In statistical terms, this implies that the internal model should have a strong bias towards a correct model, so that with very little data, a useful set of model parameters is estimated.

Priorities. One of the main tasks of the CPC may be to provide online priorities for the tasks it is expected to do. While current computers are capable of providing priorities for the tasks they have to address, these priorities are set by the software of the operating system, and all possible events for which priority should change are pre-programmed into the system. The situation is again different in the CPC case, as the number of possible combinations of inputs which should affect priorities is practically infinite and furthermore, the number of possible input representations for which the scene has to be parsed and understood is practically infinite as well. While there are some rules for the variability of the input environment, these rules are yet to be define and algorithms to automatically learn them from a small sample of the environment are not yet available. This problem is further complicated by the fact that all the factors that have been described above may be user specific, namely, the environment in which the CPC operates, the priorities, the parsing of the scene and the input/output relations of the decisions that have to be made, may all be (partially) user specific. Thus, it is difficult to use data from one CPC to train another CPC, and in fact, it is likely that data from a single person should be used for each CPC. In short, we need to better understand the process of attention and find ways to mimic it efficiently.

Fusion of Sensors. Another manifestation of the above problem is the need to fuse dynamically information arriving from several sensor modalities into a single coherent data representation that is amenable to scene analysis. Unfortunately, while the technology for sensors which are advanced enough, is available, the computational algorithms for taking full advantage of several sensors and fusing the information in a useful way is not yet available. This area is subject to intense research these days. (see the web page of the information fusion society {www.inforfusion.org}.)

The above differences pose great challenges to computer science research in the following decade. In the meantime, it is likely that wearable computers for small specific tasks will appear. One of the first domain of applications for such CPC is medical monitoring, a topic that could be addressed in the next workshop meeting.

INDEX

Acting 83, 84, 87, 103, 167, 175, 240
Active vision 197
Adaptivity 106, 164, 211
Agent 51, 123, 170, 209, 238
    autonomous 173
    bayesian 62
    cognitive 62, 178
    cycle 52
    environmet's state 52
    intelligent 89, 90, 167, 169
    Markov 58
    modelling 54
    multiagent 176, 179
    rational 51, 55
    reactive 58
        non-Markov 60
    robotic 51, 174
    with memory 63
Architectures 41, 177
    asynchronous 177
    brain-inspired 42
    hybrid 177
    network 150
    planning 127, 175
    pyramidal 203
        bin-pyramids 204
        coarse granularity 204
        fine granularity 204
        logical pyramid 205
        quad-pyramid 204
    reactive 168
    robot 178
    software 177
Artificial
    action 234
    intelligence 41, 42, 81, 84, 127, 160, 168, 170, 173, 239
    system 149, 210, 240, 237
        autonomous 85
        neuronal network 149, 150
Attention
    capacity models 199
    focus of 118, 194
    inherited 114
    interrupt theories 198
    searchlight 85
    visual
        binding problem 198
        inhibition of return 200
        lateral inhibition 200
            flat search function 200
            negative search function 200
            positive search function 200
        pre-attentive 198
            early vision 198
                pop-out theory 198
                ROI indexing 199
            interrupt theories 199
                textons 199
                onset 199
        reconstructionist approach 198
        search-light 198
        selective 199
        spatial 200
            spot-light model 200
            zoom-lens model 200
            gradient model 200
        visual search
            feature search mode 199
            singleton detection mode 199
Awareness 195, 196
Auditory system 237

Behavior 41, 84, 167, 168
    autonomous 123
    biological 162
    complex 88
    concurrent 178
    emergent 162
    intelligent 174

Behavior (*cont'd*)
- predicted 161
- rational 165
- robot 169
- rules 151
- uncertain 174

Biological
- brain 189
- diversity 191
- evolution 150, 189
- instructions 190
- neural network 150
- organism 189
- species 149, 150
- system 149

Block
- re-stacking problem 42, 45

Brain 189, 190
- biological 189
- imaging 189
- functions 192

Cerebral
- circuitry 190
- microwiring 190
- structure 190

Coding 193
Cognitive science 189, 193
Coherence 131, 134, 240
Complexity 108, 109, 123, 124, 127, 151. 153, 168
- Kolmogorov's algorithms 239, 240

Concept 215
- acquisition 217
- formation 217
- heuristic 215
- hierarchy 216

Conceptual clustering 217
Consciousness 44, 84, 86, 125, 186, 195, 196
- self 82, 84

Content retrieval 138
- query by example 138
- query by sketch 138
- of perception 86

Co-pilot computer (CPC) 239
Correctness
- verification 134, 176
- political 87

Cybernetics 211

Decision
- autonomous 167
- constructive 82
- contextual factors 81
- continuity argument 80
- control 3
- descriptive approach 77

Decision
- descriptive approach (*cont'd*)
  - prospect theory 77
    - certainty effect 78
    - utility
      - value function 78
      - framing effect 79
    - mental account 79
    - mental activity 84
- frequentist approach 73
- heuristic strategies 68
  - contextual factors
  - systematic errors (biases) 67
- making 1, 67, 159, 162, 164, 183, 193, 195
- normative approach 77
  - expected-utility theory 77
- pragmatic approach 74
  - model(-free) condition
- pragmatic information 83

DNA 192, 193
Depiction 42, 43, 44
Dynamic concept 240
Dynamic environmet 126
Dynamic Time Warping (DTW) 30

Emergence 43, 86, 109
Evolution 43, 84, 85, 150, 175, 191
- algorithms 150, 151
- artificial 159
- biological 189, 194
- computation 150
- learning 149, 151, 156
- mammalian 190
- nondirectional 191
- theory 150, 151

Expert systems 95
- knowledge revision 95

Features extraction 240, 237, 238
- Fourier descriptors 138
- space 150, 151, 152

Feedback 211
Forgetting
- factor 32
- selective 194

Fusion of sensors 239
Fuzzification 4

Generalisation 3
- f-generalisation 2
- f.g-generalization 2
- Nl-generalization 2

Genome 189, 190
Genetic
- information 193
- mutations 191

Genetic (*cont'd*)
 redundancy 193
GUI 238

Heuristic
 availability 70
 concept 239
 representativeness 68
  base rate fallacy 70
  conjunction fallacy 71
   Linda problem 72
  Gambler fallacy 70
 search 175
Hierarchical clustering 217
Hough transform 226
Human
 cloning 190
 consciousness 190
 organism 194
Human-Machine interaction 237, 238

Image 238, 239
 classification 138
 filter 123
 model 114
 processing 177
Information
 biological 193
 category 237
 compression 93
 context-dependent 237
 elaboration of 192, 194
 embedded 149
 fusion 126
 geometrical 175
 hidden 210
 imperfect 165
 information loss 192
 perception-based 1,2
 pragmatic 83
 processing 3, 161, 195
 spatial 175
 system 124
Insight 108, 172
Instability 240
Internal time 210
Intuition 105

Language
 L 238, 239
 natural 2,3
 generalised constraint 2, 20
Learning 105, 108, 125, 149
 algorithms 151, 152
 automatic 239
 autonomous system 82
 back-propagation rule 98

Learning (*cont'd*)
 cycle 151
 continuous 185
 classifiers
  neural networks 96
 distributed 185
 from mistake 90
 incremental 94
 locally receptive field networks 90
 mimicking 149
 one step 94
 open field networks 90
 Q-learning 59
 reinforcement 99
 rules 162
 supervised 151, 152, 237
 classification theory 94
  theory refinement 95
  theory revision 95
   goal 237
 concept 91, 92, 93, 149, 150
  general-to-specific 93
  mixed 94
  specific-to-general 93
  symbolic representation 93
 system 151, 153
 unsupervised 237
 vector quantization 150, 151
Law
 casual 171
 general 239
 of form 87
 of large numbers 70
 of Retaliation 125
 of small numbers 70
 taken theorem 36
Lie algebra 226, 228
Logic 86, 87
 dynamic 169
 first order
  theory revision 96
 reinforcement 59
 fuzzy 2, 16
 planning 170

Machine learning 91, 93, 95, 150, 239, 240, 237, 239
Meaning 195
Memory
 short term 194
 long term 194
Mind 42, 84, 86, 195
Model
 abstraction 238, 239
 ARX 28
 conformable polynomial regression (CPR) 114

Model (*cont'd*)
ecological 162
conceptualisation 238
distribution 150
free 59
general 82
geometric 174
Markov decision process (MDP) 58
Markov environment 54
polynomial regression (DPR) 114
Piage 82
of visual attention 114
symbolic 174
Molecular biology 189, 191

Neural networks 82, 92, 126, 161
activation functions 96
locally receptive field 96
open field 96
Hopfield energy-based analysis 43
LVQ 152
Neurophysiology 212
Neurones 189, 165
central nervous system 189
cerebral cortex 189
firing of neurones 193
Nervous
system 190, 191
central system 195
synapses 189, 190
Neuroscence 193
Nucleotides 193

Object
complex 86
geographic 239
perception 238
recognition 118
transform 83

Packages
classification
HER (Hierarchical Entropy-based Representation) 139
learning
CLINT 95
CREATOR 96
CRLS 95
DUCE 95
Ductor 95
EITHER 95
ENIGMA 96
IA-EBL 96
ID3 93
ID5 94
Induce 94
INTHELEX 96

Packages
learning (*cont'd*)
KBANN 96
k-NN 96
KRUST 96
Mitchell's Version Space 94
NEITHER 95
NTR 96
Q-learning 99
REFINER 96
RIGEL 94
RTLS 95
SEEK2 95
SHARADE 93
STAGGER 94
THESEO 97
WHY 96
representation
HER (Hierarchical Entropy-based Representation) 137
NRM (Neural Representation Modeller) 45
Parallel processes 195, 196
Phenotype 192
Perception 54, 86, 104, 174, 176, 238
faithful 55
noiseless 55
P-Operator 237
process 239, 240
reasoning 167, 195
role 239
system 240
transform 238
Planner 210
Planning 210
Plasticity 210, 211, 212, 213
Precision 82
Principle
category structure 240
cognitive economy 240
internal model 104
maxim of quantity 75
pertinence 75
puzzle 109
rationality 75
satisfying 80
universal minimality 104
Probability
estimation 2
law of 132
model 132
event 132
theory 20, 169
Process
abstraction 237, 239
cognitive 240
concurrent 169

Process (*cont'd*)
constructive 82, 85
evolutionary 152
formalisation 238
learning 149
logmap transformation 201
multiresolution 201
normalisation 238, 237
perception 239, 240
partially observable Markov Decision, 61
statistic analysis 127, 152
stochastic 176
thinking 82
Projection matrix 228, 231, 232
Propositional logic
theory revision 95
Pseudo-concepts 211
Psychology 189

Rationalization 184
Reconfigurability 211, 212
Representation
models feature graph 202
pattern tree 202
refinements grammar 202
Robotics
evolutionary 159
guided 225
RNA 192
Retina 237
(ir)Rational 186
Risk analysis 131

Scan-path 197
Searching 42
Selection
natural 191
Self
governing 149
reconfiguration 126
similarity 36
Hurst exponent 36
Sensing 105, 167, 169
action 169, 178
device 177
Standardisation 185
metadata standards 185
Set
crisp 1,4
fuzzy sets 3, 8
graph 9
logic 16
granulation 4, 5
functions 3, 18
relations 12
Single layer perceptron 161
Superior colliculus 46
Systems
ACADIA 203
adaptive 106
CLUSTER/2 217
COBWEB 217
CONDENSATION 226, 227
Cortex I 202
DISCON 217
FOVEA 202
KL-ONE 217
KLUSTER 217
natural 210
PAPIA II 205
RANSAC 227
RAPID 226, 227
SENSAR 203
visual servoing, 233, 234
UNIMEM 217

Takens theorem 36
Teaching by showing 225, 233
Theory
ecologically grounded 165
normative 160
prospect 160
of rational choice 161
Thinking 42, 82, 84, 103, 124, 194, 240
construction 82
definition 84
mechanism 84
necessary 81, 83
process 82,85
transform 83
Tracking,
rigid motion 228
edges 229

Video image 230
Virtual brains 44
Vision
early 198
Visual
attention mechanism 113
awareness 46
relevance function 116
system 123
thinking 184
tracking 226, 227, 233